Mathematical Methods of

Optimization

LARS-CHR [illegible]

Studentlitteratur

The papers and inks used in this product are eco-friendly.

Art. No 34235
ISBN 978-91-44-07075-9
Edition 1:1

www.studentlitteratur.se
Studentlitteratur AB, Lund

Cover illustration: Shutterstock
Cover design by Francisco Ortega

Printed by Holmbergs i Malmö AB, Sweden 2010

Preface

This book is an expanded version of my lecture notes on the course in *Optimization* at the Lund Institute of Technology (LTH). The aim of the course and of the book is to present a suitable blend of practical optimization methods and some central parts of the theory, in particular convexity and constrained optimization. On the matter of practical optimization, the aspiration is to give the flavour of some simple general algorithms, rather than to discuss in depth such details which might obscure the overview. Likewise, in the presentation of the theory we do not go for the most general results. Instead we prefer to conduct the discussion in a setting most suited to illuminate the concepts involved. We believe that the more advanced theory of optimization had better be deferred to a second course on the subject.

The mathematical theory is developed rigorously, with careful statements of definitions and theorems and with full proofs. However, parts of the text has been set in a small font, to indicate that the contents can be omitted without loss of continuity.

Prerequisites for the course are calculus in one and several variables, and linear algebra including some eigenvalue theory. Positive definite matrices are discussed in an appendix.

The interdependence of the different chapters is shown in the picture on page viii.

For any course in mathematics it is essential that the student solves plenty of exercises. They serve at least three different purposes: to support the theory by in simple situations indicating what it is all about, to show how the theory can be applied, and to give the student general training in problem solving, thus enforcing the understanding of the concepts involved. Most sections in this book are provided with a set of exercises. At the end of many chapters there are some further mixed exercises. Some of these — but certainly not all — might be a little more challenging to the student. Answers are given to all problems where it is meaningful. Selected exercises, marked with an S, have been provided with a full solution. — Notation like (2006–01–12, 1) at the end of a problem informs the reader that the problem was given at the exam at LTH on the indicated date, as problem number 1 in a row of six.

The exercises are, with few exceptions, intended to be solved with pencil and paper and a minicomputer. In some cases it might be advantageous to use Matlab in the calculations. — However, in a course like the present one, the student should also be given opportunity to do some problems in many variables on a computer, in order to obtain a proper feeling for the issues involved.

This course was developed in Lund many years ago, by Gunnar Sparr and later Stefan Diehl. Their influence on the content of the book is therefore profound, although they have not set eyes on the text itself. — Another source of input has been the experience of teachers who, beside myself, have taught the course in recent years at LTH: Andrey Ghulchak, Anders Holst, Carl Olsson, Jörg Schmeling, Lars Vretare. In particular, Anders Holst and Lars Vretare have helped to clear up many obscure points. Finally, response from students to the earlier lecture notes have of course made an impact on the final version. — My thanks to them all.

Lund in August 2010

Lars-Christer Böiers

Contents

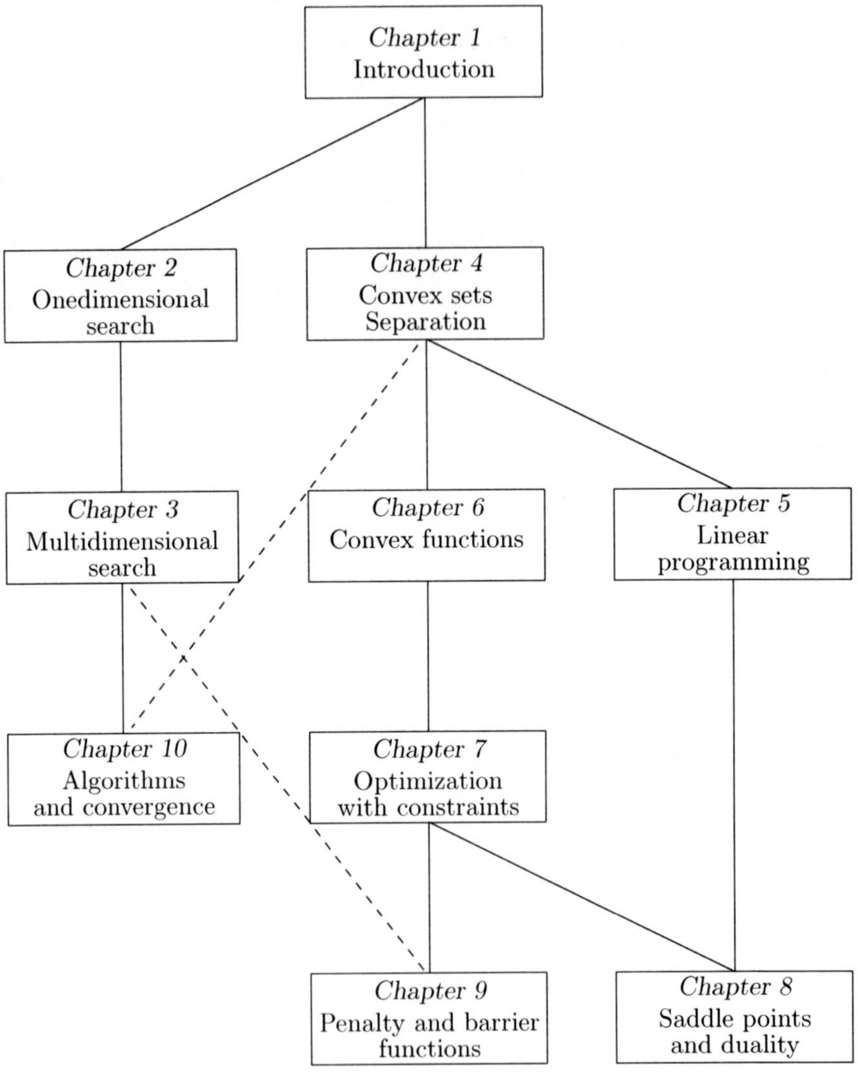
Chapter 1
Introduction
Chapter 2
Onedimensional search
Chapter 4
Convex sets
Separation
Chapter 3
Multidimensional search
Chapter 6
Convex functions
Chapter 5
Linear programming
Chapter 10
Algorithms and convergence
Chapter 7
Optimization with constraints
Chapter 9
Penalty and barrier functions
Chapter 8
Saddle points and duality

Chapter 1

INTRODUCTION

In this chapter we give a general overview and some illustrative examples of the concept of optimization. We also review some important topics from multidimensional analysis. However, the reader who is unfamiliar with the differential calculus of $\mathbf{R}^n$ must consult other literature for a full discussion.

We start with some presumably well-known notation and then give the formal definitions of a minimum and a maximum.

1.1 Some notation and terminology

1.1.1 The space $\mathbf{R}^n$

We are going to work in the space $\mathbf{R}^n$ of n-tuples $\boldsymbol{x} = (x_1, \ldots, x_n)$ of real numbers. Often we use geometrical language, calling $\boldsymbol{x}$ a *point* in $\mathbf{R}^n$. Sometimes it is convenient to consider these elements as *vectors* and even as *column matrices*.

For the elements of $\mathbf{R}^n$ we define **addition** and **multiplication with a scalar** (a real number) in the usual way:

$$\begin{aligned}(x_1, \ldots, x_n) + (y_1, \ldots, y_n) &= (x_1 + y_1, \ldots, x_n + y_n) \\ \lambda(x_1, \ldots, x_n) &= (\lambda x_1, \ldots, \lambda x_n).\end{aligned}$$

We also have the **scalar product**

$$\boldsymbol{x}^T \boldsymbol{y} = x_1 y_1 + \ldots + x_n y_n$$

(where $\boldsymbol{x}^T$ denotes the *transpose* of the column vector $\boldsymbol{x}$, i.e. a *row* vector).

The **norm** or **length** of $\boldsymbol{x} \in \mathbf{R}^n$ is denoted by $\|\boldsymbol{x}\|$ and defined as the number

$$\|\boldsymbol{x}\| = \sqrt{x_1^2 + \ldots + x_n^2}.$$

The **distance** between two points $\boldsymbol{x}$ and $\boldsymbol{y}$ in $\mathbf{R}^n$ is the number $\|\boldsymbol{x}-\boldsymbol{y}\| = \|\boldsymbol{y}-\boldsymbol{x}\|$.

For $\boldsymbol{a} \in \mathbf{R}^n$ and $\varepsilon > 0$ the set

$$S_\varepsilon(\boldsymbol{a}) = \{\, \boldsymbol{x} \in \mathbf{R}^n\,;\, \|\boldsymbol{x} - \boldsymbol{a}\| = \varepsilon \,\}$$

is called a **sphere** with center $\boldsymbol{a}$ and radius ε, and the set

$$B_\varepsilon(\boldsymbol{a}) = \{\, \boldsymbol{x} \in \mathbf{R}^n\,;\, \|\boldsymbol{x} - \boldsymbol{a}\| < \varepsilon \,\}$$

is called an open **ball** with center $\boldsymbol{a}$ and radius ε. If equality is allowed we have a *closed* ball.

We may on occasion need the *Cauchy-Schwarz inequality*

$$|\boldsymbol{x}^T\boldsymbol{y}| \le \|\boldsymbol{x}\|\,\|\boldsymbol{y}\|$$

or its consequence the *triangle inequality*

$$\|\boldsymbol{x} + \boldsymbol{y}\| \le \|\boldsymbol{x}\| + \|\boldsymbol{y}\|.$$

Neither of these will be proven here.

1.1.2 Optimization

Consider a function f defined on a subset S of $\mathbf{R}^n$.

DEFINITION 1. *The function f is said to have a (global)* **minimum** *at a point $\bar{\boldsymbol{x}}$ in S if*

$$f(\boldsymbol{x}) \ge f(\bar{\boldsymbol{x}}) \qquad \text{for all} \quad \boldsymbol{x} \in S.$$

DEFINITION 2. *The function f is said to have a local* **minimum** *at a point $\bar{\boldsymbol{x}}$ in S if there is a neighbourhood $\|\boldsymbol{x} - \bar{\boldsymbol{x}}\| < \delta$ of $\bar{\boldsymbol{x}}$ such that*

$$f(\boldsymbol{x}) \ge f(\bar{\boldsymbol{x}}) \qquad \text{for all } \ \boldsymbol{x} \in S \ \text{ with } \|\boldsymbol{x} - \bar{\boldsymbol{x}}\| < \delta\ .$$

In this situation we call $\bar{x}$ a (local) **minimum point** and $f(\bar{x})$ a (local) **minimum value**; the word *minimum* may refer to the point or the value depending on the context.

Note that there might be more than one global minimum point; the global minimum value is of course unique.

A global or local minimum is called **strict** if there is strict inequality above except when $x = \bar{x}$.

A typical minimization problem may have the following appearance:

Find the minimum of $f(x)$ subject to

$$\begin{aligned} &g_i(x) \le 0, \quad i = 1, \dots, m, \\ &h_j(x) = 0, \quad j = 1, \dots, l, \\ &x \in X. \end{aligned} \tag{1}$$

Here $g_1, \dots, g_m$ and $h_1, \dots, h_l$ are real-valued functions on $\mathbf{R}^n$, and X is some subset of $\mathbf{R}^n$. We call f the **objective function**. The conditions $g_i(x) \le 0$ and $h_j(x) = 0$ are called **constraints**. — Of course any inequality $g(x) \ge 0$ can be transformed to the opposite inequality by multiplication by -1. Hence it is no restriction to assume that all inequality constraints are of the form $g(x) \le 0$.

The conditions in (1) together define the set S used in the definitions above.

The purpose of the set X in (1) is to incorporate conditions which might not be explicitly given as constraints, or which one for some reason prefers to handle independently in some optimization algorithm.

The regularity of the functions f, g_i, h_j may vary; they may be continuous, once differentiable, twice differentiable, ...

Note the following terminology: a point $\bar{x}$ in X satisfying the conditions (1) is called a **feasible solution** to the problem. The minimization problem can thus be rephrased as to find a feasible solution $\bar{x}$ such that

$$f(x) \ge f(\bar{x}) \quad \text{for all feasible } x.$$

We call a constraint $g_i(x) \le 0$ **active** at a feasible point $\bar{x}$ if $g_i(\bar{x}) = 0$. Equality constraints are of course by nature always active.

Example 1. Here is an example in the case of two variables: find the minimum of $(x-3)^2+(y-4)^2$ subject to

$$y \geq \ln x, \qquad xy \leq 5, \qquad x > 0, \qquad y = x^3.$$

In this problem there are two inequality constraints,

$$g_1(x,y) = \ln x - y \leq 0, \qquad\qquad g_2(x,y) = xy - 5 \leq 0,$$

and one equality constraint,

$$h_1(x,y) = x^3 - y = 0.$$

The set X is in this case $X = \{\,(x,y)\,;\, x > 0\,\}$, necessary whether the condition $x > 0$ is explicitly stated or not because of the logarithm in one of the constraints.

The objective function is

$$f(x,y) = (x-3)^2 + (y-4)^2.$$

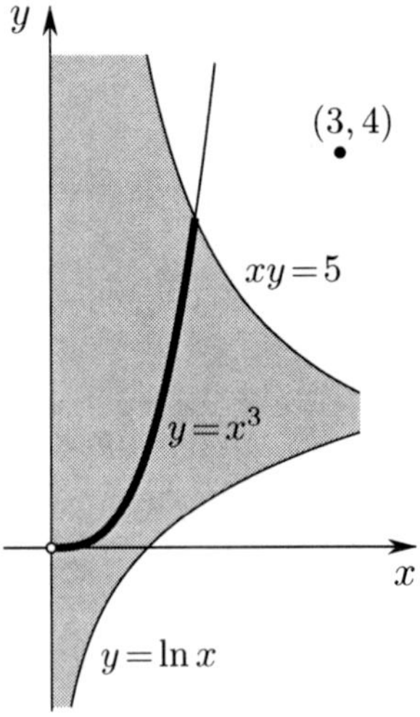

The set S which is considered to be the domain of f in this problem is the heavy piece of curve in the figure. At the uppermost point of the curve the constraint $xy \leq 5$ is active.

Considering the geometrical interpretation of the objective function as the square of a distance, the reader should have no difficulty in finding the answer to this problem.[1]

□

The various definitions of a maximum are of course analogous to those of a minimum.

Definition 3. *The function f is said to have a (global)* **maximum** *at a point $\bar{\boldsymbol{x}}$ in S if*

$$f(\boldsymbol{x}) \leq f(\bar{\boldsymbol{x}}) \qquad \textit{for all} \quad \boldsymbol{x} \in S.$$

[1] The minimum point is $(\sqrt[4]{5},\, \sqrt[4]{5^3})$, the minimum value is ≈ 2.6947.

Definition 4. *The function f is said to have a local* **maximum** *at a point $\bar{x}$ in S if there is a neighbourhood $\|x - \bar{x}\| < \delta$ of $\bar{x}$ such that*

$$f(x) \leq f(\bar{x}) \qquad \text{for all } x \in S \text{ with } \|x - \bar{x}\| < \delta .$$

The terminology in connection with maxima corresponds to that for minima above. The term *optimum* refers to either a minimum or a maximum. The name *extreme point* is used for either a minimum point or a maximum point.

A maximum problem can always be turned into a minimum problem, since

$$\max f(x) = -\min(-f(x)).$$

Consequently we will henceforth restrict the discussion to minimum problems only, except in a few particular cases.

There are two complementary issues involved in the discussion of a minimization problem.

- Practical optimization methods, where the goal is to find some algorithm which will compute a certain minimum with prescribed accuracy in a reasonable amount of time.

- The theory of optimization: existence, and perhaps uniqueness, of a minimum. Conditions (necessary or sufficient) for a point $\bar{x}$ to be a minimum point. Properties of minima. Connections between different optimization problems, so called *duality theory*.

Both these aspects are relevant for any optimization problem. Except in the simplest cases the theory produces equations for the minimum point which are so complicated that a numerical treatment is necessary at the end. On the other hand, without a theory to support the numerical methods applied, there is no guarantee that the results obtained have any relevance to the problem posed.

Exercise

1.1 Consider the following optimization problem:

$$\begin{aligned}
\text{Minimize} \quad & (x_1 - 3)^2 + (x_2 - 3)^2 \\
\text{subject to} \quad & 4x_1^2 + 9x_2^2 \le 36 \\
& x_1^2 + 3x_2 = 3 \\
& (x_1, x_2) \in X = \{\, (x_1, x_2)\,;\, x_1 \ge -1 \,\}.
\end{aligned}$$

Sketch the feasible region defined by the constraints, and the level curves of the objective function. Use the graphical interpretation to find the minimum point.

1.2 Examples of optimization problems

Example 2. We are going to consider a so called *neural network.*[2] The network consists of *input nodes* (numbered $1, \dots, A$), *hidden nodes* (numbered $1, \dots, B$) and *output nodes* (numbered $1, \dots, C$).[3] Every input node a receives an input signal x_a.

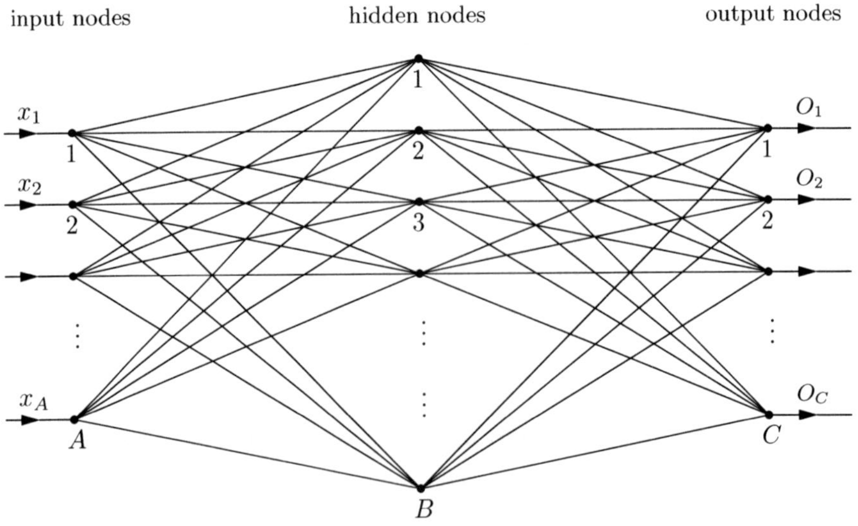

[2]This is a simplified and incomplete version of methods used in some Master's Theses at the Department of Mathematics, LTH.

[3]In one case $A \approx 260$, $B \approx 150$, $C \approx 120$.

These input signals are transmitted to the hidden nodes in such a way that hidden node b receives the signal

$$h_b = \sum_{a=1}^{A} w_{1,a,b} x_a.$$

Here $w_{1,a,b}$ are positive numbers (*weights*) to be decided. Each hidden node transforms its input signal to an output signal

$$H_b = g(h_b),$$

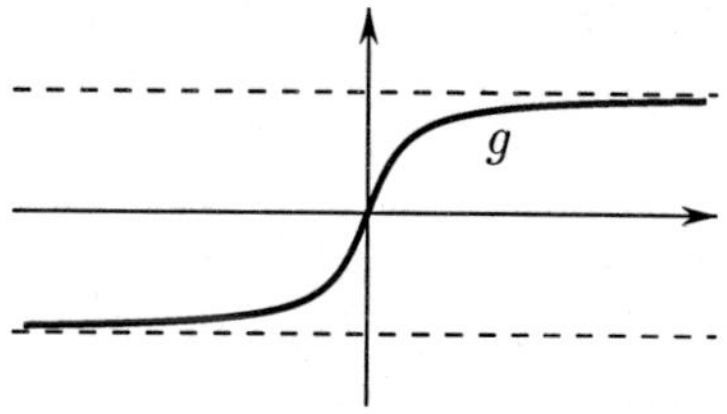

using some prescribed function g of one variable. This function should have the properties indicated in the figure.

The outputs from the hidden nodes are transmitted to the output nodes, making output node c receive the input signal

$$o_c = \sum_{b=1}^{B} w_{2,b,c} H_b.$$

The numbers $w_{2,b,c}$ are some new weights to be decided. The output signals from the output nodes are

$$O_c = g(o_c),$$

where g may or may not be the same function as above.

For some given input signals $x_1, \ldots, x_A$ the *desired* output signals are $y_1, \ldots, y_C$. The goal is to choose the weights to meet this requirement. Since this normally can not be done exactly we have to be satisfied with less: to find weights $w_{1,a,b}$, $w_{2,b,c}$ which minimize the *square error*

$$E(\boldsymbol{w}) = \sum_{c=1}^{C} (y_c - O_c)^2.$$

Note that the variables in this function are the weights; the number of them is $AB + BC$.[4]

We do not intend to discuss methods for solving this problem just now. However, we will make some comments on it after the next example.

□

[4] About 50 000 in one case.

Example 3. Suppose that we wish to make a tent consisting of two rectangular walls and two gables in the form of isosceles triangles, and no bottom. For a given volume V of the tent the dimensions are to be chosen so as to minimize the material required, i.e., to minimize the area.

Denote by l the length of the tent, and by b and h the base length and height of the triangles. The total tent area is the sum of two rectangle areas and two triangle areas and is equal to

$$f(l,b,h) = 2l\sqrt{\frac{b^2}{4} + h^2} + bh.$$

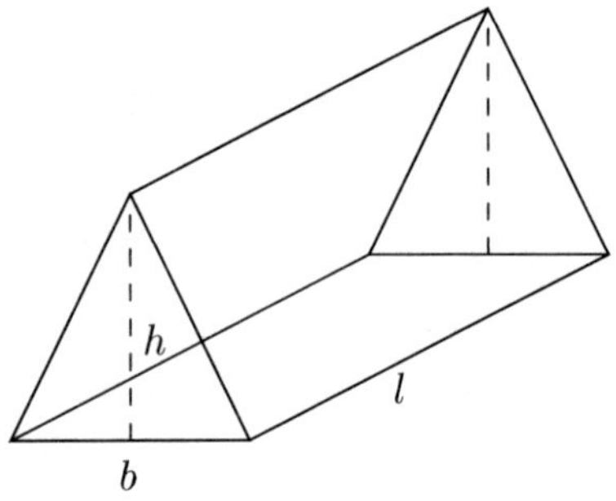

The problem is to minimize this function subject to the constraint on the volume,

$$\frac{bhl}{2} = V.$$

The existence of this minimum is of course not obvious.

The reader should be able to solve this problem using previous knowledge of optimization. (Solve the constraint condition for l and then minimize a function of the two variables b and h with no constraint.[5]) Our purpose in presenting the problem here is to contrast it with the problem in example 2. □

The two previous examples both exhibit minimization problems, but with very different characteristics. The reader should contemplate the span of the problems one may encounter under the subject of optimization. Some comments may be in order:

1. The number of variables in the tent-making problem is 3 (or 2 if you eliminate l), whereas in the neural network problem the number is several thousands. The answer in the latter case is hardly likely to be written down on paper.

2. The second problem has a geometrical interpretation, and can in some sense be understood on geometrical grounds. The answer, whether it is computed exactly or by the numerical solving of some

[5] The answer is $h = (V/\sqrt{2})^{1/3}$, $b = 2(V/\sqrt{2})^{1/3}$, $l = \sqrt{2}\,(V/\sqrt{2})^{1/3}$.

system of equations, can be checked against "common sense". Errors resulting in gravely wrong answers will be detected. You can also check the answer by insertion of the results into the original equations.

These qualities are absent from the first problem. You probably have no sense at all for the magnitude of the minimum point or the minimum value. You have to put your trust in the numerical optimization method used. This stresses the importance of knowing what one is doing.

In problems like example 2 it is essential that you try out the proposed solution method on some small test problems (of the same nature as the real problem), of which you know the answer beforehand. This will give you the opportunity to check the behaviour of the algorithm, and perhaps detect some possible errors.

3. Also note that the second problem is a "once only" problem. When you have found the solution, and is satisfied that it is correct, you will probably never have reason to return to the problem again.

   In contrast to this, the neural network problem is intended to be solved again and again, using different sets of input signals. The iterations will be handled by a computer, and no human being will set eyes on the intermediate results. This again emphasizes that you must be acquainted with the theoretical background of the methods used, so that you are aware of what you are doing.

   The fact that you intend to iterate a minimization procedure many times also has the consequence that issues like running time and memory storage requirements become important.

4. Another difference: in the first example the minimum *value* is probably of no interest at all, whereas the minimum value of the second problem corresponds to the least amount of material possible, and thus has a concrete meaning.

5. Finally, note that the second example exhibits a problem where the answer to the question posed is the main issue. In contrast, the optimization problem in example 2 appears as a small subproblem inside some actual problem of interest.

We give some further examples, engineering applications somewhere in the middle between the two extreme ones above. The reader should consider how the comments above apply to these.

Example 4. We are going to study a simple model for the manoeuvring of a satellite in space.

Consider a twodimensional (for simplicity) rigid body with center of gravity at the origin. By applying forces at a number of points (x_i, y_i), $i = 1, \ldots, n$, we wish to bring about a prescribed total force $\boldsymbol{F} = (F_x, F_y)$ and a prescribed torque T on the body. The forces to be applied will act in directions with angles θ_i, $i = 1, \ldots, n$, to the positive x-axis (supposed to be given beforehand), and we denote their magnitudes by f_i.

The total force requirement can be stated as

$$\begin{aligned} \sum_{i=1}^{n} f_i \cos\theta_i &= F_x \\ \sum_{i=1}^{n} f_i \sin\theta_i &= F_y, \end{aligned} \tag{2}$$

and the condition on the torque is

$$\sum_{i=1}^{n} (x_i f_i \sin\theta_i - y_i f_i \cos\theta_i) = T. \tag{3}$$

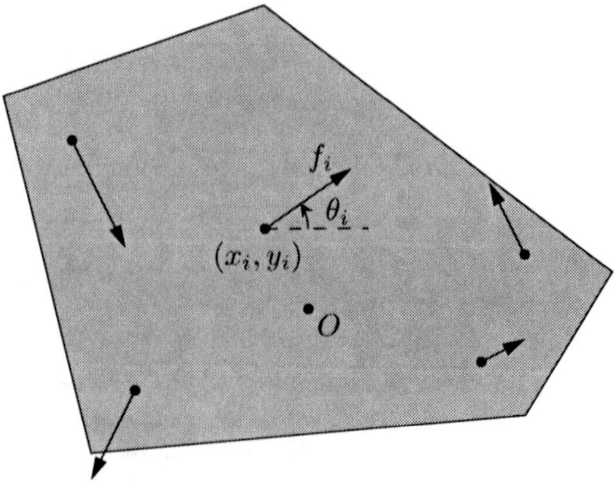

There is an upper limit on the magnitudes of the forces. After normalization we may assume

$$0 \le f_i \le 1, \qquad i = 1, \dots, n. \tag{4}$$

The optimization problem is to accomplish the desired total force and torque with a minimum of fuel consumption. The energy required to achieve a certain force is assumed to be proportional to its magnitude. Thus we arrive at the following optimization problem: to minimize

$$f_1 + f_2 + \dots + f_n$$

subject to the constraints (2)–(4).

This turns out to be a so called linear programming (LP) problem. The principles for its solution will be discussed later in the course. □

Example 5. In one application in image analysis two threedimensional vectors $\boldsymbol{a}$ and $\boldsymbol{b}$ are computed, which are supposed to be orthogonal. Due to noise in the input data the resulting vectors in practice deviate from orthogonality. To rectify this one wishes to substitute two truly orthogonal vectors $\boldsymbol{x}$ and $\boldsymbol{y}$ which in some sense are close to $\boldsymbol{a}$ and $\boldsymbol{b}$ respectively. Such vectors can be found as solutions to the problem of minimizing

$$\|\boldsymbol{x} - \boldsymbol{a}\|^2 + \|\boldsymbol{y} - \boldsymbol{b}\|^2$$

subject to the constraint $\boldsymbol{x}^T\boldsymbol{y} = 0$.

This is a problem in six variables $x_1, x_2, x_3, y_1, y_2, y_3$. Written out in full the objective function is

$$(x_1 - a_1)^2 + (x_2 - a_2)^2 + (x_3 - a_3)^2 + (y_1 - b_1)^2 + (y_2 - b_2)^2 + (y_3 - b_3)^2,$$

and the constraint is

$$x_1y_1 + x_2y_2 + x_3y_3 = 0.$$

However, it is preferable to retain the vector notation as long as possible when trying to solve this problem. We will present the solution in example 7.23, page 285. □

Example 6. Here is an application in *experimental design.* Let

$$x_1,\ x_2,\ \ldots,\ x_n$$

denote input variables (quantities of ingredients, temperature, pressure, ...) in some chemical reaction, and let y denote some output variable (quantity of some resulting compound, purity, ...). We wish to optimize y by a suitable choice of $x_1, \ldots, x_n$. Due to lack of a theoretical model, or the inadequacy of one, this must be done by experiments.

Thus we wish to design a sequence of inputs $\boldsymbol{x}^{(1)}$, $\boldsymbol{x}^{(2)}$, ... aiming at finding an $\boldsymbol{x} = (x_1, \ldots, x_n)$ which gives an optimal output y. The choice of the input $\boldsymbol{x}^{(k)}$ is to decided from the outcomes of the preceding experiments. For obvious reasons (minimization of cost or of time) the choices should be made so as to reach a reasonable approximation of the optimal variables $\boldsymbol{x}$ in as few experiments as possible. Problem: what strategy should be used for the choices of $\boldsymbol{x}^{(k)}$, $k = 1, 2, \ldots$? □

1.3 Review of some concepts from n-dimensional analysis

For an open subset S of $\mathbf{R}^n$ we denote by $\mathcal{C}^k(S)$ the set of all k times differentiable functions on S with continuous derivatives.

1.3.1 The gradient

Let f be a differentiable function on some subset of $\mathbf{R}^n$. We denote by $\boldsymbol{\nabla} f(\boldsymbol{x})$ the *gradient vector* at $\boldsymbol{x}$, i.e., the vector whose components are the partial derivatives of f computed at $\boldsymbol{x}$. In this text we always consider $\boldsymbol{\nabla} f(\boldsymbol{x})$ as a *column* vector (column matrix)

$$\boldsymbol{\nabla} f(\boldsymbol{x}) = \begin{pmatrix} \dfrac{\partial f}{\partial x_1}(\boldsymbol{x}) \\ \vdots \\ \dfrac{\partial f}{\partial x_n}(\boldsymbol{x}) \end{pmatrix}.$$

The gradient vector enjoys some properties with which the reader is supposed to be familiar:

- For the *directional derivative* at $\boldsymbol{x}$ in direction $\boldsymbol{d}$ we have the formula
$$f'(\boldsymbol{x};\boldsymbol{d}) = \boldsymbol{\nabla} f(\boldsymbol{x})^T \boldsymbol{d},$$
assuming that $||\boldsymbol{d}|| = 1$.

- The gradient vector indicates the direction from $\boldsymbol{x}$ in which f has its maximal growth. (This is a consequence of the formula for the directional derivative above and the Cauchy-Schwarz inequality.) As a consequence, $-\boldsymbol{\nabla} f(\boldsymbol{x})$ points in the direction of steepest descent.

- The vector $\boldsymbol{\nabla} f(\boldsymbol{x})$ is orthogonal to the *level surface* $f(\boldsymbol{x}) = C$ (C a constant) through $\boldsymbol{x}$.

The reader should prove these facts as an exercise, or consult other literature on the subject.

A vector $\boldsymbol{d} \neq \boldsymbol{0}$ is called a **direction of descent** from $\boldsymbol{x}$ if there is a positive number δ such that

$$f(\boldsymbol{x} + \lambda \boldsymbol{d}) < f(\boldsymbol{x}) \qquad \text{when} \quad 0 < \lambda < \delta.$$

From the formula for the directional derivative it follows that $\boldsymbol{d}$ is a descent direction if $\boldsymbol{\nabla} f(\boldsymbol{x})^T \boldsymbol{d} < 0$.

1.3.2 The Jacobian

Let $\boldsymbol{g}$ be a function from $\mathbf{R}^n$ to $\mathbf{R}^m$, i.e., a vector-valued function of n variables. Denote the components of $\boldsymbol{g}$ by $g_1, \ldots, g_m$. Then the *Jacobian* of $\boldsymbol{g}$ is the matrix

$$\begin{pmatrix} \dfrac{\partial g_1}{\partial x_1} & \cdots & \dfrac{\partial g_1}{\partial x_n} \\ \vdots & & \vdots \\ \dfrac{\partial g_m}{\partial x_1} & \cdots & \dfrac{\partial g_m}{\partial x_n} \end{pmatrix}.$$

Note that row i contains the components of the gradient $\boldsymbol{\nabla} g_i$.

For example, the linear function $\boldsymbol{x} \mapsto \boldsymbol{g}(\boldsymbol{x}) = \boldsymbol{A}\boldsymbol{x}$, where $\boldsymbol{A}$ is an $m \times n$ matrix, has the Jacobian $\boldsymbol{A}$.

1.3.3 The Hessian

Now assume that $f\colon \mathbf{R}^n \to \mathbf{R}$ is twice differentiable. The *Hessian* $\boldsymbol{H}(\boldsymbol{x})$ of f at $\boldsymbol{x}$ is the quadratic matrix whose entries are the second order derivatives of f:

$$\boldsymbol{H}(\boldsymbol{x}) = \left(\frac{\partial^2 f}{\partial x_i \partial x_k}(\boldsymbol{x}) \right).$$

Under very weak assumptions (which we will always assumed fulfilled) the Hessian is a *symmetric* matrix. — Another notation for the Hessian is $\boldsymbol{\nabla}^2 f(\boldsymbol{x})$.

We remind the reader of *Taylor's formula*. In matrix notation it says that close to a point $\boldsymbol{a} \in \mathbf{R}^n$

$$\begin{aligned} &f(\boldsymbol{x}) = \\ (5) \qquad &= f(\boldsymbol{a}) + \boldsymbol{\nabla} f(\boldsymbol{a})^T(\boldsymbol{x}-\boldsymbol{a}) + \frac{1}{2}(\boldsymbol{x}-\boldsymbol{a})^T \boldsymbol{H}(\boldsymbol{a})(\boldsymbol{x}-\boldsymbol{a}) + O(\|\boldsymbol{x}-\boldsymbol{a}\|^3). \end{aligned}$$

We use $O(t)$ to denote any function with the property that $\dfrac{O(t)}{t}$ is bounded for t close to 0.

1.3.4 Quadratic functions

If we neglect the error term $O(\|\boldsymbol{x}-\boldsymbol{a}\|^3)$ in (5) we obtain a quadratic function of $\boldsymbol{x}-\boldsymbol{a}$. In general a *quadratic function* is a real-valued function on $\mathbf{R}^n$ of the form

$$q(\boldsymbol{x}) = \frac{1}{2}\boldsymbol{x}^T \boldsymbol{H}\boldsymbol{x} + \boldsymbol{c}^T\boldsymbol{x} + d,$$

with $\boldsymbol{H}$ a (constant) symmetric $n \times n$ matrix, $\boldsymbol{c}$ an $n \times 1$ matrix (column vector) and d a number.[6]

The reader should prove (important exercise!) that $\boldsymbol{\nabla} q(\boldsymbol{x}) = \boldsymbol{H}\boldsymbol{x} + \boldsymbol{c}$ and that the Hessian of q is $\boldsymbol{H}$. These facts will be used extensively during the course.

With only second order terms present a quadratic function is called a *quadratic form*:

$$\frac{1}{2}\boldsymbol{x}^T \boldsymbol{H}\boldsymbol{x}.$$

---

[6] We will often omit d in the future, since it does not affect the minimum point and only adds a constant to the minimum value.

The reader is supposed to be acquainted with the various ways of, by a change of coordinates, writing a quadratic form in *diagonal form*, that is, with square terms only and no cross products. In this form $\boldsymbol{H}$ is a diagonal matrix.

1.3.5 Miscellanous on optimization

It is a well known fact that if a differentiable function f on $\mathbf{R}^n$ has a local minimum at an inner point $\bar{\boldsymbol{x}}$ of its domain then

$$\nabla f(\bar{\boldsymbol{x}}) = \boldsymbol{0}.$$

This is also true if f has a local maximum at $\bar{\boldsymbol{x}}$. However, the gradient may become zero in many other situations as well, for instance at a saddle point in the case of two variables. Hence $\nabla f(\bar{\boldsymbol{x}}) = \boldsymbol{0}$ is a *necessary* but *not sufficient* condition for a local optimum at $\bar{\boldsymbol{x}}$.

A point $\bar{\boldsymbol{x}}$ satisfying $\nabla f(\bar{\boldsymbol{x}}) = \boldsymbol{0}$ is called a **stationary point.**

It is important to note the condition above that $\bar{\boldsymbol{x}}$ should be an inner point of the domain of f. For a general optimization problem with domain defined by conditions (1) we can not expect a minimum to occur at an inner point. In fact, in the presence of an equality constraint $h(\boldsymbol{x}) = 0$ the domain contains no inner points at all!

The conclusion of these observations is that one can not hope to solve optimization problems by the simple procedure of putting the gradient equal to zero and solve the resulting equations. More sophisticated methods are required. This explains the need for courses like the present one.

Now might be a good moment to warn the reader not to expect methods that are general enough to work for any optimization problem. Rather, most methods we discuss are geared at problems having some particular properties, will work well on those but will behave poorly or maybe not succeed at all on others.

Consider now the Taylor expansion (5) of f at a stationary point $\bar{\boldsymbol{x}}$. Since the gradient is zero we have

$$f(\boldsymbol{x}) = f(\bar{\boldsymbol{x}}) + \frac{1}{2}(\boldsymbol{x} - \bar{\boldsymbol{x}})^T \boldsymbol{H}(\bar{\boldsymbol{x}})(\boldsymbol{x} - \bar{\boldsymbol{x}}) + O(\|\boldsymbol{x} - \bar{\boldsymbol{x}}\|^3).$$

If the remainder term can be neglected (this is not always the case), the behaviour of f near $\bar{\boldsymbol{x}}$ is governed by the Hessian $\boldsymbol{H}(\bar{\boldsymbol{x}})$ at $\bar{\boldsymbol{x}}$. The reader should recognize from earlier courses that if this matrix is positive definite[7] then f has a local minimum at $\bar{\boldsymbol{x}}$. This provides a *sufficient* but *not necessary* condition for a local minimum.

The reader should check on the proofs of all the statements above him/herself.

Example 7. Consider in particular a quadratic function

$$q(\boldsymbol{x}) = \frac{1}{2}\boldsymbol{x}^T\boldsymbol{H}\boldsymbol{x} + \boldsymbol{c}^T\boldsymbol{x} + d.$$

We assume no constraints, i.e., the function q has domain $\mathbf{R}^n$. Putting the gradient equal to zero we obtain

$$\boldsymbol{H}\boldsymbol{x} + \boldsymbol{c} = \boldsymbol{0} \quad \Longleftrightarrow \quad \boldsymbol{H}\boldsymbol{x} = -\boldsymbol{c}.$$

From this system of equations we obtain all stationary points of q.

Suppose that $\boldsymbol{H}$ is positive definite. Then it is invertible, and there is only one stationary point

$$\bar{\boldsymbol{x}} = -\boldsymbol{H}^{-1}\boldsymbol{c}.$$

This is a local minimum by the assumption that $\boldsymbol{H}$ is positive definite; in fact it is easily seen to be a *global* minimum point. □

Remark. Using the stationary point $\bar{\boldsymbol{x}}$ we can rewrite $q(\boldsymbol{x})$ in the form

$$q(\boldsymbol{x}) = \frac{1}{2}(\boldsymbol{x} - \bar{\boldsymbol{x}})^T\boldsymbol{H}(\boldsymbol{x} - \bar{\boldsymbol{x}}) + d',$$

where d' is a new constant (in fact $d' = q(\bar{\boldsymbol{x}})$). This is a consequence of Taylor's formula but can be proved directly by simple matrix computations. This is left as an exercise for the reader.

A consequence of this rewriting is that the introduction of new variables $\boldsymbol{y} = \boldsymbol{x} - \bar{\boldsymbol{x}}$ results in a quadratic form

$$\frac{1}{2}\boldsymbol{y}^T\boldsymbol{H}\boldsymbol{y},$$

if we ignore the constant d'.

---

[7]See the appendix for a full discussion of positive definite matrices.

The content of the above remark will be needed later. For future reference we state the result as a lemma.

LEMMA 1. *Let* $q(\boldsymbol{x}) = \frac{1}{2}\boldsymbol{x}^T\boldsymbol{H}\boldsymbol{x} + \boldsymbol{c}^T\boldsymbol{x} + d$ *be a quadratic form, and let* $\bar{\boldsymbol{x}}$ *be a stationary point of* q. *Then* $q(\boldsymbol{x})$ *can be rewritten*

$$q(\boldsymbol{x}) = \frac{1}{2}(\boldsymbol{x} - \bar{\boldsymbol{x}})^T\boldsymbol{H}(\boldsymbol{x} - \bar{\boldsymbol{x}}) + d', \qquad \text{where } d' = q(\bar{\boldsymbol{x}}).$$

From example 7 we know all there is to know about optimization of quadratic functions. In spite of that they will occur frequently in the future. Taylor's formula says somewhat vaguely that near an inner minimum point the behaviour of the function f is very much like that of a certain quadratic function. As a consequence we expect that an algorithm for finding the minimum of a general function in particular should be able to find the minimum of a quadratic one. Hence all methods suggested for finding an optimum should be analysed as to their behaviour on a quadratic function. In fact it is often such analyses that lie behind the inventions of new optimization methods and improvements of existing ones.

Exercises

1.2 a) Find the gradient $\nabla f(\boldsymbol{x})$ of the function $f(\boldsymbol{x}) = \boldsymbol{c}^T\boldsymbol{x}$. (Here $\boldsymbol{c}$ is a constant column matrix.)

b) Find the Jacobian of $\boldsymbol{g}(\boldsymbol{x}) = \boldsymbol{A}\boldsymbol{x}$, where $\boldsymbol{A}$ is a constant $m \times n$-matrix. In particular, what is the Jacobian of $\boldsymbol{x}$?

c) Write down the matrix $\nabla(\nabla^T f)(\boldsymbol{x})$ for a general twice differentiable function $f\colon \mathbf{R}^n \to \mathbf{R}$. Verify that it is symmetric. This matrix is the *Hessian* of f, denoted by $\nabla^2 f(\boldsymbol{x})$.

d) Find the gradient and the Hessian of $f(\boldsymbol{x}) = \boldsymbol{x}^T\boldsymbol{A}\boldsymbol{x}$, where $\boldsymbol{A}$ is a constant (not necessarily symmetric) $n \times n$-matrix.

e) Find the gradient and the Hessian of $f(\boldsymbol{x}) = \frac{1}{2}\boldsymbol{x}^T\boldsymbol{H}\boldsymbol{x} + \boldsymbol{c}^T\boldsymbol{x}$, where $\boldsymbol{H}$ is a constant symmetric $n \times n$-matrix and $\boldsymbol{c}$ is a constant column matrix.

f) Find the gradient and Hessian of $f(\boldsymbol{x}) = \|\boldsymbol{x}\|^2 = \boldsymbol{x}^T\boldsymbol{x}$.

g) For functions $\boldsymbol{g}$ and $\boldsymbol{h}$ from $\mathbf{R}^n$ to $\mathbf{R}^n$, prove the formula

$$\nabla(\boldsymbol{g}^T\boldsymbol{h}) = (\nabla\boldsymbol{g}^T)\boldsymbol{h} + (\nabla\boldsymbol{h}^T)\boldsymbol{g}.$$

(Note the order.) This can be used to give another solution to problems d) and e) above.

1.3 Let $f(\boldsymbol{x}) \in \mathcal{C}^2(\mathbf{R}^n)$ be a scalar function and

$$\boldsymbol{x}(\alpha) = (x_1(\alpha),\ x_2(\alpha),\ \ldots,\ x_n(\alpha))^T$$

a vector valued function whose components are differentiable functions defined on $\mathbf{R}$. Let $\varphi(\alpha) = f(\boldsymbol{x}(\alpha))$.

a) Give an expression for $\varphi'(\alpha)$.

b) Assume in particular that $\boldsymbol{x}(\alpha) = \hat{\boldsymbol{x}} + \alpha\boldsymbol{d}$ (a straight line in $\mathbf{R}^n$). Compute $\varphi'(0)$ and $\varphi''(0)$.

S 1.4 Let $f \in \mathcal{C}^1(\mathbf{R}^n)$ and consider the problem of minimizing f along a line through $\boldsymbol{x}$ with direction $\boldsymbol{d} \neq \boldsymbol{0}$. Put

$$F(\lambda) = f(\boldsymbol{x} + \lambda\boldsymbol{d}).$$

Assume that F has a minimum for $\lambda = \lambda^*$ and let $\boldsymbol{x}^* = \boldsymbol{x} + \lambda^*\boldsymbol{d}$. Prove that $\nabla f(\boldsymbol{x}^*)$ is orthogonal to $\boldsymbol{d}$, i.e., that $\boldsymbol{d}$ is tangent to the level surface of f through $\boldsymbol{x}^*$.

1.5 a) Consider the quadratic function

$$q(\boldsymbol{x}) = \frac{1}{2}\boldsymbol{x}^T\boldsymbol{H}\boldsymbol{x} + \boldsymbol{c}^T\boldsymbol{x} + d,$$

where $\boldsymbol{H}$ is symmetric. Show that the stationary points are given by the solutions to the system of equations

$$\boldsymbol{H}\boldsymbol{x} + \boldsymbol{c} = \boldsymbol{0}.$$

b) Show that $q(\boldsymbol{x})$ can be written

$$q(\boldsymbol{x}) = \frac{1}{2}(\boldsymbol{x} - \bar{\boldsymbol{x}})^T\boldsymbol{H}(\boldsymbol{x} - \bar{\boldsymbol{x}}) + d'$$

where $\bar{\boldsymbol{x}}$ is a stationary point of q. What is d'?

1.6 Let $q(\boldsymbol{x})$ be a quadratic function with Hessian $\boldsymbol{H}$. Denote by $\boldsymbol{\delta}$ the difference $\boldsymbol{x}_1 - \boldsymbol{x}_2$ between any two points and by $\boldsymbol{\gamma}$ the corresponding difference in gradients. Show that $\boldsymbol{\gamma} = \boldsymbol{H}\boldsymbol{\delta}$.

1.7 Let $f\colon \mathbf{R}^n \to \mathbf{R}$, and let $\boldsymbol{g}(\boldsymbol{x}) = \nabla f(\boldsymbol{x})$. Write down the Taylor expansion of the gradient $\boldsymbol{g}(\boldsymbol{a} + \boldsymbol{\delta})$ around $\boldsymbol{a}$, neglecting terms of order $\|\boldsymbol{\delta}\|^2$.

1.8 Determine the character (positive definite, negative definite, ...) of the quadratic forms

a) $x_1^2 + 6x_2^2 + 23x_3^2 + 4x_1x_2 - 6x_1x_3 - 20x_2x_3$,

b) $3x_1^2 + x_2^2 + x_3^2 + 3x_1x_2 + 2x_2x_3$.

S 1.9 For

$$f(\boldsymbol{x}) = 2x_1^3 - 3x_1^2 - 6x_1x_2(x_1 - x_2 - 1),$$

compute the stationary points, and find out whether or not they are local extreme points.

1.10 Compute the stationary points and determine their character for

a) $f(\boldsymbol{x}) = 2x_1^2 + x_1x_2 + x_2^2 + x_2x_3 + x_3^2 - 6x_1 - 7x_2 - 8x_3 + 9$,

b) $f(\boldsymbol{x}) = x_1^4 + 4x_1^2x_2^2 - 2x_1^2 + 2x_2^2 - 1$.

1.11 Let

$$f(\boldsymbol{x}) = (x_1^2 + x_2^2)^2.$$

Verify that $(0, 0)$ is a stationary point and a global minimizer for f, but that $\boldsymbol{\nabla}^2 f(0, 0)$ is not positive definite.

1.12 At a point $\hat{\boldsymbol{x}}$ for which $\hat{\boldsymbol{g}} = \boldsymbol{\nabla} f(\hat{\boldsymbol{x}}) \neq \boldsymbol{0}$, show that f has its maximal slope in direction $\boldsymbol{d} = \hat{\boldsymbol{g}}/\|\hat{\boldsymbol{g}}\|$, over all vectors for which $\|\boldsymbol{d}\| = 1$. (The *steepest ascent* direction.)

Hint: Use the directional derivative and the Cauchy-Schwarz inequality.

1.13 The set $S = \{\, \boldsymbol{x} \in \mathbf{R}^n \,;\, f(\boldsymbol{x}) = C \,\}$, where $C \in \mathbf{R}$, is called a *level set* or *level surface* of f. Let $\hat{\boldsymbol{x}} \in S$. Show that $\boldsymbol{\nabla} f(\hat{\boldsymbol{x}})$ is orthogonal to S and to its tangent plane at $\hat{\boldsymbol{x}}$.

Hint: Let $\boldsymbol{x} = \boldsymbol{x}(\alpha)$ be an arbitrary (differentiable) curve which lies in S and passes through the point $\hat{\boldsymbol{x}}$, i.e., $\boldsymbol{x}(\alpha_0) = \hat{\boldsymbol{x}}$ for some $\alpha_0 \in \mathbf{R}$. Then $f(\boldsymbol{x}(\alpha)) = C$ for all α in a neighbourhood of α_0.

Chapter 2

ONEDIMENSIONAL SEARCH

Most numerical methods for the minimization of a function f of n variables use a strategy somewhat like the following: find a sequence $\boldsymbol{x}_1$, $\boldsymbol{x}_2$, ... of points in $\mathbf{R}^n$ such that

$$f(\boldsymbol{x}_1) \geq f(\boldsymbol{x}_2) \geq \dots ,$$

and stop when convenient. More precisely, they operate by iteration of the following scheme:

- Given a point $\boldsymbol{x}_k$, choose a direction $\boldsymbol{d}_k$.
- Search from $\boldsymbol{x}_k$ in direction $\boldsymbol{d}_k$. Choose a suitable point

  $$\boldsymbol{x}_{k+1} = \boldsymbol{x}_k + \lambda \boldsymbol{d}_k$$

  such that $f(\boldsymbol{x}_{k+1}) \leq f(\boldsymbol{x}_k)$.

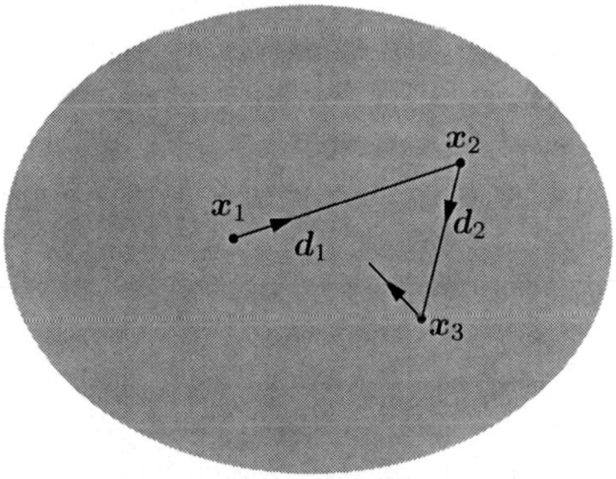

There are at least three problems in connection with this approach.

1. How to choose the search directions $\boldsymbol{d}_1, \boldsymbol{d}_2, \ldots$.

2. How to choose the number λ which appears in the formula for $\boldsymbol{x}_{k+1}$.

3. How to choose a suitable stop criterion.

The second problem is concerned with the minimization of a function of *one* variable $F(\lambda) = f(\boldsymbol{x}_k + \lambda \boldsymbol{d}_k)$ (*onedimensional search* or *line search*). It is to this issue that we first turn our attention. Thus, in this chapter we study various ways of finding the minimum of a one variable function. General multidimensional search will be discussed in chapter 3.

2.1 Line search without the use of derivatives

Let F be a function of one variable λ, defined on a compact interval $a \leq \lambda \leq b$. On F we make the following assumptions:

- F has a minimum at some λ^* in $[a, b]$.

- F is (weakly) decreasing when $a \leq \lambda \leq \lambda^*$, and (weakly) increasing when $\lambda^* \leq \lambda \leq b$.

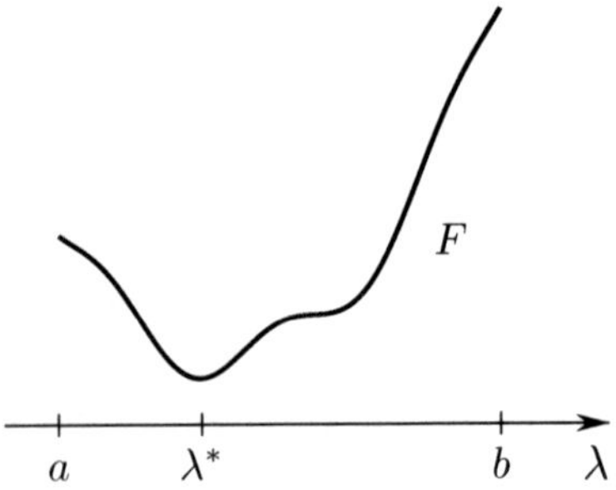

Such a function F is sometimes called *unimodal*.

Our goal is to find an approximation to λ^* by a (hopefully clever) search among the values of F. In this context we introduce the term *interval of uncertainty*, an interval to which λ^* belongs. The initial interval of uncertainty is $[a, b]$, and our objective is to find a much smaller one.

To achieve our goal we must compute the values of F at a number of points. In this context the evaluation of a function at a point is considered an expensive operation, requiring much computer time. Since we expect to have to perform this operation many times during the process, the number of function evaluations required in a certain minimization method is important when judging its merits. The fewer evaluations needed to achieve a certain degree of accuracy the better.

To begin with we study methods which do not use derivatives. There are several reasons for the interest in those. First of all, our objective function F might in fact not be differentiable, or we do not know whether it is differentiable or not. Secondly, even though F is known to be differentiable its derivative might be difficult or impossible to compute. Maybe the function is given by a list of function values (perhaps found by experiments), and we must resort to numerical differentiation. For a third reason, the formula for F' might be so complicated that we prefer not to use it for evaluation.

We begin with a simple example of how one might proceed. Better methods will be discussed afterwards.

Example 1. (*Uniform search.*) Divide the interval $[a, b]$ into subintervals using N equidistant grid points

$$a < \lambda_1 < \lambda_2 < \ldots < \lambda_N < b.$$

Compute the values of F at these points, and find the grid point $\overline{\lambda}$ with the least value. Then one of the two subintervals with endpoint $\overline{\lambda}$ must contain λ^*. Together they constitute a new interval of uncertainty.

Each subinterval of the partition has length $(b-a)/(N+1)$. Thus we

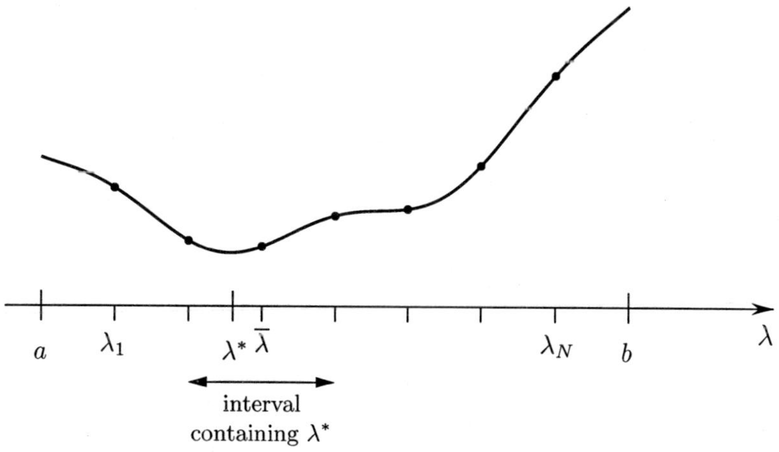

have found a new interval of uncertainty with length

$$L_N = \frac{2}{N+1}(b-a) \tag{1}$$

at the cost of N function evaluations. □

In example 1 we carried out a so called *simultaneous search*; the grid points were chosen beforehand, and then the objective function was evaluated at all points. We now turn our attention to *sequential search*, where we choose points one at a time, and let the outcome of one function evaluation influence the choices that follow.

The methods to be discussed all rely on the following fundamental idea. Compute the values of F at two points λ and μ with

$$a < \lambda < \mu < b.$$

Now distinguish between two cases:

1. Suppose $F(\lambda) > F(\mu)$. Then $[\lambda, b]$ is a new interval of uncertainty.

2. Suppose $F(\lambda) \leq F(\mu)$. Then a new interval of uncertainty is $[a, \mu]$.

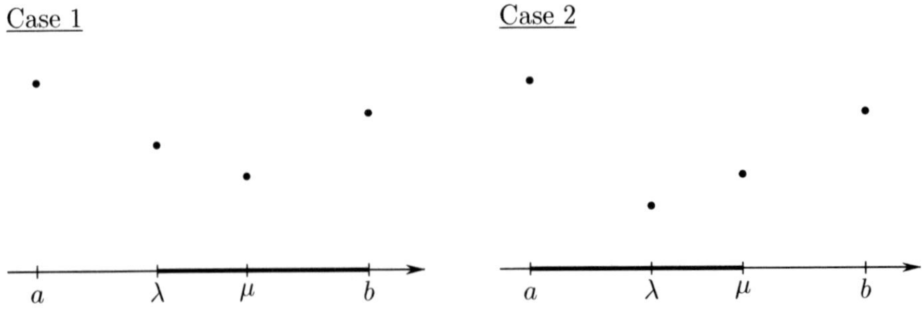

We iterate the procedure outlined above, obtaining a sequence of diminishing intervals of uncertainty $[a_k, b_k]$. Different methods will be characterized by different ways of choosing the numbers λ_k, μ_k in $[a_k, b_k]$.

2.1.1 Dichotomous search

Here we choose both λ_k and μ_k close to the midpoint of $[a_k, b_k]$:

$$\lambda_k = \frac{a_k + b_k}{2} - \varepsilon, \qquad \mu_k = \frac{a_k + b_k}{2} + \varepsilon,$$

where ε is a small number chosen in advance. Ignoring the effect of ε, the length of the interval of uncertainty is roughly halved at every iteration. However, each iteration requires two function evaluations. Hence, if L_N is the length of the uncertainty interval obtained after N function evaluations, then

$$L_N \approx \frac{1}{2^{N/2}}(b_1 - a_1). \tag{2}$$

Note that N must be an even number here.

The next method shows that we can do better.

2.1.2 Golden section search

This method relies on two ideas. The first is that we require the lengths $L_k = b_k - a_k$ of the intervals of uncertainty to decrease at a constant rate:

$$L_k = \alpha L_{k-1}$$

for some constant α, $0 < \alpha < 1$. The second idea is to reuse one of the function evaluations from the latest iteration step, resulting in only one function evaluation being required at each step except the first. To achieve this we must have

in case 1: $\lambda_{k+1} = \mu_k$

in case 2: $\mu_{k+1} = \lambda_k$.

See the figure on the next page, which shows the effect of one iteration step.

The requirement that the length of the interval be reduced by the factor α is equivalent to

$$\begin{aligned} \lambda_k &= a_k + (1-\alpha)(b_k - a_k) \\ \mu_k &= a_k + \alpha\,(b_k - a_k). \end{aligned} \tag{3}$$

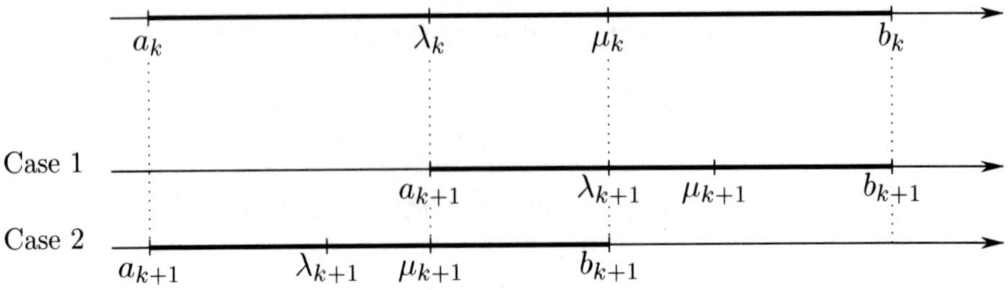

Using these formulas and considering case 1, the condition $\lambda_{k+1} = \mu_k$ can be rewritten as

$$a_{k+1} + (1-\alpha)(b_{k+1} - a_{k+1}) = a_k + \alpha(b_k - a_k).$$

Here $a_{k+1} = \lambda_k$, which is given by (3). Furthermore, $b_{k+1} - a_{k+1} = \alpha(b_k - a_k)$. Hence we obtain

$$a_k + (1-\alpha)(b_k - a_k) + (1-\alpha)\alpha(b_k - a_k) = a_k + \alpha(b_k - a_k).$$

After some cancellations we arrive at the following equation for α:

$$(1-\alpha) + (1-\alpha)\alpha = \alpha,$$

which simplifies to

$$\alpha^2 + \alpha - 1 = 0,$$

a second degree equation with solutions $\alpha = \dfrac{-1 \pm \sqrt{5}}{2}$. We choose the positive root

$$\alpha = \frac{\sqrt{5} - 1}{2} \approx 0.618.$$

This is the famous *golden section number*, from which the method derives its name.

It is left to the reader to check that case 2 gives the same result.

To summarize, the iteration step in the golden section method consists of choosing numbers λ_k and μ_k according to (3), evaluating F at these points and choosing the new interval of uncertainty by the result of the evaluations. The merit of the scheme is that one function evaluation is inherited from earlier iterations, except at the first step when of course two function values must be computed. Thus we have the formula

$$L_N = (b_1 - a_1)\alpha^{N-1} \tag{4}$$

for the length L_N of the uncertainty interval after N function evaluations.

Example 2. Let

$$F(\lambda) = \lambda^2 + e^{-\lambda},$$

and let $[0,1]$ be the initial interval of uncertainty. The table shows the computations when 8 iterations are performed. The resulting interval of uncertainty is $[0.339, 0.361]$.

We might mention that the use of a routine from Matlab's optimization toolbox results in the value 0.3517 for the minimum point.

k	a_k	b_k	λ_k	μ_k	$F(\lambda_k)$	$F(\mu_k)$	Comments
1	0	1	**0.3820**	**0.6180**	**0.8284**	**0.9210**	Since $F(\lambda_1) < F(\mu_1)$ the new interval of uncertainty is $[a_1, \mu_1]$.
2	0	0.6180	**0.2361**	0.3820	**0.8455**	0.8284	New interval of uncertainty is $[\lambda_2, b_2]$. New evaluations are printed in boldface.
3	0.2361	0.6180	0.3820	**0.4721**	0.8284	**0.8466**	New interval of uncertainty is $[a_3, \mu_3]$.
4	0.2361	0.4721	**0.3262**	0.3820	**0.8281**	0.8284	
5	0.2361	0.3820	**0.2918**	0.3262	**0.8321**	0.8281	
6	0.2918	0.3820	0.3262	**0.3475**	0.8281	**0.827208**	
7	0.3262	0.3820	0.3475	**0.3607**	0.827208	**0.827292**	
8	0.3262	0.3607	**0.3394**	0.3475	**0.827390**	0.827208	Final interval of uncertainty is $[\lambda_8, b_8]$.
9	0.3394	0.3607					

□

2.1.3 Fibonacci search

We now consider a variant of the golden section method.

The *Fibonacci numbers* $\{F_k\}_{k=0}^{\infty}$ are defined recursively by the formulas

$$\begin{cases} F_k = F_{k-1} + F_{k-2}, & k \geq 2 \\ F_0 = F_1 = 1. \end{cases} \tag{5}$$

The first ten Fibonacci numbers are

$$1, \quad 1, \quad 2, \quad 3, \quad 5, \quad 8, \quad 13, \quad 21, \quad 34, \quad 55.$$

We use these numbers to modify the algorithm of the golden section method. In (3) we allow the number α to depend on k, by fixing in advance a positive integer n and putting

$$\alpha = \alpha(n,k) = \frac{F_{n-k}}{F_{n-k+1}}.$$

By the recursion formula (5) this makes $1-\alpha = \dfrac{F_{n-k-1}}{F_{n-k+1}}$. Hence (3) is replaced with

$$\begin{aligned} \lambda_k &= a_k + \frac{F_{n-k-1}}{F_{n-k+1}}(b_k - a_k) \\ \mu_k &= a_k + \frac{F_{n-k}}{F_{n-k+1}}(b_k - a_k). \end{aligned} \tag{6}$$

We have now waived the requirement that the width of the intervals diminish at a constant rate. However, the second characteristic of the golden section method, that only one function evaluation is required at each step, is nevertheless valid. In other words we still have

$$\begin{aligned} &\text{in case 1:} \quad \lambda_{k+1} = \mu_k \\ &\text{in case 2:} \quad \mu_{k+1} = \lambda_k. \end{aligned}$$

We give a proof in the first case. (The reader should check the second one.) By (6)

$$\lambda_{k+1} = a_{k+1} + \frac{F_{n-k-2}}{F_{n-k}}(b_{k+1} - a_{k+1}).$$

Here $a_{k+1} = \lambda_k$, which is defined in (6), and

$$b_{k+1} - a_{k+1} = \alpha(n,k)\,(b_k - a_k).$$

Using this we obtain

$$\lambda_{k+1} = a_k + \frac{F_{n-k-1}}{F_{n-k+1}}(b_k - a_k) + \frac{F_{n-k-2}}{F_{n-k}} \frac{F_{n-k}}{F_{n-k+1}}(b_k - a_k) =$$

$$= a_k + \frac{F_{n-k-1} + F_{n-k-2}}{F_{n-k+1}}(b_k - a_k) = a_k + \frac{F_{n-k}}{F_{n-k+1}}(b_k - a_k) = \mu_k.$$

The last but one equality is due to the recursion formula (5) for the Fibonacci numbers.

Example 3. Let $n = 9$. Then

$$\alpha(9, k) = \frac{F_{9-k}}{F_{10-k}}, \qquad 1 - \alpha(9, k) = \frac{F_{8-k}}{F_{10-k}}.$$

The Fibonacci numbers F_k for $k = 0, 1, \ldots, 9$ are given above. The coefficients to be used at the successive iteration steps (6) are as follows.

	$1 - \alpha(9, k)$	$\alpha(9, k)$
$k = 1$	$\frac{F_7}{F_9} = \frac{21}{55}$	$\frac{F_8}{F_9} = \frac{34}{55}$
$k = 2$	$\frac{F_6}{F_8} = \frac{13}{34}$	$\frac{F_7}{F_8} = \frac{21}{34}$
$k = 3$	$\frac{F_5}{F_7} = \frac{8}{21}$	$\frac{F_6}{F_7} = \frac{13}{21}$
$k = 4$	$\frac{F_4}{F_6} = \frac{5}{13}$	$\frac{F_5}{F_6} = \frac{8}{13}$
$k = 5$	$\frac{F_3}{F_5} = \frac{3}{8}$	$\frac{F_4}{F_5} = \frac{5}{8}$
$k = 6$	$\frac{F_2}{F_4} = \frac{2}{5}$	$\frac{F_3}{F_4} = \frac{3}{5}$
$k = 7$	$\frac{F_1}{F_3} = \frac{1}{3}$	$\frac{F_2}{F_3} = \frac{2}{3}$
$k = 8$	$\frac{F_0}{F_2} = \frac{1}{2}$	$\frac{F_1}{F_2} = \frac{1}{2}$

In the last step the coefficients are equal, and hence $\lambda_8 = \mu_8$. At this stage one usually substitutes $\mu_8 + \varepsilon$ for μ_8, where ε is a small number (compare Dichotomous search), and then carries out one further iteration. □

The artificial device in the last step in example 3 is often used, in the general case resulting in that $n-1$ iterations being executed. Consequently the objective function is evaluated n times. In fact it is from this observation that we derive the value to choose for n. (Note that this must be done in advance.)

The lengths of the intervals of uncertainty fulfil

$$b_{k+1} - a_{k+1} = \frac{F_{n-k}}{F_{n-k+1}} (b_k - a_k).$$

From this recursion formula it follows that $b_n - a_n = \dfrac{1}{F_n}(b_1 - a_1)$. If we use the number N of function evaluations as a measure we get

$$L_N = \frac{1}{F_N} (b_1 - a_1). \tag{7}$$

Example 4. Consider again the function $F(\lambda) = \lambda^2 + e^{-\lambda}$ from example 2. To carry out 8 iterations in the Fibonacci method we must choose $n = 9$.

k	a_k	b_k	λ_k	μ_k	$F(\lambda_k)$	$F(\mu_k)$	Comments
1	0	1	**0.3818**	**0.6182**	**0.8284**	**0.9211**	Since $F(\lambda_1) < F(\mu_1)$ the new interval of uncertainty is $[a_1, \mu_1]$.
2	0	0.6182	**0.2364**	0.3818	**0.8454**	0.8284	New interval of uncertainty is $[\lambda_2, b_2]$. New evaluations are printed in boldface.
3	0.2364	0.6182	0.3818	**0.4727**	0.8284	**0.8468**	New interval of uncertainty is $[a_3, \mu_3]$.
4	0.2364	0.4727	**0.3273**	0.3818	**0.8280**	0.8284	
5	0.2364	0.3818	**0.2909**	0.3273	**0.8322**	0.8280	
6	0.2909	0.3818	0.3273	**0.3455**	0.8280	**0.827237**	
7	0.3273	0.3818	0.3455	**0.3636**	0.827237	**0.827375**	
8	0.3273	0.3636	0.3455	**0.3455**+ε	0.827237	**0.827222**	To get a separation in the last iteration we add a small number ε to μ_8. Here we chose $\varepsilon = 0.001$.
9	0.3455	0.3636					This is $[\lambda_8, b_8]$.

The weights to be used are those in example 3. The results obtained are exhibited in the table. The resulting interval of uncertainty is $[0.345, 0.364]$. □

2.1.4 Comparison between methods

Suppose that we require a certain ratio between the length $b_1 - a_1$ of the initial interval of uncertainty and the length L of the final one. Then the number N of function evaluations required in the different methods considered above follow from conditions (1), (2), (4) and (7) respectively, and are given by:

Uniform search $$\frac{N+1}{2} \geq \frac{b_1 - a_1}{L}.$$

Dichotomous search $$\sqrt{2}^N \geq \frac{b_1 - a_1}{L}, \qquad N \text{ even}.$$

Golden section search $$\left(\frac{1}{\alpha}\right)^{N-1} \geq \frac{b_1 - a_1}{L}.$$

Fibonacci search $$F_N \geq \frac{b_1 - a_1}{L}.$$

It can be proved that to the nearest integer

$$F_n \approx \frac{1}{\sqrt{5}} \left(\frac{1+\sqrt{5}}{2}\right)^{n+1} = \frac{3+\sqrt{5}}{2\sqrt{5}} \left(\frac{1}{\alpha}\right)^{n-1} \qquad \text{for large } n,$$

where α is the golden section number. Consequently Fibonacci search behaves somewhat better than the golden section method, but they are asymptotically equal. Both are superior to the other two methods when we compare the number of function evaluations required to achieve a certain accuracy.

Exercises

S 2.1 Find the minimum of the function $F(\lambda) = 1 - \lambda e^{-\lambda}$

a) using golden section search with 4 iterations,

b) using Fibonacci search with 4 iterations.

In both cases start with the interval $[0, 2]$.

2.2 a) Show that $f(x) = 2x^2 + 3e^{-x}$ has exactly one minimum between 0 and 1. Compute an interval of uncertainty of length ≤ 0.1 using the golden section search method.

b) How many function value evaluations did you have to carry out above? How many would have been required if the dichotomous search method had been used?

(1997–12–17, 5)

2.3 Find the minimum of $F(\lambda) = e^{-\lambda} + \lambda^2$ by each of the following procedures.

a) Golden section method. Start from the interval $[-1, 1]$ and carry out 5 iterations.

b) Dichotomous search. Start from the same interval, choose $\varepsilon = 0.01$, carry out 5 iterations. — Now we get a smaller interval than by the golden section method. Why is this no contradiction to the results in subsection 2.1.4?

2.4 Consider the function f of two variables defined by

$$f(\boldsymbol{x}) = (x_1^3 + x_2)^2 + 2(x_2 - x_1 - 4)^4.$$

Given a point $\hat{\boldsymbol{x}}$ and a nonzero direction $\boldsymbol{d}$, let $F(\lambda) = f(\hat{\boldsymbol{x}} + \lambda \boldsymbol{d})$.

a) Obtain an explicit expression for $F(\lambda)$.

b) For $\hat{\boldsymbol{x}} = (0, 0)$ and $\boldsymbol{d} = (1, 1)$, using the Fibonacci method, find the value of λ that solves the problem to minimize $F(\lambda)$ subject to $\lambda \in \mathbf{R}$. (Start with the interval $[-2, 2]$ and choose 5 iterations.)

c) For $\hat{\boldsymbol{x}} = (4, 5)$ and $\boldsymbol{d} = (1, -2)$, using the golden section method, find the value of λ that solves the problem to minimize $F(\lambda)$ subject to $\lambda \in \mathbf{R}$. (Start with the interval $[-2, 2]$. Carry out 5 iterations.)

2.5 Let q denote the ratio of the length of the final interval of uncertainty to the length of the initial interval of uncertainty.

For 1° the uniform search method, 2° the dichotomous search method, 3° the golden section method, and 4° the Fibonacci search method, compute the number of function evaluations required for

$$q = 0.1, \quad 0.01, \quad 0.001 \quad \text{and} \quad 0.0001.$$

2.2 Line search using derivatives

We are now going to discuss two line search methods in which the derivatives of the objective function F come into use.

2.2.1 The bisection method

We assume that F belongs to class $\mathcal{C}^1$, with F' monotone. In the iteration step, starting from an interval of uncertainty $[a_k, b_k]$, we compute the midpoint $\lambda_k = \frac{a_k + b_k}{2}$ and evaluate F' at λ_k. If $F'(\lambda_k) > 0$ we conclude that the minimum point belongs to the left half of the interval, otherwise to the right half. Thus we distinguish two cases:

1. Suppose $F'(\lambda_k) \leq 0$. Then $[\lambda_k, b_k]$ is a new interval of uncertainty.

2. Suppose $F'(\lambda_k) > 0$. Then a new interval of uncertainty is $[a_k, \lambda_k]$.

The figure shows the effect of one iteration step.

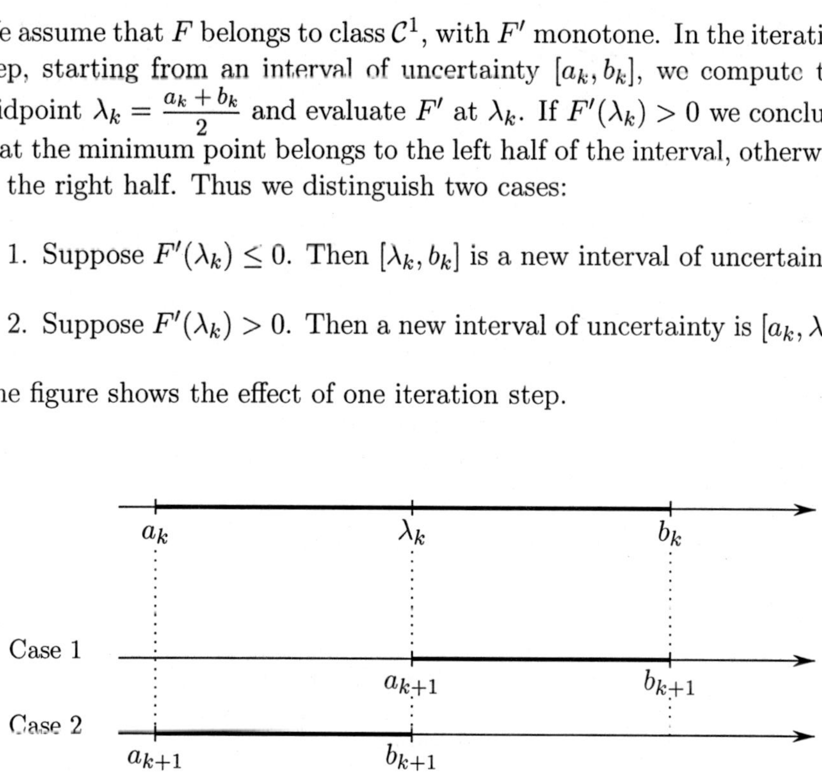

Obviously the length of the interval of uncertainty is halved at each iteration step. Every iteration step requires one function evaluation (of the derivative). Hence the length L_N after N function evaluations is

$$L_N = \frac{1}{2^N}\,(b_1 - a_1).$$

This is better than the non-derivative methods above.

2.2.2 Newton's method

In contrast to the previous ones, Newton's method does not work by construction of a sequence of diminishing intervals of uncertainty. Instead we generate a sequence $\{\lambda_k\}_1^\infty$ of numbers which is expected to converge to the minimum point λ^*.

Assume that the function F is twice continuously differentiable on $\mathbf{R}$. Then, by Taylor's formula, near λ_k we can approximate $F(\lambda)$ by the second degree polynomial

$$q(\lambda) = F(\lambda_k) + F'(\lambda_k)(\lambda - \lambda_k) + \frac{1}{2}F''(\lambda_k)(\lambda - \lambda_k)^2.$$

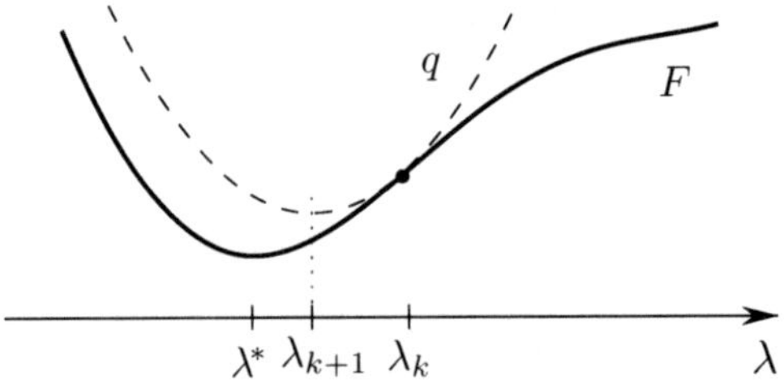

The fundamental idea is to minimize q instead of F. This can be done simply by setting the derivative of q equal to zero, resulting in the first degree equation

$$F'(\lambda_k) + F''(\lambda_k)(\lambda - \lambda_k) = 0$$

for λ. We define λ_{k+1} to be the solution of this equation, i.e.

$$\lambda_{k+1} = \lambda_k - \frac{F'(\lambda_k)}{F''(\lambda_k)}. \tag{8}$$

(Assume that $F''(\lambda_k) \neq 0$.) Starting from an initial point λ_1 a sequence $\{\lambda_k\}_1^\infty$ is generated by iteration of (8).

The reader might recognize that this procedure is in fact the well known *Newton-Raphson method* for solving the equation $F'(\lambda) = 0$.

The methods discussed previously obviously have the property that they converge to a minimum in the long run. In Newton's method there is no guarantee for that. In fact, starting at a discretionary point, you may well obtain a sequence tending to infinity, or oscillating between two points, or converging to another solution to $F'(\lambda) = 0$ than the one of interest. Also, if there is a limit it might not be a minimum point. This kind of behaviour is well known from the Newton-Raphson method for solving equations.

We will now prove that if the initial point λ_1 is chosen *sufficiently close* to a minimum λ^* the generated sequence will in fact converge to λ^*. We begin by studying the effect of one iteration step.

LEMMA 1. *Assume that $F \in C^3(\mathbf{R})$.*[1] *Let λ^* be a minimum point for F, and assume that $F''(\lambda^*) \neq 0$. Choose the number ε so that there are constants A and B making*

$$\frac{1}{|F''(\lambda)|} \leq A \quad \text{when } |\lambda - \lambda^*| < \varepsilon$$

$$|F'''(\lambda)| \leq 2B \quad \text{when } |\lambda - \lambda^*| < \varepsilon.$$

Suppose that λ_k belongs to the interval $|\lambda - \lambda^*| < \varepsilon$, and let λ_{k+1} be defined by the recursion formula (8). Then

$$|\lambda_{k+1} - \lambda^*| \leq AB|\lambda_k - \lambda^*|^2. \tag{9}$$

PROOF. Since λ^* is a minimum point $F'(\lambda^*) = 0$. Consequently, using the recursion formula (8),

$$|\lambda_{k+1} - \lambda^*| = \left| \lambda_k - \frac{F'(\lambda_k)}{F''(\lambda_k)} - \lambda^* + \frac{F'(\lambda^*)}{F''(\lambda_k)} \right|.$$

This in turn can be rewritten as

$$|\lambda_{k+1} - \lambda^*| = \frac{1}{|F''(\lambda_k)|} \left| F'(\lambda^*) - F'(\lambda_k) - F''(\lambda_k)(\lambda^* - \lambda_k) \right|.$$

We now make use of Taylor's formula for the derivative F' at λ_k:

$$F'(\lambda) = F'(\lambda_k) + F''(\lambda_k)(\lambda - \lambda_k) + \frac{1}{2} F'''(\xi)(\lambda - \lambda_k)^2,$$

---

[1] It is unnecessarily restrictive to assume that F is three times differentiable. However, this assumption will make the proof more transparent. We refer the reader to other literature for a more general statement. Also compare with the multidimensional version of the theorem, page 64.

for some ξ between λ and λ_k. Putting $\lambda = \lambda^*$ and inserting above we obtain

$$|\lambda_{k+1} - \lambda^*| = \frac{1}{|F''(\lambda_k)|}\frac{1}{2}|F'''(\xi)|\,|\lambda_k - \lambda^*|^2.$$

Now (9) follows from the assumptions of the lemma. □

THEOREM 1. *We use the notation and assumptions of lemma 1. If the initial point λ_1 is sufficiently close to λ^* then the sequence $\{\lambda_k\}_1^\infty$ generated by (8) converges to λ^*.*

PROOF. By choosing ε in lemma 1 small enough we may assume that $c = AB\varepsilon < 1$. From (9) it in particular follows that if $|\lambda_k - \lambda^*| < \varepsilon$ then

$$|\lambda_{k+1} - \lambda^*| \le AB\varepsilon|\lambda_k - \lambda^*| = c|\lambda_k - \lambda^*| < |\lambda_k - \lambda^*|.$$

Hence, if λ_1 belongs to the interval $|\lambda - \lambda^*| < \varepsilon$ then all subsequent elements of the sequence belong to this interval. Furthermore we have by iteration that

$$|\lambda_k - \lambda^*| < c^{k-1}|\lambda_1 - \lambda^*| \to 0 \quad \text{as } k \to \infty.$$

□

Remark. Consider again inequality (9). With the square in the right hand side it states that λ_k converges *quadratically* to λ^*. For the previous methods the reader may verify that we can prove *linear convergence* only:

$$|\lambda_{k+1} - \lambda^*| \le C|\lambda_k - \lambda^*|, \qquad C \text{ some constant.}$$

Convergence rate is discussed more thoroughly in appendix 1 to chapter 3, page 103.

The quadratic convergence of Newton's method implies very fast convergence indeed in the vicinity of a minimum point. Only a few iterations are needed to achieve an accurate estimate of the limit.

Example 5. Consider once more the function

$$F(\lambda) = \lambda^2 + e^{-\lambda}.$$

The derivatives are

$$F'(\lambda) = 2\lambda - e^{-\lambda}, \qquad F''(\lambda) = 2 + e^{-\lambda}.$$

The recursion formula (8) in this case becomes

$$\lambda_{k+1} = \lambda_k - \frac{2\lambda_k - e^{-\lambda_k}}{2 + e^{-\lambda_k}}.$$

Starting at $\lambda_1 = 0$ we successively obtain the figures in the table.

k	1	2	3	4
λ	0	0.3333	0.3517	0.3517
$F(\lambda_k)$	1	0.8276	0.8272	0.8272

After three iterations we have arrived at the minimum point 0.3517. □

Example 6. Here is an interesting example which shows one thing that might occur if you start too far from the minimum.

Let

$$F(\lambda) = 25\lambda^4 - 25\lambda^3 - 14\lambda^2 + 21\lambda.$$

Then

$$F'(\lambda) = 100\lambda^3 - 75\lambda^2 - 28\lambda + 21, \qquad F''(\lambda) = 300\lambda^2 - 150\lambda - 28,$$

and the recursion formula takes the form

$$\lambda_{k+1} = \lambda_k - \frac{100\lambda_k^3 - 75\lambda_k^2 - 28\lambda_k + 21}{300\lambda_k^2 - 150\lambda_k - 28}.$$

Starting at $\lambda_1 = 0$ we obtain $\lambda_2 = \frac{3}{4}$. In the next iteration step we find that $F'(\frac{3}{4}) = 0$, making $\lambda_3 = \lambda_2$. We have reached a stationary point at λ_2, and the process has terminated. Since $F''(\lambda_2) = \frac{113}{4} > 0$ the function has a local minimum at λ_2.

However, this minimum is not global. In fact we have moved *upwards* from λ_1, since

$$F(0) = 0 \quad \text{and} \quad F(\tfrac{3}{4}) = \tfrac{1341}{256} > 0.$$

Note also that $F'(0) = 21 > 0$. The true global minimum of F is to the left of λ_1, at $\lambda = -\sqrt{7}/5$, as the reader might verify.

The figure on the next page shows the graph of the function, and that of the approximating second degree polynomial at $\lambda_1 = 0$. □

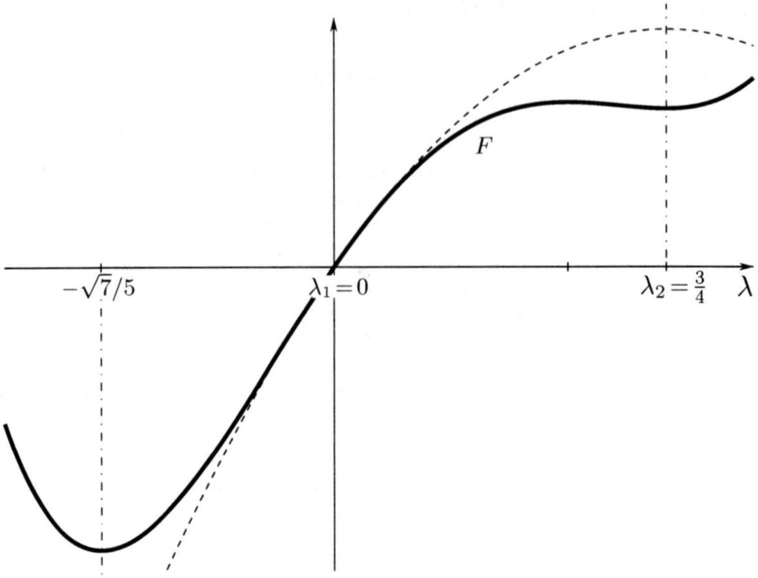

Exercises

2.6 Find the minimum of $F(\lambda) = e^{-\lambda} + \lambda^2$ by each of the following procedures.

a) Newton's method. Start from $\lambda = 1$.

b) Bisecting search method. Start from the interval $[-1, 1]$ and carry out 5 iterations.

Compare with the results of exercises 2.3.

S 2.7 Use Newton's method to find the minimum point of

$$F(\lambda) = e^{\lambda^2} - 2\lambda.$$

Start at $\lambda = 0$.

2.3 Inexact line search

The most frequently used methods of those mentioned above seem to be the golden section search and the bisection method.

When the underlying problem is to minimize a function of several variables, and line search is used as a subroutine in another optimization algorithm, it is often not worth while to carry out all iterations required to obtain convergence (i.e., to be very close to the optimum). The price to pay for the many function evaluations is too high, and it is better to stop early, choose another search direction and start on a new line search. In these cases a rather loose stop criterion might be used, perhaps founded on the relative closeness to the minimum point.

Extending this idea, one sometimes takes a single step only in the search direction. Then you have to choose the one step length only. This choice of course requires some care; it should not be too long, running the risk of overstepping the minimum point, and it should not be unnecessarily small, since after all the goal is to make a substantial step towards the minimum point.

We describe one popular method by Armijo for the choice of a suitable step length for a one-step search method.

Armijo's rule

The method uses two parameters, $0 < \varepsilon < 1$ and $\alpha > 1$, which are to be chosen in advance.

Let F be a differentiable function of one variable, and assume that $F'(0) < 0$. Starting at $\lambda = 0$ we wish to choose a point $\overline{\lambda}$ to the right as a (coarse) approximation of a minimum point.

Put

$$T(\lambda) = F(0) + \varepsilon\lambda F'(0).$$

If $\varepsilon = 1$ this function would represent the tangent line at 0 to the curve $y = F(\lambda)$. However $0 < \varepsilon < 1$, which means that there is a slight overshoot. If the curve meets the line $y = T(\lambda)$ somewhere, we may or may not have passed a minimum. In any case, we consider this intersection an upper limit for the choice of $\overline{\lambda}$. More precisely, we consider those $\overline{\lambda}$ acceptable as a step length for which

$$F(\overline{\lambda}) \leq T(\overline{\lambda}). \tag{10}$$

However, $\overline{\lambda}$ should not be chosen too small. This we achieve by the further requirement that

$$F(\alpha\overline{\lambda}) \geq T(\alpha\overline{\lambda}). \tag{11}$$

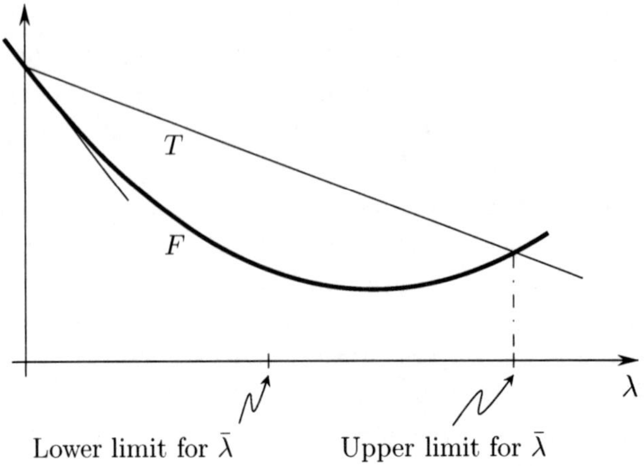

The figure shows the case when $\alpha = 2$.

Armijo's rule consists of choosing a number $\overline{\lambda}$ satisfying these two requirements.

In practical implementations an initial step length $\overline{\lambda}$ is given beforehand. If this length satisfies (10) and (11) it is accepted and chosen as the step to take. If $\overline{\lambda}$ proves to be too large, that is, if (10) is not satisfied, it is divided by 2 (assuming $\alpha = 2$) and tested for acceptance anew. If it is still not satisfactory a new division by 2 is made, and so on. If the initial step length is too small, so that (11) is not satisfied, $\overline{\lambda}$ is doubled iteratively until an acceptable value is arrived at.

Chapter 3

MULTIDIMENSIONAL SEARCH

3.1 Introduction

The reader should be aware that the problem of finding a minimum of a function of several variables is a difficult one. Several local minima may exist (or none); neither of them might be a global one. Questions of this kind can of course be answered only by a careful analysis of the actual problem at hand. Even disregarding complications of this nature it is only in simple cases that the minimum can be found exactly. Hence one usually has to resort to numerical methods. From a general viewpoint these often consist of some scheme for wandering about in the domain of the objective function, trying to find smaller and smaller function values. Of course there are more or less clever ways of doing this, which is why some theory is needed to support the different methods. Also, different applications require different kinds of cleverness. Sometimes it is a question of computer storage, sometimes running time is the issue, sometimes other properties are the crucial ones. There is no general method that will work for all kinds of problems; rather, most methods are designed to work well in certain applications but may be useless in other cases. Commercially available programs often have the ability to switch between different methods when circumstances demand it.

It should perhaps also be emphasized that the choice of an optimization method for a certain application should be supported not only by theory but also by careful testing of realistic problems.

In the following sections some optimization methods will be presented. The aim is not to describe in detail any of the more complicated methods used in practice. Instead we wish to promote an understanding of how such methods work by the study of some simple schemes whose characteristics are not obscured by too many messy details.

The various methods in existence to handle the numerical problems which arise in the carrying out of computations constitute an important subject in itself, and will not be touched upon here.

The optimization methods we discuss will (with one exception) all work by iteration of the following steps.

- Given the current point $\boldsymbol{x}_k$, choose a direction $\boldsymbol{d}_k$.
- Search from $\boldsymbol{x}_k$ in direction $\boldsymbol{d}_k$. Choose a suitable

$$\boldsymbol{x}_{k+1} = \boldsymbol{x}_k + \lambda_k \boldsymbol{d}_k$$

  such that $f(\boldsymbol{x}_{k+1}) \leq f(\boldsymbol{x}_k)$.

The differences between various methods lie essentially in the way the search directions $\boldsymbol{d}_k$ are chosen.

The value of λ_k is often decided by minimizing f along the line from $\boldsymbol{x}_k$ in direction $\boldsymbol{d}_k$. In one case (Newton's method) a unit step in direction $\boldsymbol{d}_k$ is taken, i.e., we choose $\lambda_k = 1$. Some ways of minimizing functions of one variable have been discussed in chapter 2. In this chapter we are going to assume that an *exact* line search is carried out, that is, a λ_k is chosen which solves the problem

$$\min_{\lambda} f(\boldsymbol{x}_k + \lambda \boldsymbol{d}_k)$$

exactly. While this of course is not realistic in practice[1], it serves the double purpose of concentrating the discussion to the more pertinent question of the pure multidimensional properties of the methods, and to simplify the theoretical discussion.

We will in general not discuss the convergence properties of the methods that follow. These issues will be taken up in chapter 10 in connection with the theory of algorithms.

The following consequence of an exact line search is important.

[1] It is perhaps not even desirable; see the discussion of Armijo's rule, section 2.3.

LEMMA 1. *Assume that f is a differentiable function of n variables. Let $F(\lambda) = f(\boldsymbol{x}+\lambda\boldsymbol{d})$, where $\boldsymbol{x}$ is a point in the domain of f and $\boldsymbol{d} \neq \boldsymbol{0}$ a direction in $\mathbf{R}^n$. Assume that $\bar{\boldsymbol{x}} = \boldsymbol{x} + \overline{\lambda}\boldsymbol{d}$ belongs to the domain of f. If $F'(\overline{\lambda}) = 0$ (in particular, if F has a local minimum at $\overline{\lambda}$) the vectors $\nabla f(\bar{\boldsymbol{x}})$ and $\boldsymbol{d}$ are orthogonal.*

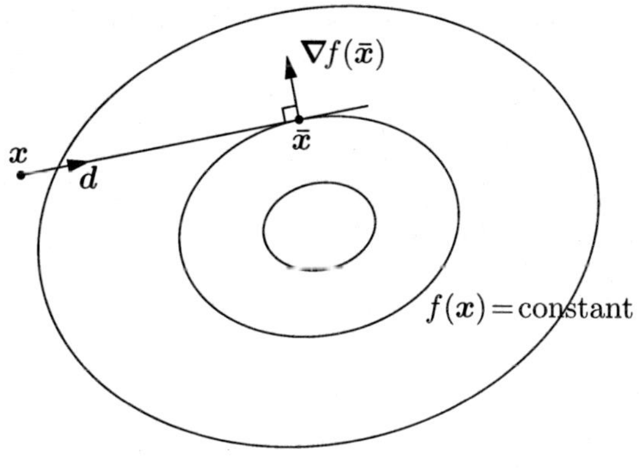

PROOF. By the chain rule

$$F'(\lambda) = \nabla f(\boldsymbol{x}+\lambda\boldsymbol{d})^T\boldsymbol{d}.$$

From the assumption that $F'(\overline{\lambda}) = 0$ it follows that $\nabla f(\bar{\boldsymbol{x}})^T\boldsymbol{d} = 0$, which was to be proved. □

One implication of the lemma is that an exact line search along a direction $\boldsymbol{d}$ ends up at a point $\bar{\boldsymbol{x}}$ where $\boldsymbol{d}$ is tangent to the level surface of f.

Exercise

S 3.1 Consider the problem of minimizing a quadratic function

$$q(\boldsymbol{x}) = \frac{1}{2}\boldsymbol{x}^T\boldsymbol{H}\boldsymbol{x} + \boldsymbol{c}^T\boldsymbol{x}$$

($\boldsymbol{H}$ is positive definite). Suppose that we search along the line from $\boldsymbol{x}_k$ in direction $\boldsymbol{d}_k$. Find that λ_k which minimizes $\lambda \mapsto q(\boldsymbol{x}_k + \lambda\boldsymbol{d}_k)$ exactly.

3.2 Search without using derivatives

The reasons for not using derivatives are the same as in the onedimensional case, section 2.1; it might not be clear whether they exist, the formulas for the derivatives may not be available, or they may be too complicated to be of any use.

In the absence of any information from the derivatives, starting from a point $\boldsymbol{x}_1$ in $\mathbf{R}^n$ there is nothing to suggest some direction in which it might be profitable to search for a minimum. Then we might as well search parallel to a coordinate axis, i.e., vary only one coordinate at a time. This is how the method of *cyclic coordinates* works.

The method of cyclic coordinates

Let

$$\boldsymbol{e}_1 = (1, 0, \ldots, 0), \quad \boldsymbol{e}_2 = (0, 1, 0, \ldots, 0), \quad \ldots, \quad \boldsymbol{e}_n = (0, \ldots, 0, 1)$$

denote the canonical basis vectors in $\mathbf{R}^n$. In the algorithm of cyclic coordinates there is one inner loop and one outer loop. The inner loop consists of successively using these n vectors as search directions. The outer loop repeats this procedure until the decision to terminate is taken.

More precisely, assume that an initial point $\boldsymbol{x}_1$ is chosen. The algorithm of cyclic coordinates runs as follows.

For $k := 1$ *until termination do*

- *Put* $\boldsymbol{y}_1 := \boldsymbol{x}_k$.
- *For* $j := 1$ *until* n *do*
  - *Let* λ_j *be the solution to the onedimensional problem*

    $$\min_{\lambda} f(\boldsymbol{y}_j + \lambda \boldsymbol{e}_j).$$

  - *Put* $\boldsymbol{y}_{j+1} := \boldsymbol{y}_j + \lambda_j \boldsymbol{e}_j$.
- *Let* $\boldsymbol{x}_{k+1} := \boldsymbol{y}_{n+1}$.
- *Decide whether to terminate or continue.*

We will now see how this works out in practice.

Example 1. We demonstrate the behaviour of the algorithm on the objective function of two variables

$$f(x_1, x_2) = 2(x_1 - 2)^4 + (2x_1 - 3x_2)^2.$$

The exact minimum point obviously is $(2, \frac{4}{3})$, and the minimum value is 0.

We choose the initial point $(0, 2.5)$ and carry out 8 iterations (each with 2 inner iterations). The numerical results are displayed in the following table. The path taken by the sequence $x_1, \ldots, x_9$ in the plane is shown in the following figure together with the level curves of f.[2] Note the zigzagging movement. It requires about 75 (outer) iterations to reach a point within the distance of 0.1 from the true minimum point $(2, \frac{4}{3})$. For the line search a Matlab minimization routine was used.

k	x_k	$f(x_k)$	j	e_j	y_j	$f(y_j)$
1	(0 , 2.5000)	88.2500	1	$(1,0)$	(2.9343, 2.5000)	4.1854
			2	$(0,1)$	(2.9343, 1.9562)	1.5241
2	(2.9343, 1.9562)	1.5241	1	$(1,0)$	(2.6543, 1.9562)	0.6802
			2	$(0,1)$	(2.6543, 1.7695)	0.3665
3	(2.6543, 1.7695)	0.3665	1	$(1,0)$	(2.5165, 1.7695)	0.2183
			2	$(0,1)$	(2.5165, 1.6777)	0.1423
4	(2.5165, 1.6777)	0.1423	1	$(1,0)$	(2.4345, 1.6777)	0.0982
			2	$(0,1)$	(2.4345, 1.6230)	0.0713
5	(2.4345, 1.6230)	0.0713	1	$(1,0)$	(2.3797, 1.6230)	0.0536
			2	$(0,1)$	(2.3797, 1.5865)	0.0416
6	(2.3797, 1.5865)	0.0416	1	$(1,0)$	(2.3403, 1.5865)	0.0330
			2	$(0,1)$	(2.3403, 1.5602)	0.0268
7	(2.3403, 1.5602)	0.0268	1	$(1,0)$	(2.3104, 1.5602)	0.0221
			2	$(0,1)$	(2.3104, 1.5403)	0.0186
8	(2.3104, 1.5403)	0.0186	1	$(1,0)$	(2.2868, 1.5403)	0.0158
			2	$(0,1)$	(2.2868, 1.5245)	0.0135
9	(2.2868, 1.5245)	0.0135				

[2] The intervals of the level curves are approximately logarithmic.

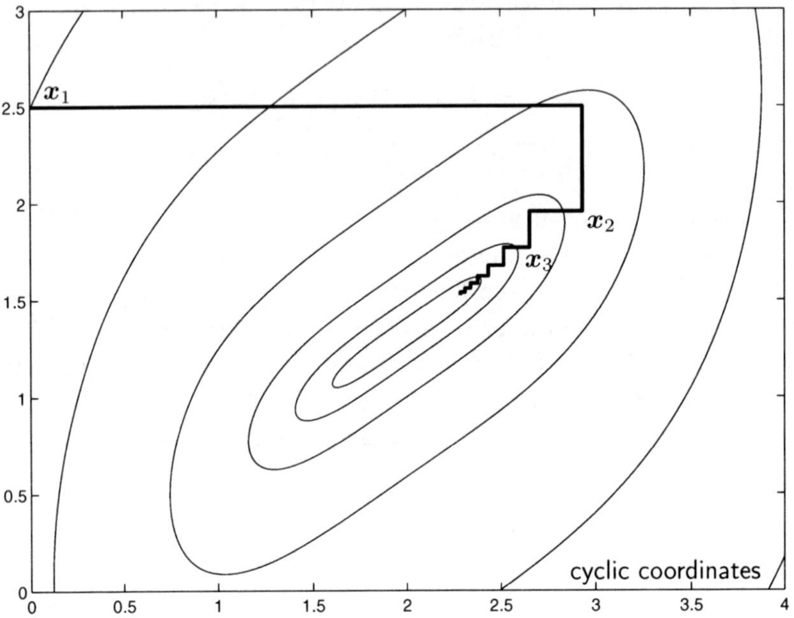

□

It can be shown that if f is differentiable the generated sequence $\{x_k\}_1^\infty$ converges to a stationary point under some general assumptions. However, if f is not differentiable the search may end at a non-stationary point. The figure below shows level curves of a function f of two vari-

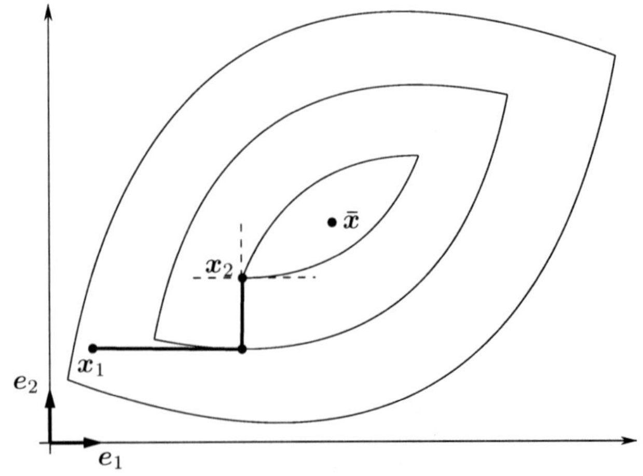

ables, and how the generated sequence gets stuck at the non-optimal point x_2.

The difficulty here could of course be overcome by the introduction of a new search direction. Even in the differentiable case a change of search direction (called an *acceleration step* or a *pattern search step*) is often introduced regularly in the cyclic coordinate method. We give some examples, without going into details.

Example 2. In the *method of Hooke and Jeeves*, from x_2 we search in direction $x_2 - x_1$. Call the resulting point z_2. After that another cycle of search along the basis directions follows, starting from z_2 and resulting in a point x_3. The next step is taken in the direction $x_3 - x_2$, and is followed by another sequence of cyclic coordinates search, and so on.

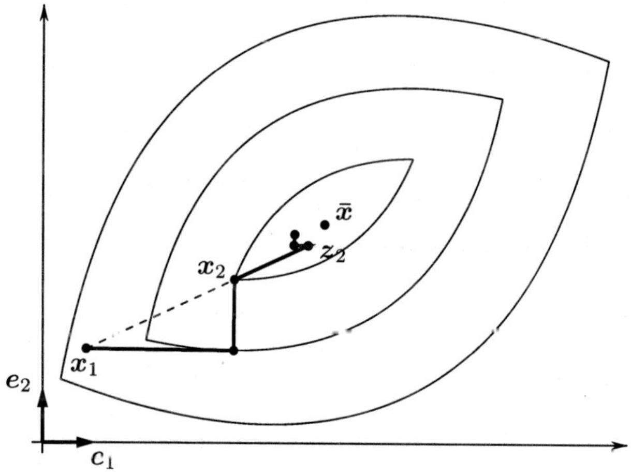

□

Here is another variant.

Example 3. In the *method of Rosenbrock* a new set of basis vectors is computed[3] after each cycle, with the direction $x_{k+1} - x_k$ as the first vector. The new basis is used as search directions in the next cycle.

---

[3]This can be done by the *Gram–Schmidt orthogonalization process.* See a course in *Linear algebra.*

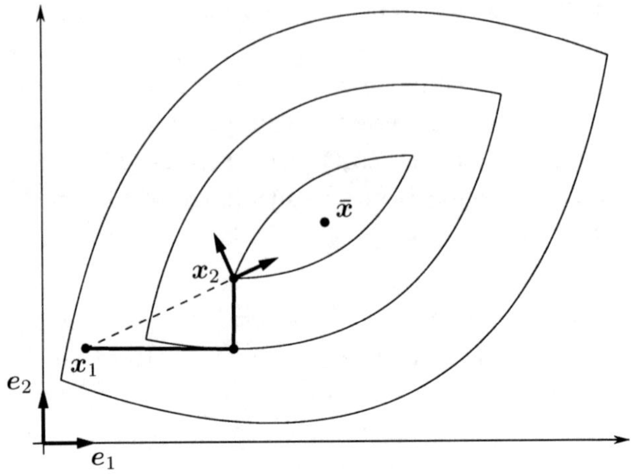

□

3.3 The steepest descent method

From now on we assume that the function f to be minimized is of class $\mathcal{C}^1$, that is, continuously differentiable. It is a well known fact that at a given point $\boldsymbol{x}$ the gradient vector $\boldsymbol{\nabla} f(\boldsymbol{x})$ indicates the direction from $\boldsymbol{x}$ in which f has its largest growth. We call it the direction of *steepest*

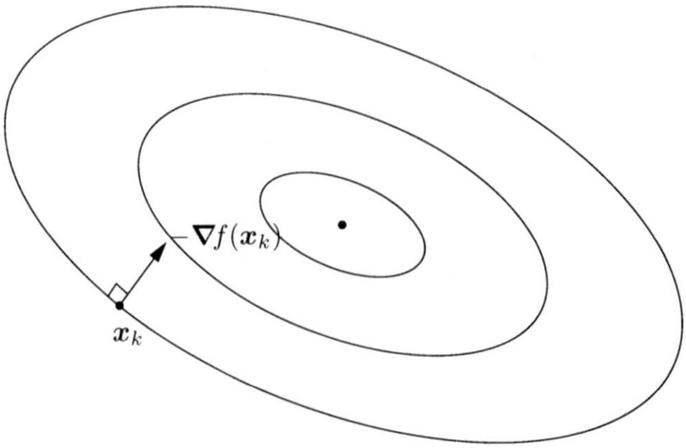

ascent. In consequence, $-\nabla f(\boldsymbol{x})$ is the direction in which f has its maximal descent (the direction of *steepest descent*). This is of course an obvious choice for a search direction in a minimizing algorithm. The *steepest descent method* (SD for short) works according to this principle, by iteration of

- choose $\boldsymbol{d}_k = -\nabla f(\boldsymbol{x}_k)$,
- let $\boldsymbol{x}_{k+1} = \boldsymbol{x}_k + \lambda_k \boldsymbol{d}_k$, where λ_k solves $\min_{\lambda} f(\boldsymbol{x}_k + \lambda \boldsymbol{d}_k)$.

However, in spite of its apparent attractiveness the method has certain drawbacks, as we shall see in a moment. Nevertheless it is used in some applications. Also, most existing methods are in some sense a refinement of it. Thus, understanding of the method is important, and we shall devote some time to its study.

Example 4. We use the steepest descent method on the same objective function as in example 1, with the same initial point $(0, 2.5)$. The numerical results are shown in the table on the next page, and the figure shows the movement in the x_1x_2-plane. After 8 iterations we reach the point (2.1663, 1.4439).

The convergence is slow except for the first step, and can not be seen on this scale. A magnification (next figure, page 51) shows a zigzagging motion towards the minimum which is typical of the steepest descent method. In the case of two variables this movement can be understood directly from lemma 1. A consequence of the result there is that only two search directions will appear, and that these are orthogonal. We will explain this in greater detail in the next example.

To obtain a point with a distance less than 0.1 to the exact minimum point $(2, \frac{4}{3})$ would require around 130 iterations in this case. However, the minimum *value* is rather close to zero already after the first 8 iterations.

For the problem of finding the minimum point numerically this function f is in fact considered a more than normally difficult case, due to the long flat valley that can be discerned in the level curves.

k	$\boldsymbol{x}_k$	$f(\boldsymbol{x}_k)$	$\boldsymbol{d}_k$
1	(0 , 2.5)	88.2500	(94 , −45)
2	(2.1799, 1.4564)	0.0022	(−0.0083, −0.0574)
3	(2.1793, 1.4525)	0.0021	(−0.0510, 0.0074)
4	(2.1752, 1.4531)	0.0020	(−0.0077, −0.0530)
5	(2.1746, 1.4494)	0.0019	(−0.0472, 0.0068)
6	(2.1708, 1.4499)	0.0018	(−0.0071, −0.0491)
7	(2.1703, 1.4465)	0.0017	(−0.0437, 0.0063)
8	(2.1667, 1.4470)	0.0016	(−0.0066, −0.0457)
9	(2.1663, 1.4439)	0.0015	

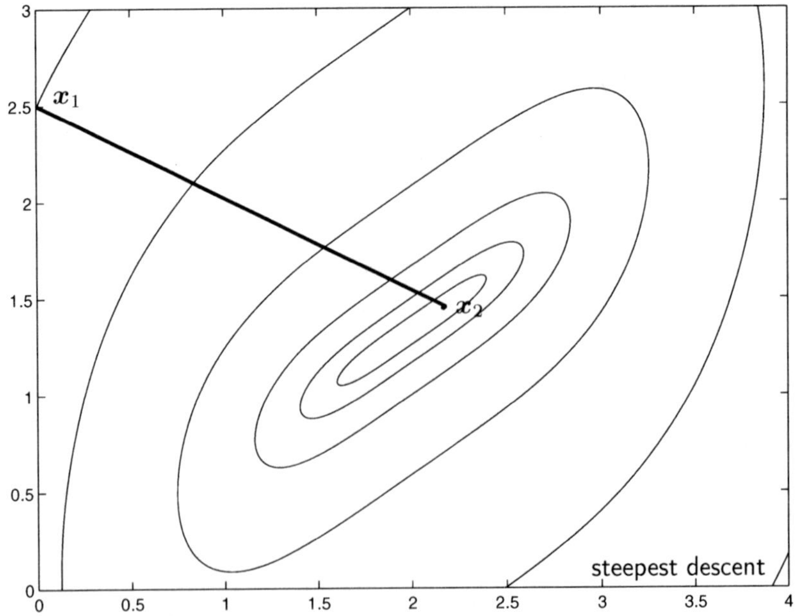

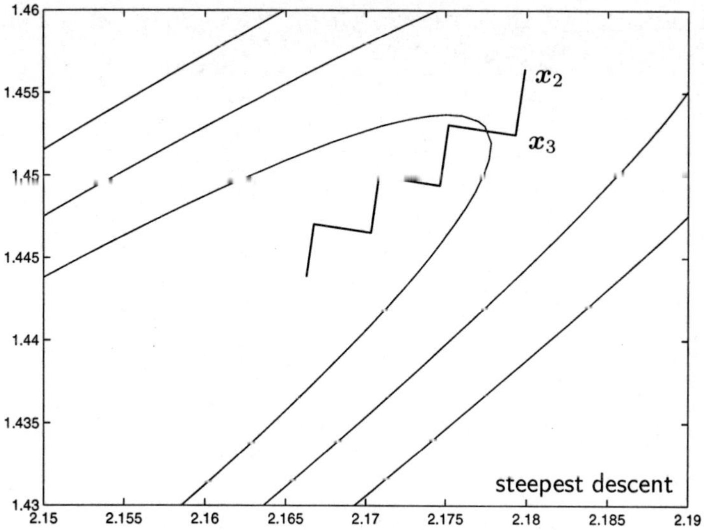

□

In the next example we work through the computations in a simple case.

Example 5. Let

$$f(x_1, x_2) = \frac{1}{2}(x_1^2 + \alpha x_2^2), \qquad \alpha > 1. \tag{1}$$

For values of α close to 1 the level curves of f are roughly circular, and when α is large they are eccentric ellipses. Obviously the minimum of f occurs at $(0, 0)$.

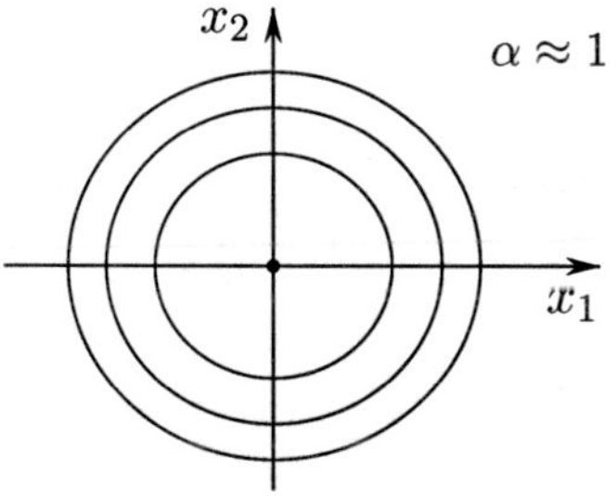

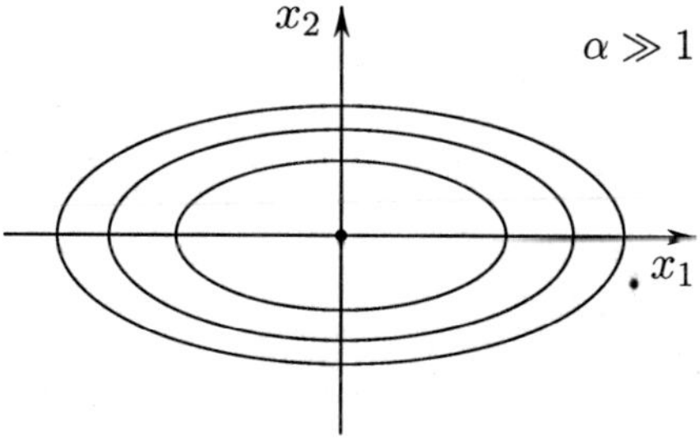

Assume that we start the search for a minimum at a point $\boldsymbol{x} = (x_1, x_2)$ with $x_1 \neq 0$, $x_2 \neq 0$. Put

$$\boldsymbol{d} = -\boldsymbol{\nabla} f(\boldsymbol{x}) = \begin{pmatrix} -x_1 \\ -\alpha x_2 \end{pmatrix}$$

and define

$$F(\lambda) = f(\boldsymbol{x} + \lambda \boldsymbol{d}) = \frac{1}{2}\left((x_1 - \lambda x_1)^2 + \alpha(x_2 - \lambda \alpha x_2)^2\right).$$

After some simplification

$$F(\lambda) = \frac{1}{2}x_1^2\,(1-\lambda)^2 + \frac{1}{2}\alpha x_2^2\,(1-\alpha\lambda)^2.$$

We wish to minimize this function, which is of the second degree in λ. Differentiating we get

$$F'(\lambda) = -x_1^2\,(1-\lambda) - \alpha^2 x_2^2\,(1-\alpha\lambda),$$

and setting the derivative equal to zero gives after some calculation the solution

$$\lambda = \frac{x_1^2 + \alpha^2 x_2^2}{x_1^2 + \alpha^3 x_2^2}.$$

Using this value of λ we obtain the result of the line search

$$\boldsymbol{x}_{\text{new}} = \boldsymbol{x} + \lambda \boldsymbol{d} = \ldots = (\alpha - 1)\frac{x_1 x_2}{x_1^2 + \alpha^3 x_2^2}\begin{pmatrix} \alpha^2 x_2 \\ -x_1 \end{pmatrix}. \tag{2}$$

Note that the quotient of the second and the first coordinate of $\boldsymbol{x}_{\text{new}}$ is

$$\frac{(x_{\text{new}})_2}{(x_{\text{new}})_1} = -\frac{1}{\alpha^2}\frac{x_1}{x_2}. \tag{3}$$

From this observation we conclude the following. Assume that $\{\boldsymbol{x}^{(k)}\}_1^\infty$ is a sequence generated by the steepest descent algorithm. Denote by γ the quotient

$$\gamma = \frac{x_2^{(1)}}{x_1^{(1)}}.$$

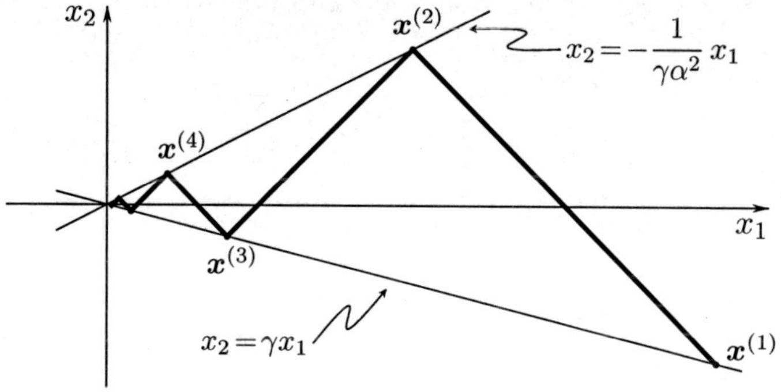

Then the points $\boldsymbol{x}^{(k)}$, $k = 1, 2, \ldots$, belong alternatingly to the lines

$$x_2 = \gamma x_1, \qquad x_2 = -\frac{1}{\gamma\alpha^2}\, x_1.$$

Any quadratic function of two variables can by a change of coordinates be put in the form (1). Hence this phenomenon occurs for any such function. Furthermore, close to a local minimum point any twice differentiable function can be approximated by a quadratic function. We should thus expect the algorithm to behave similarly when applied to any function of two variables near a minimum. Cf. example 4.

We now turn our attention to the sequence of function *values* $\{f_k\}_1^\infty$, where we for shortness of notation put $f_k = f(\boldsymbol{x}_k)$. We have

$$\frac{f_{k+1}}{f_k} = \frac{\left(x_1^{(k+1)}\right)^2 + \alpha\left(x_2^{(k+1)}\right)^2}{\left(x_1^{(k)}\right)^2 + \alpha\left(x_2^{(k)}\right)^2}. \tag{4}$$

Put

$$b_k = \frac{x_1^{(k)}}{x_2^{(k)}}.$$

From (2) we obtain

$$\begin{cases} x_1^{(k+1)} = (\alpha - 1)\dfrac{b_k}{b_k^2 + \alpha^3}\, \alpha^2 x_2^{(k)} \\[2ex] x_2^{(k+1)} = -(\alpha - 1)\dfrac{b_k}{b_k^2 + \alpha^3}\, x_1^{(k)}. \end{cases}$$

Insertion of this in (4) results after some simplification in

$$\frac{f_{k+1}}{f_k} = \frac{\alpha(\alpha-1)^2 b_k^2}{(b_k^2+\alpha)(b_k^2+\alpha^3)}.$$

Now, the function of one variable $x \mapsto \dfrac{x}{(x+\alpha)(x+\alpha^3)}$ has its maximum equal to

$$\frac{\alpha^2}{(\alpha^2+\alpha)(\alpha^2+\alpha^3)} = \frac{1}{\alpha(\alpha+1)^2},$$

occuring for $x = \alpha^2$. (The reader should verify this.) Thus

$$\frac{f_{k+1}}{f_k} \le \frac{(\alpha-1)^2}{(\alpha+1)^2},$$

with equality when $|b_k| = \alpha$. Denote the right hand side constant by C_α. By iteration we obtain

$$f_{k+1} \le C_\alpha^k f_1,$$

again with equality if the initial point $\boldsymbol{x}^{(1)}$ satisfies $|b_1| = \alpha$. (If $|b_1| = \alpha$ then $|b_k| = \alpha$ for all k by (3).)

Note that $C_\alpha \approx 1$ if $\alpha \gg 1$. This implies that for such an α the convergence to $f_{\min} = 0$ is very slow. □

In example 5 the Hessian of the objective function f is a diagonal matrix

$$\begin{pmatrix} 1 & 0 \\ 0 & \alpha \end{pmatrix}$$

with eigenvalues 1 and α. For a general quadratic form $q(\boldsymbol{x}) = \frac{1}{2}\boldsymbol{x}^T\boldsymbol{H}\boldsymbol{x}$ on $\mathbf{R}^n$, with $\boldsymbol{H}$ positive definite we define

$$\alpha = \frac{\text{largest eigenvalue of } \boldsymbol{H}}{\text{smallest eigenvalue of } \boldsymbol{H}}$$

(the *condition number*). We now state results corresponding to the example in this general case.

THEOREM 1. *Let* $\{\boldsymbol{x}^{(k)}\}_1^\infty$ *be a sequence generated by the steepest descent algorithm for the quadratic form* $q(\boldsymbol{x}) = \frac{1}{2}\boldsymbol{x}^T\boldsymbol{H}\boldsymbol{x}$*, and let* $q_k = q(\boldsymbol{x}^{(k)})$.

a) Then for $k = 1, 2, \ldots$

$$q_{k+1} \le C_\alpha^k q_1, \qquad \text{where } C_\alpha = \frac{(\alpha-1)^2}{(\alpha+1)^2}, \tag{5}$$

with equality occuring for some choice of initial point.

b) Put $e(\boldsymbol{x}) = \sqrt{\boldsymbol{x}^T \boldsymbol{H} \boldsymbol{x}}$ *(the error function). For* $k = 1, 2, \ldots$ *we have*

$$e(\boldsymbol{x}^{(k+1)}) \le C_\alpha^{k/2} e(\boldsymbol{x}^{(1)}). \tag{6}$$

Remark. For a general quadratic function $q(\boldsymbol{x}) = \frac{1}{2}\boldsymbol{x}^T \boldsymbol{H} \boldsymbol{x} + \boldsymbol{c}^T \boldsymbol{x} + d$ the inequality in (5) should read $q_{k+1} - q_{\min} \le C_\alpha^k (q_1 - q_{\min})$. In (6) distances $e(\boldsymbol{x} - \bar{\boldsymbol{x}})$ should be considered. Here $\bar{\boldsymbol{x}}$ is the minimum point and $q_{\min} = q(\bar{\boldsymbol{x}})$.

Remark. Inequality (5) is equivalent to

$$q_{k+1} \le C_\alpha q_k, \qquad k = 1, 2, \ldots .$$

The constant C_α is less than 1. Thus the convergence of the sequence $\{q_k\}_1^\infty$ is *linear*.[4]

The proof of theorem 1 is rather lengthy, and we postpone it until the end of the section.

Instead we now indicate some ways of modifying the steepest descent method in order to make it more efficient. One modification is to deflect the gradient by multiplying it with a suitably chosen positive definite matrix. This means choosing a search direction of the form

$$\boldsymbol{d}_k = -\boldsymbol{B}_k \nabla f(\boldsymbol{x}_k).$$

(The reader might prove that for a positive definite $\boldsymbol{B}_k$ this is still a direction of descent.) We will encounter this idea later with different choices of $\boldsymbol{B}_k$ in Newton's method and in quasi-Newton methods. — Another variant is to modify the gradient by the addition of a correction vector, i.e., choosing a search direction of the form

$$\boldsymbol{d}_k = -\nabla f(\boldsymbol{x}_k) + \boldsymbol{b}_k$$

for some suitable $\boldsymbol{b}_k$. This is what is done in the conjugate gradient method.

[4] See appendix 1, page 103, for the terminology used in connection with convergence rate.

Proof of theorem 1

This proof is intended for the interested reader only, and should anyway be passed over at the first reading of the text.

We begin by proving the following lemma on positive definite matrices.

LEMMA 2. (KANTOROVICH) *Let $\boldsymbol{H}$ be a positive definite matrix with eigenvalues*

$$0 < \lambda_1 \leq \lambda_2 \leq \ldots \leq \lambda_n.$$

Then

$$1 \leq \frac{\boldsymbol{x}^T \boldsymbol{H} \boldsymbol{x}}{\boldsymbol{x}^T \boldsymbol{x}} \cdot \frac{\boldsymbol{x}^T \boldsymbol{H}^{-1} \boldsymbol{x}}{\boldsymbol{x}^T \boldsymbol{x}} \leq \frac{(\lambda_1 + \lambda_n)^2}{4\lambda_1 \lambda_n} \qquad \text{for all } \boldsymbol{x} \neq \boldsymbol{0}. \tag{7}$$

There is equality in the right inequality of (7) when $\boldsymbol{x}$ is parallel to $\boldsymbol{v}_1 + \boldsymbol{v}_n$ or $\boldsymbol{v}_1 - \boldsymbol{v}_n$, where $\boldsymbol{v}_1$ and $\boldsymbol{v}_n$ are normalized orthogonal eigenvectors of $\boldsymbol{H}$ with eigenvalues λ_1 and λ_n respectively.

PROOF. We first simplify (7) by a change of coordinates.

Since $\boldsymbol{H}$ is symmetric a diagonal matrix $\boldsymbol{D}$ (with the eigenvalues as entries) and an orthogonal matrix $\boldsymbol{Q}$ exist such that

$$\boldsymbol{H} = \boldsymbol{Q}\boldsymbol{D}\boldsymbol{Q}^T \quad \text{and} \quad \boldsymbol{H}^{-1} = \boldsymbol{Q}\boldsymbol{D}^{-1}\boldsymbol{Q}^T.$$

(See theorem A.2, page 348.) Then

$$\begin{aligned}
\boldsymbol{x}^T \boldsymbol{H} \boldsymbol{x} &= (\boldsymbol{Q}^T \boldsymbol{x})^T \boldsymbol{D} (\boldsymbol{Q}^T \boldsymbol{x}), \\
\boldsymbol{x}^T \boldsymbol{H}^{-1} \boldsymbol{x} &= (\boldsymbol{Q}^T \boldsymbol{x})^T \boldsymbol{D}^{-1} (\boldsymbol{Q}^T \boldsymbol{x}), \\
\boldsymbol{x}^T \boldsymbol{x} &= (\boldsymbol{Q}^T \boldsymbol{x})^T (\boldsymbol{Q}^T \boldsymbol{x}).
\end{aligned}$$

Putting $\boldsymbol{y} = \boldsymbol{Q}^T \boldsymbol{x}$ we find that (7) is equivalent to

$$1 \leq \frac{\boldsymbol{y}^T \boldsymbol{D} \boldsymbol{y}}{\boldsymbol{y}^T \boldsymbol{y}} \cdot \frac{\boldsymbol{y}^T \boldsymbol{D}^{-1} \boldsymbol{y}}{\boldsymbol{y}^T \boldsymbol{y}} \leq \frac{(\lambda_1 + \lambda_n)^2}{4\lambda_1 \lambda_n} \qquad \text{for all } \boldsymbol{y} \neq \boldsymbol{0}. \tag{8}$$

Since (8) is homogeneous in $\boldsymbol{y}$ we may normalize it to

$$1 \leq \boldsymbol{y}^T \boldsymbol{D} \boldsymbol{y} \cdot \boldsymbol{y}^T \boldsymbol{D}^{-1} \boldsymbol{y} \leq \frac{(\lambda_1 + \lambda_n)^2}{4\lambda_1 \lambda_n} \qquad \text{for all } \boldsymbol{y} \neq \boldsymbol{0} \text{ such that } \boldsymbol{y}^T \boldsymbol{y} = 1.$$

Here

$$\boldsymbol{y}^T \boldsymbol{D} \boldsymbol{y} = \sum_{i=1}^n \lambda_i y_i^2, \qquad \boldsymbol{y}^T \boldsymbol{D}^{-1} \boldsymbol{y} = \sum_{i=1}^n \frac{1}{\lambda_i} y_i^2, \qquad \boldsymbol{y}^T \boldsymbol{y} = \sum_{i=1}^n y_i^2.$$

Changing notation, what we need to prove is that

(9)
$$1 \le \left(\sum_{i=1}^{n} \lambda_i \theta_i\right)\left(\sum_{i=1}^{n} \frac{1}{\lambda_i}\theta_i\right) \le \frac{(\lambda_1+\lambda_n)^2}{4\lambda_1\lambda_n} \qquad \text{when all } \theta_i \ge 0 \text{ and } \sum_{1}^{n}\theta_i = 1.$$

We now introduce the two functions of one variable

$$\varphi(t) = \frac{1}{t}, \qquad \psi(t) = \frac{1}{\lambda_1\lambda_n}(\lambda_1+\lambda_n-t)$$

They both have the same values at $t = \lambda_1$ and $t = \lambda_n$. The function φ is convex[5] and ψ is affine, hence

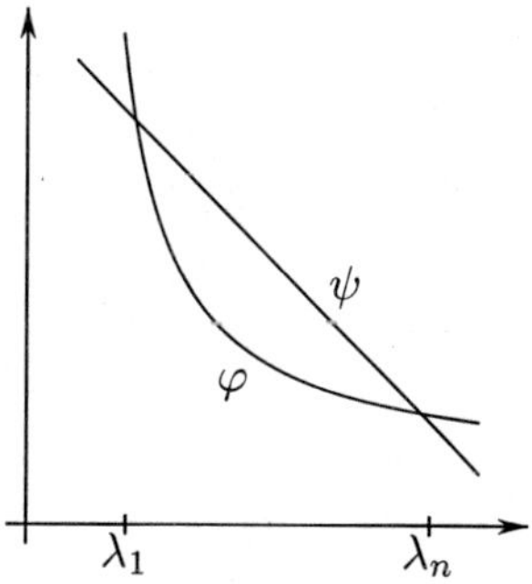

(10) $\varphi(t) \le \psi(t)$ when $\lambda_1 \le t \le \lambda_n$.

Furthermore, the convexity of φ implies that for $\theta_i \ge 0$ and $\sum_1^n \theta_i = 1$

$$\varphi\Big(\sum_{1}^{n}\theta_i\lambda_i\Big) \le \sum_{i=1}^{n}\theta_i\varphi(\lambda_i).$$

By (10) and the facts that ψ has the form $\psi(t) = At + B$ and that $\sum_1^n \theta_i = 1$

$$\sum_{i=1}^{n}\theta_i\varphi(\lambda_i) \le \sum_{i=1}^{n}\theta_i\psi(\lambda_i) = \psi\Big(\sum_{i=1}^{n}\theta_i\lambda_i\Big).$$

These inequalities imply that

$$1 \le \frac{\sum_1^n \theta_i\varphi(\lambda_i)}{\varphi(\sum_1^n\theta_i\lambda_i)} \le \frac{\psi\Big(\sum_1^n\theta_i\lambda_i\Big)}{\varphi\Big(\sum_1^n\theta_i\lambda_i\Big)},$$

where since $\varphi(t) = \dfrac{1}{t}$ the middle term is equal to

$$\left(\sum_{i=1}^{n}\lambda_i\theta_i\right)\left(\sum_{i=1}^{n}\frac{1}{\lambda_i}\theta_i\right).$$

The left inequality in (9) is now proved. To prove the right one it is enough to prove that

$$\frac{\psi\Big(\sum_1^n\theta_i\lambda_i\Big)}{\varphi\Big(\sum_1^n\theta_i\lambda_i\Big)} \le \frac{(\lambda_1+\lambda_n)^2}{4\lambda_1\lambda_n}.$$

[5] Convex functions will be discussed extensively in chapter 6.

This is done as follows. The function

$$t \longmapsto \frac{\psi(t)}{\varphi(t)} = \frac{1}{\lambda_1\lambda_n}\Big(t(\lambda_1+\lambda_n) - t^2\Big), \qquad \lambda_1 \le t \le \lambda_n,$$

is a second degree polynomial. It is easily seen to have its maximum at $t = \dfrac{\lambda_1+\lambda_n}{2}$; the maximal value is

$$\frac{1}{\lambda_1\lambda_n}\cdot\frac{(\lambda_1+\lambda_n)^2}{4}.$$

Thus inequality (7) is fully proved.

To verify the statement on equality we merely observe that if $\boldsymbol{x} = \boldsymbol{v}_1 \pm \boldsymbol{v}_n$ (where $\boldsymbol{v}_1^T\boldsymbol{v}_n = 0$, $\|\boldsymbol{v}_1\| = 1$, $\|\boldsymbol{v}_n\| = 1$) then

$$\boldsymbol{x}^T\boldsymbol{H}\boldsymbol{x} = \lambda_1+\lambda_n, \qquad \boldsymbol{x}^T\boldsymbol{H}^{-1}\boldsymbol{x} = \frac{1}{\lambda_1}+\frac{1}{\lambda_n}, \qquad \boldsymbol{x}^T\boldsymbol{x} = 2,$$

and consequently

$$\frac{\boldsymbol{x}^T\boldsymbol{H}\boldsymbol{x}}{\boldsymbol{x}^T\boldsymbol{x}}\cdot\frac{\boldsymbol{x}^T\boldsymbol{H}^{-1}\boldsymbol{x}}{\boldsymbol{x}^T\boldsymbol{x}} = \frac{1}{4}\Big(\lambda_1+\lambda_n\Big)\Big(\frac{1}{\lambda_1}+\frac{1}{\lambda_n}\Big) = \frac{(\lambda_1+\lambda_n)^2}{4\lambda_1\lambda_n}.$$

□

For the proof of theorem 1 we introduce some short notation for the gradients of q at the points $\boldsymbol{x}^{(k)}$, putting

$$\boldsymbol{g}_k = \boldsymbol{\nabla} q(\boldsymbol{x}^{(k)}) = \boldsymbol{H}\boldsymbol{x}^{(k)}.$$

For later reference we observe that

$$\boldsymbol{g}_k^T\boldsymbol{H}^{-1}\boldsymbol{g}_k = \boldsymbol{x}^{(k)^T}\boldsymbol{H}\boldsymbol{x}^{(k)} = 2q_k. \tag{11}$$

In the iteration step we have put[6]

$$\boldsymbol{x}^{(k+1)} = \boldsymbol{x}^{(k)} - \mu_k\boldsymbol{g}_k,$$

where μ_k is chosen to minimize $q(\boldsymbol{x}^{(k)} - \mu\boldsymbol{g}_k)$. First we must find μ_k. A simple computation gives

$$\begin{aligned} q(\boldsymbol{x}^{(k)} - \mu\boldsymbol{g}_k) &= \frac{1}{2}(\boldsymbol{x}^{(k)} - \mu\boldsymbol{g}_k)^T\boldsymbol{H}\,(\boldsymbol{x}^{(k)} - \mu\boldsymbol{g}_k) = \\ &= \frac{1}{2}\boldsymbol{x}^{(k)^T}\boldsymbol{H}\boldsymbol{x}^{(k)} - \mu\boldsymbol{g}_k^T\boldsymbol{H}\boldsymbol{x}^{(k)} + \frac{1}{2}\mu^2\boldsymbol{g}_k^T\boldsymbol{H}\boldsymbol{g}_k. \end{aligned} \tag{12}$$

[6] We use the variable μ in the line search here, since the notation $\lambda_1,\ldots,\lambda_n$ is reserved for the eigenvalues of $\boldsymbol{H}$.

Differentiating with respect to μ and setting the derivative equal to zero gives

$$-\boldsymbol{g}_k^T \boldsymbol{H}\boldsymbol{x}^{(k)} + \mu \boldsymbol{g}_k^T \boldsymbol{H}\boldsymbol{g}_k = 0.$$

Remembering that $\boldsymbol{H}\boldsymbol{x}^{(k)} = \boldsymbol{g}_k$ we obtain

$$\mu_k = \frac{\boldsymbol{g}_k^T \boldsymbol{g}_k}{\boldsymbol{g}_k^T \boldsymbol{H}\boldsymbol{g}_k}. \tag{13}$$

We now compute $q_{k+1} = q(\boldsymbol{x}^{(k+1)})$. Starting from the calculation in (12) we have

$$\begin{aligned} q_{k+1} &= \frac{1}{2}\boldsymbol{x}^{(k)T} \boldsymbol{H}\boldsymbol{x}^{(k)} - \mu_k \boldsymbol{g}_k^T \boldsymbol{H}\boldsymbol{x}^{(k)} + \frac{1}{2}\mu_k^2 \boldsymbol{g}_k^T \boldsymbol{H}\boldsymbol{g}_k = \\ &= q_k - \mu_k \boldsymbol{g}_k^T \boldsymbol{g}_k + \frac{1}{2}\mu_k \boldsymbol{g}_k^T \boldsymbol{g}_k = \\ &= q_k - \frac{1}{2}\mu_k \boldsymbol{g}_k^T \boldsymbol{g}_k. \end{aligned}$$

Using (11) this can be written

$$q_{k+1} = q_k - q_k \mu_k \frac{\boldsymbol{g}_k^T \boldsymbol{g}_k}{\boldsymbol{g}_k^T \boldsymbol{H}^{-1} \boldsymbol{g}_k} = q_k \Big(1 - \mu_k \frac{\boldsymbol{g}_k^T \boldsymbol{g}_k}{\boldsymbol{g}_k^T \boldsymbol{H}^{-1} \boldsymbol{g}_k}\Big).$$

Finally, inserting the value of μ_k from (13) we obtain

$$q_{k+1} = q_k \Big(1 - \frac{\boldsymbol{g}_k^T \boldsymbol{g}_k}{\boldsymbol{g}_k^T \boldsymbol{H}\boldsymbol{g}_k} \cdot \frac{\boldsymbol{g}_k^T \boldsymbol{g}_k}{\boldsymbol{g}_k^T \boldsymbol{H}^{-1} \boldsymbol{g}_k}\Big).$$

Now we invoke the Kantorovich lemma. From that we obtain the estimate

$$q_{k+1} \le q_k \Big(1 - \frac{4\lambda_1\lambda_n}{(\lambda_1+\lambda_n)^2}\Big) = q_k \frac{(\lambda_1-\lambda_n)^2}{(\lambda_1+\lambda_n)^2} = C_\alpha q_k. \tag{14}$$

From this inequality (5) follows by iteration, as in example 5.

We now turn to the question of when there is equality in (5). From lemma 2 it follows that equality in (14) occurs if $\boldsymbol{g}_k$ is parallel to either of $\boldsymbol{v}_1 \pm \boldsymbol{v}_n$ (notation as in the lemma). Assume that $\boldsymbol{g}_k = t(\boldsymbol{v}_1 + \boldsymbol{v}_n)$. Then by (13)

$$\mu_k = \frac{2}{\lambda_1 + \lambda_n}$$

and it follows that

$$\begin{aligned} \boldsymbol{g}_{k+1} &= \boldsymbol{H}\boldsymbol{x}^{(k+1)} = \boldsymbol{H}\boldsymbol{x}^{(k)} - \mu_k \boldsymbol{H}\boldsymbol{g}_k = \boldsymbol{g}_k - \mu_k \boldsymbol{H}\boldsymbol{g}_k = \\ &= t\Big(\boldsymbol{v}_1 + \boldsymbol{v}_n - \frac{2}{\lambda_1+\lambda_n}(\lambda_1\boldsymbol{v}_1 + \lambda_n\boldsymbol{v}_n)\Big) = t\,\frac{\lambda_n - \lambda_1}{\lambda_n + \lambda_1}(\boldsymbol{v}_1 - \boldsymbol{v}_n). \end{aligned}$$

Thus, if $\boldsymbol{g}_k$ is parallel to $\boldsymbol{v}_1 + \boldsymbol{v}_n$ then $\boldsymbol{g}_{k+1}$ is parallel to $\boldsymbol{v}_1 - \boldsymbol{v}_n$. The reader may check that the corresponding statement is valid when $\boldsymbol{g}_k$ is parallel to $\boldsymbol{v}_1 - \boldsymbol{v}_n$.

The consequence of this observation is that if the initial point is such that $\boldsymbol{g}_1 = \boldsymbol{H}^{-1}\boldsymbol{x}^{(1)}$ is parallel to either of $\boldsymbol{v}_1 \pm \boldsymbol{v}_n$ then all consecutive points $\boldsymbol{x}^{(k)}$ will have the same property, the gradients alternating between the two directions. Then there is equality in (14) for all k, hence equality in (5)

Finally, to prove part b) of the theorem we observe that $e(\boldsymbol{x}) = \sqrt{2q(\boldsymbol{x})}$. Hence we obtain from a) that

$$e(\boldsymbol{x}^{(k+1)})^2 = 2q_{k+1} \leq 2C_\alpha^k q_1 = C_\alpha^k e(\boldsymbol{x}^{(1)})^2,$$

which is equivalent to (6). □

Exercises

S 3.2 *(Steepest descent for a quadratic function.)* Let

$$q(\boldsymbol{x}) = \frac{1}{2}\boldsymbol{x}^T\boldsymbol{H}\boldsymbol{x} + \boldsymbol{c}^T\boldsymbol{x}$$

where $\boldsymbol{H}$ is positive definite.

a) Prove that $\lambda_k = \dfrac{\boldsymbol{g}_k^T\boldsymbol{g}_k}{\boldsymbol{g}_k^T\boldsymbol{H}\boldsymbol{g}_k}$, when exact line searches are used in the SD algorithm. Here $\boldsymbol{g}_k = \boldsymbol{\nabla} q(\boldsymbol{x}_k)$.

b) Verify by means of a matrix computation that $\boldsymbol{g}_{k+1} \perp \boldsymbol{g}_k$.

c) Show that in two dimensions there are only two possible search directions, and that these are orthogonal.

3.3 Consider $q(\boldsymbol{x}) = \frac{1}{2}\boldsymbol{x}^T\boldsymbol{H}\boldsymbol{x}+\boldsymbol{c}^T\boldsymbol{x}$, where $\boldsymbol{H} = \begin{pmatrix} 4 & 1 \\ 1 & 4 \end{pmatrix}$, $\boldsymbol{c} = \begin{pmatrix} -7 \\ 2 \end{pmatrix}$.

a) Find the exact point of minimum $\bar{\boldsymbol{x}}$.

b) When minimizing using the steepest descent method the iterations as you know follow a zigzag pattern. Find the two occuring directions when starting at

$$\boldsymbol{x}^{(1)} = (2, 3).$$

c) Give an estimate of the number of iterations required to make

$$q(\boldsymbol{x}^{(k)}) - q(\bar{\boldsymbol{x}}) \leq \frac{1}{2} \cdot 10^{-4},$$

when starting at $\boldsymbol{x}^{(1)}$ above.

(1997–04–04, 4)

3.4 Consider the minimization of the function $f(x_1, x_2) = x_1^2 + 2x_1x_2 + 7x_2^2$.

a) The figure below shows the level curves of f. Draw the result of the first five SD iterations, starting from (1.3, 0.4).

b) Compute the first two SD iterations starting from (1.3, 0.4). Use the exact line search formulas from exercise 3.1 or 3.2.

c) Check that the two search directions used in b) are orthogonal.

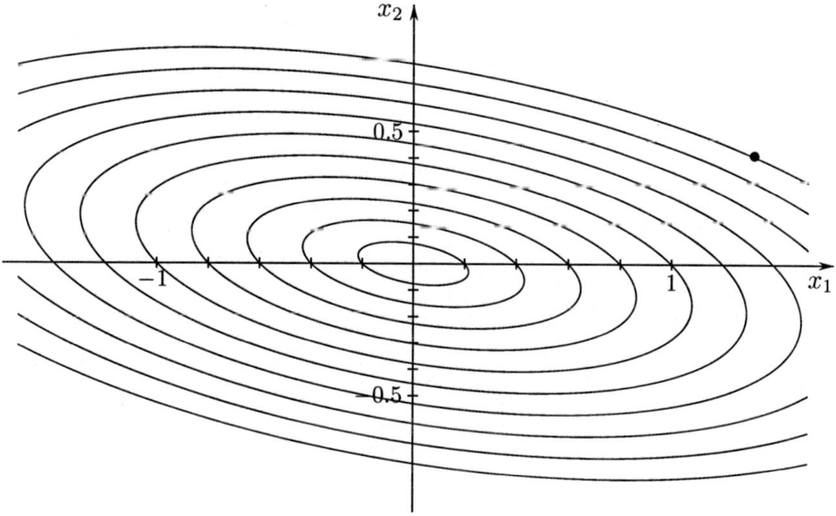

3.5 If $\boldsymbol{H}$ has condition number 10 ($= \lambda_{max}/\lambda_{min}$), how many SD iterations are needed to guarantee an error $\leq 5 \cdot 10^{-7}$ in the minimum value of

$$f(\boldsymbol{x}) = \frac{1}{2}(\boldsymbol{x} - \bar{\boldsymbol{x}})^T \boldsymbol{H}(\boldsymbol{x} - \bar{\boldsymbol{x}}),$$

if $f(\boldsymbol{x}_1) = 1$?

3.6 a) Sketch some level curves of the function

$$f(\boldsymbol{x}) = x_1^2 + 4x_2^2 - 4x_1 - 8x_2.$$

Find the point $\bar{\boldsymbol{x}}$ which minimizes f.

b) Show in a figure some SD iterations from $\boldsymbol{x}_1 = (0, 0)$.

c) Show that if the method of steepest descent is started from $\boldsymbol{x}_1 = (0, 0)$ it cannot converge to $\bar{\boldsymbol{x}}$ in a finite number of steps. Are there any initial points $\boldsymbol{x}_1$ for which there will be convergence in a finite number of steps?

3.4 Newton's method and modifications

3.4.1 Newton's method

Assume that the function of n variables f is of class $\mathcal{C}^2$. Then, by Taylor's formula, near a point $\boldsymbol{x}_k$ we can approximate $f(\boldsymbol{x})$ by the quadratic function

$$q(\boldsymbol{x}) = f(\boldsymbol{x}_k) + \nabla f(\boldsymbol{x}_k)^T(\boldsymbol{x}-\boldsymbol{x}_k) + \frac{1}{2}(\boldsymbol{x}-\boldsymbol{x}_k)^T \boldsymbol{H}(\boldsymbol{x}_k)(\boldsymbol{x}-\boldsymbol{x}_k).$$

Here $\boldsymbol{H}(\boldsymbol{x}_k)$ denotes the Hessian of f at $\boldsymbol{x}_k$. If this matrix is positive definite q has a minimum where $\nabla q(\boldsymbol{x}) = \boldsymbol{0}$. By differentiation of q we obtain

$$\nabla q(\boldsymbol{x}) = \boldsymbol{0} \iff \nabla f(\boldsymbol{x}_k) + \boldsymbol{H}(\boldsymbol{x}_k)(\boldsymbol{x}-\boldsymbol{x}_k) = \boldsymbol{0} \iff$$
$$\iff \boldsymbol{x} = \boldsymbol{x}_k - \boldsymbol{H}(\boldsymbol{x}_k)^{-1}\nabla f(\boldsymbol{x}_k).$$

If $\boldsymbol{x}_k$ is close to a minimum point $\bar{\boldsymbol{x}}$ of f it seems natural to use this formula as a basis for the iteration step, choosing the minimum point of q as our next approximation of $\bar{\boldsymbol{x}}$. This means letting

$$(15) \qquad \boldsymbol{x}_{k+1} = \boldsymbol{x}_k - \boldsymbol{H}(\boldsymbol{x}_k)^{-1}\nabla f(\boldsymbol{x}_k).$$

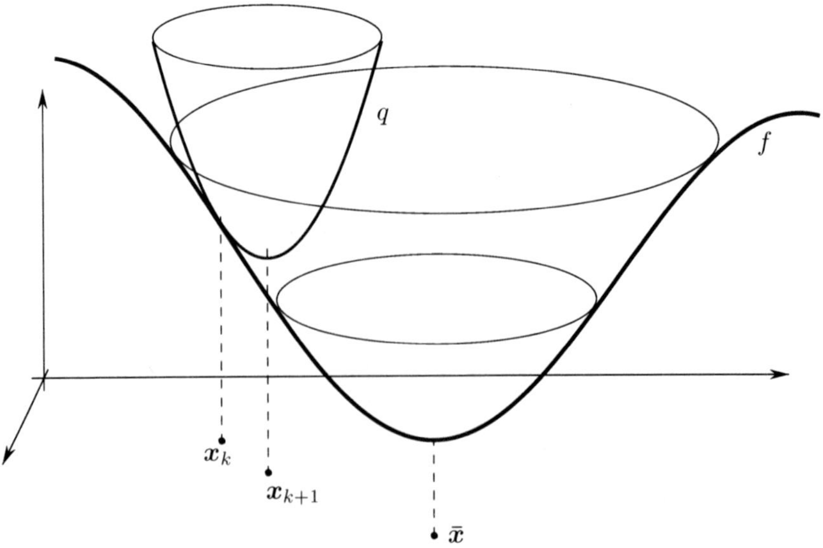

Newton's method is defined as the iteration of formula (15) starting from some chosen initial point $\boldsymbol{x}_1$. In the general scheme described in the introduction to this chapter Newton's method implies picking the search directions

$$\boldsymbol{d}_k = -\boldsymbol{H}(\boldsymbol{x}_k)^{-1}\nabla f(\boldsymbol{x}_k)$$

and taking a *unit step* in that direction (thus dispensing with the line search). The reader should compare with the one variable version of Newton's method in section 2.2.

Note that if $\bar{\boldsymbol{x}}$ is a local minimum point then $\boldsymbol{H}(\bar{\boldsymbol{x}})$ is positive semidefinite. If it is in fact positive definite then $\boldsymbol{H}(\boldsymbol{x})$ is by continuity positive definite when $\boldsymbol{x}$ is close to $\bar{\boldsymbol{x}}$. Then in particular the matrices $\boldsymbol{H}(\boldsymbol{x}_k)$ are invertible. — If $\boldsymbol{H}(\bar{\boldsymbol{x}})$ is non-invertible Newton's method does not work very well. The assumptions in theorem 2 below are not fulfilled. (Cf. the onedimensional case.)

Example 6. Consider the objective function from examples 1 and 4. We use the same initial point $(0, 2.5)$ as before. The table shows the

k	$\boldsymbol{x}_k$	$f(\boldsymbol{x}_k)$	$\boldsymbol{H}(\boldsymbol{x}_k)$	$\boldsymbol{d}_k$
1	(0 , 2.5)	88.2500	$\begin{pmatrix} 104 & -12 \\ -12 & 18 \end{pmatrix}$	(0.7273, −2.0152)
2	(0.7273, 0.4848)	5.2477	$\begin{pmatrix} 44.6942 & -12 \\ -12 & 18 \end{pmatrix}$	(0.4495, 0.2996)
3	(1.1767, 0.7845)	0.9187	$\begin{pmatrix} 25.6801 & -12 \\ -12 & 18 \end{pmatrix}$	(0.2525, 0.1683)
4	(1.4292, 0.9528)	0.2123	$\begin{pmatrix} 19.2528 & -12 \\ -12 & 18 \end{pmatrix}$	(0.1322, 0.0881)
5	(1.5614, 1.0409)	0.0740	$\begin{pmatrix} 17.1078 & -12 \\ -12 & 18 \end{pmatrix}$	(0.0741, 0.0494)
6	(1.6355, 1.0903)	0.0353	$\begin{pmatrix} 16.2725 & -12 \\ -12 & 18 \end{pmatrix}$	(0.0468, 0.0312)
7	(1.6823, 1.1216)	0.0204	$\begin{pmatrix} 15.8805 & -12 \\ -12 & 18 \end{pmatrix}$	(0.0325, 0.0217)
8	(1.7149, 1.1433)	0.0132	$\begin{pmatrix} 15.6701 & -12 \\ -12 & 18 \end{pmatrix}$	(0.0242, 0.0161)
9	(1.7391, 1.1594)	0.0093		

results of 8 iterations with Newtons method, after which we reached the point (1.7391, 1, 1594). The convergence is shown in the figure below. To obtain a point within a circle of radius 0.1 around the exact minimum we must carry out almost 70 iterations. The slow convergence in this case can be explained by the failure of the objective function to resemble a quadratic function in the region of interest. See the level curves.

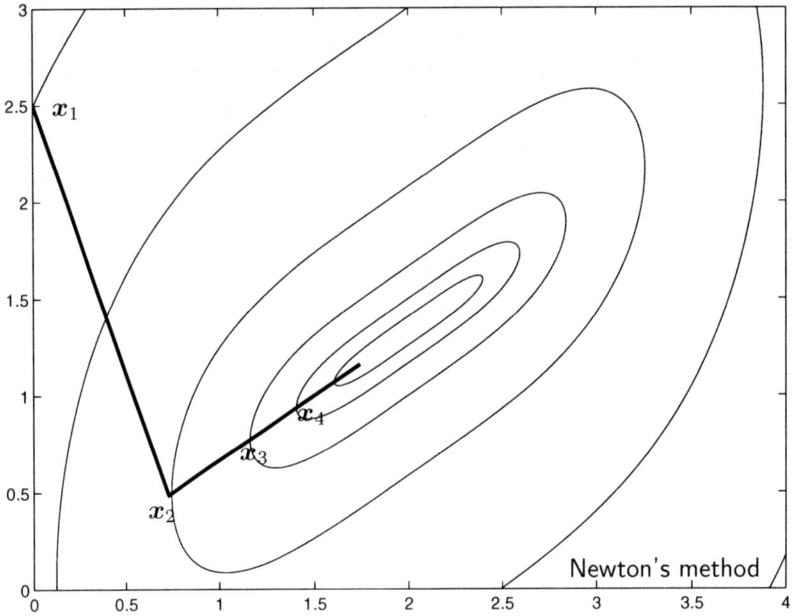

□

As in the onedimensional case

$$\boldsymbol{x}_k \to \bar{\boldsymbol{x}} \quad \text{as} \quad k \to \infty$$

if the initial point $\boldsymbol{x}_1$ is close enough to the minimum point $\bar{\boldsymbol{x}}$ of f. This is the content of the following theorem.

THEOREM 2. *Assume that* $f \in \mathcal{C}^2(\mathbf{R}^n)$. *Let* $\bar{\boldsymbol{x}}$ *be a point satisfying* $\nabla f(\bar{\boldsymbol{x}}) = \mathbf{0}$, *and assume that the Hessian* $\boldsymbol{H}(\bar{\boldsymbol{x}})$ *of* f *at* $\bar{\boldsymbol{x}}$ *is invertible. Suppose that there are numbers* A *and* B *such that when* $\boldsymbol{x}$ *is sufficiently close to* $\bar{\boldsymbol{x}}$ [7]

$$\|\boldsymbol{H}(\boldsymbol{x})^{-1}\| \leq A,$$

$$\|\nabla f(\bar{\boldsymbol{x}}) - \nabla f(\boldsymbol{x}) - \boldsymbol{H}(\boldsymbol{x})(\bar{\boldsymbol{x}} - \boldsymbol{x})\| \leq B\|\boldsymbol{x} - \bar{\boldsymbol{x}}\|^2.$$

Let the sequence $\{\boldsymbol{x}_k\}_1^\infty$ *be generated by iteration of (15) with the initial point* $\boldsymbol{x}_1$ *close enough to* $\bar{\boldsymbol{x}}$. *Then* $\boldsymbol{x}_k$ *tends to* $\bar{\boldsymbol{x}}$ *as* $k \to \infty$ *with (at least) quadratic convergence rate.*

The proof is an adaption of the one variable version to the context of several variables. It can be skipped without loss of continuity.

PROOF. We have by (15)

$$\begin{aligned} \boldsymbol{x}_{k+1} - \bar{\boldsymbol{x}} &= \boldsymbol{x}_k - \bar{\boldsymbol{x}} - \boldsymbol{H}(\boldsymbol{x}_k)^{-1}\Big(\nabla f(\boldsymbol{x}_k) - \nabla f(\bar{\boldsymbol{x}})\Big) = \\ &= \boldsymbol{H}(\boldsymbol{x}_k)^{-1}\Big(\nabla f(\bar{\boldsymbol{x}}) - \nabla f(\boldsymbol{x}_k) - \boldsymbol{H}(\boldsymbol{x}_k)(\bar{\boldsymbol{x}} - \boldsymbol{x}_k)\Big). \end{aligned}$$

Taking norms we get

$$\|\boldsymbol{x}_{k+1} - \bar{\boldsymbol{x}}\| \leq \|\boldsymbol{H}(\boldsymbol{x}_k)^{-1}\| \, \|\nabla f(\bar{\boldsymbol{x}}) - \nabla f(\boldsymbol{x}_k) - \boldsymbol{H}(\boldsymbol{x}_k)(\bar{\boldsymbol{x}} - \boldsymbol{x}_k)\| \leq AB\|\boldsymbol{x}_k - \bar{\boldsymbol{x}}\|^2. \quad (16)$$

Hence, if $AB\|\boldsymbol{x}_k - \bar{\boldsymbol{x}}\| < 1$,

$$\|\boldsymbol{x}_{k+1} - \bar{\boldsymbol{x}}\| < \|\boldsymbol{x}_k - \bar{\boldsymbol{x}}\|.$$

From this inequality we conclude that if $AB\|\boldsymbol{x}_1 - \bar{\boldsymbol{x}}\| < 1$ then the same inequality is valid for all $\boldsymbol{x}_k$. Furthermore, (16) implies that $\|\boldsymbol{x}_{k+1} - \bar{\boldsymbol{x}}\| \leq c^k\|\boldsymbol{x}_1 - \bar{\boldsymbol{x}}\|$ for some constant $c < 1$. Consequently $\boldsymbol{x}_k \to \bar{\boldsymbol{x}}$ when $k \to \infty$.

That the convergence is quadratic follows from (16). □

3.4.2 Modified Newton methods

By theorem 2 Newton's method works very well when we start close to the minimum point. If we do not some trouble may (or may not) appear.

- The Hessian $\boldsymbol{H}(\boldsymbol{x}_k)$ might not be invertible.
- The vector $\boldsymbol{d}_k = -\boldsymbol{H}(\boldsymbol{x}_k)^{-1}\nabla f(\boldsymbol{x}_k)$ might not be a descent direction for f.

[7] In the first condition we use the *norm* of a matrix, defined by $\|\boldsymbol{A}\| = \sup_{\|\boldsymbol{x}\| \neq 0} \frac{\|\boldsymbol{A}\boldsymbol{x}\|}{\|\boldsymbol{x}\|}$. From this definition the inequality $\|\boldsymbol{A}\boldsymbol{x}\| \leq \|\boldsymbol{A}\| \, \|\boldsymbol{x}\|$ follows directly.

- The "choice" of $\lambda_k = 1$ in (15) might result in too large a step in direction $\boldsymbol{d}_k$.

A remedy for the last problem is of course to, at some cost, actually carry out a line search in the direction $\boldsymbol{d}_k$. The other problems could be overcome by using another matrix $\boldsymbol{B}_k$ than the inverse Hessian to deflect the negative gradient, i.e., by putting

$$\boldsymbol{d}_k = -\boldsymbol{B}_k \nabla f(\boldsymbol{x}_k).$$

Here $\boldsymbol{B}_k$ should be a symmetric, positive definite matrix. Then it is invertible, and it can be proved that $\boldsymbol{d}_k$ is a descent direction. (Exercise: provide this proof.) The following approach leads to a suitable matrix.

We modify the Hessian of f by adding a multiple of the identity matrix. This implies the addition of the same constant to all eigenvalues of $\boldsymbol{H}(\boldsymbol{x}_k)$. Thus the matrix $\boldsymbol{B}_k$ is chosen as

$$\boldsymbol{B}_k = (\varepsilon_k \boldsymbol{I} + \boldsymbol{H}(\boldsymbol{x}_k))^{-1},$$

where the number ε_k is the least number making all eigenvalues of $\varepsilon_k \boldsymbol{I} + \boldsymbol{H}(\boldsymbol{x}_k)$ larger than some specified positive constant.

The convergence of a thus generated sequence will be discussed in chapter 10. — If $\boldsymbol{H}(\bar{\boldsymbol{x}})$ is positive definite at the minimum point $\bar{\boldsymbol{x}}$, by continuity $\boldsymbol{H}(\boldsymbol{x})$ is positive definite when $\boldsymbol{x}$ is close to $\bar{\boldsymbol{x}}$. Consequently, if $\boldsymbol{x}_k$ has $\bar{\boldsymbol{x}}$ as its limit point ε_k should tend to zero. Hence this scheme becomes in fact Newton's method when we get close to the minimum point.

In implementations of the algorithm one never actually computes an inverse matrix. The iteration step

$$\boldsymbol{x}_{k+1} = \boldsymbol{x}_k - \boldsymbol{B}_k \nabla f(\boldsymbol{x}_k)$$

can be formulated as a system of linear equations

$$\Big(\varepsilon_k \boldsymbol{I} + \boldsymbol{H}(\boldsymbol{x}_k)\Big)(\boldsymbol{x}_{k+1} - \boldsymbol{x}_k) = -\nabla f(\boldsymbol{x}_k).$$

This system is solved by use of a so called *Cholesky factorization*[8]

$$\varepsilon_k \boldsymbol{I} + \boldsymbol{H}(\boldsymbol{x}_k) = \boldsymbol{L}\boldsymbol{L}^T,$$

[8]See appendix A.3, page 360.

with $\boldsymbol{L}$ a lower triangular matrix. The choice of ε_k can be made as follows. Start with some initial value. If no Cholesky factorization can be found, multiply the trial value with some constant (typically 4) and start anew. If still unsuccessful continue in the same way until a factorization is found.

Note that too large an ε_k is not wanted. Then $\boldsymbol{B}_k$ would be nearly a constant times an identity matrix, and the method resembles steepest descent, with slow convergence at the beginning as a consequence.

Exercises

3.7 Let

$$f(\boldsymbol{x}) = \frac{1}{2}(x_1^2 - x_2)^2 + \frac{1}{2}(1 - x_1)^2.$$

a) Minimize $f(\boldsymbol{x})$ analytically.

b) Carry out one iteration step in the Newton algorithm, starting at $\boldsymbol{x}_1 = (2, 2)$. Compute and compare the function values in $\boldsymbol{x}_1$ and $\boldsymbol{x}_2$.

3.8 Show that, for a quadratic function $q(\boldsymbol{x}) = \frac{1}{2}\boldsymbol{x}^T\boldsymbol{H}\boldsymbol{x} + \boldsymbol{c}^T\boldsymbol{x} + d$, Newton's method converges in one iteration step, whatever the starting point $\boldsymbol{x}_1$. Assume that $\boldsymbol{H}$ is positive definite.

3.9 The function

$$f(\boldsymbol{x}) - x_1^3 + x_1x_2 + x_1^2x_2^2 - 3x_1$$

has the stationary points

$$\begin{aligned} &(0,3) \text{ and } (-1, \tfrac{1}{2}), \quad \text{saddlepoints,} \\ &(1, -\tfrac{1}{2}), \quad \text{local minimum.} \end{aligned}$$

a) Verify this.

b) Compute the Hessian $\boldsymbol{H}(1,1)$, and decide whether it is positive definite. Compute one Newton step from $\boldsymbol{x}_1 = (1,1)$ and compare $f(\boldsymbol{x}_1)$ and $f(\boldsymbol{x}_2)$.

c) Replace $\boldsymbol{H}(1,1)$ by $\boldsymbol{H}(1,1) + 2\boldsymbol{I}$ and investigate the definiteness. Compute the modified Newton step and $f(\boldsymbol{x}_2)$.

3.10 Show that if $\boldsymbol{B}$ is a positive definite matrix then $-\boldsymbol{B}\nabla f(\boldsymbol{x})$ is a direction of descent for f at $\boldsymbol{x}$, unless $\boldsymbol{x}$ is a stationary point.

S 3.11 Explain why Newton's method cannot be applied satisfactorily to the problem of minimizing the function

$$f(\boldsymbol{x}) = x_1^4 + x_1 x_2 + (1 + x_2)^2,$$

starting at $\boldsymbol{x}_1 = (0,0)$. Show that for $\boldsymbol{d} = -\boldsymbol{H}^{-1}\nabla f(\boldsymbol{x}_1)$ neither of the directions $\boldsymbol{d}$ or $-\boldsymbol{d}$ is downhill. — If a modified Newton method uses the matrix $\boldsymbol{H} + \varepsilon \boldsymbol{I}$, what range of ε makes $\boldsymbol{H} + \varepsilon \boldsymbol{I}$ positive definite? What search direction is obtained by using $\varepsilon = 1$? (Do not carry out a line search.)

3.12 Let

$$f(\boldsymbol{x}) = \frac{1}{24} x_1^2 (x_1^2 - 12) + x_2\big(2x_1 + \frac{3}{2} x_2 + 3x_3 + 2\big) + 4(x_3 + 1)^2,$$

and let $\boldsymbol{H}(\boldsymbol{x})$ denote its Hessian. Let $\boldsymbol{x}^{(1)} = (0,0,0)$.

a) Is $\boldsymbol{H}(\boldsymbol{x}^{(1)})$ positive definite?

b) Prove that $\boldsymbol{H}(\boldsymbol{x}^{(1)}) + 2\boldsymbol{I}$ is positive definite. Compute its Cholesky factorization (see theorem A.9, page 361) $\boldsymbol{L}\boldsymbol{L}^T$, where $\boldsymbol{L}$ is lower triangular.

c) Use the Cholesky factorization to solve

$$\Big(\boldsymbol{H}(\boldsymbol{x}^{(1)}) + 2\boldsymbol{I}\Big)\,(\boldsymbol{x}^{(2)} - \boldsymbol{x}^{(1)}) = -\nabla f(\boldsymbol{x}^{(1)}).$$

Find the modified Newton step $\boldsymbol{x}^{(2)}$. (No line search.)

(1998–12–14, 5)

3.5 Conjugate direction methods

3.5.1 Conjugate directions

Let $\boldsymbol{H}$ be a symmetric $n \times n$ matrix.

DEFINITION 1. *Vectors $\boldsymbol{d}_1, \boldsymbol{d}_2, \ldots, \boldsymbol{d}_k$ in $\mathbf{R}^n$ are called **$\boldsymbol{H}$-conjugate** (or **conjugate with respect to $\boldsymbol{H}$**) if they are linearly independent and if*

(17) $$\boldsymbol{d}_i^T \boldsymbol{H} \boldsymbol{d}_j = 0 \qquad \text{when } i \neq j.$$

Remark. The reader might recognize that in the case when $\boldsymbol{H}$ is positive definite condition (17) in fact requires the vectors to be orthogonal with respect to the scalar product

$$\langle \boldsymbol{x}, \boldsymbol{y} \rangle = \boldsymbol{x}^T \boldsymbol{H} \boldsymbol{y}.$$

However, the term $\boldsymbol{H}$-conjugate is used also when $\boldsymbol{H}$ is not positive definite.

We leave as an exercise for the reader to show that in the positive definite case vectors satisfying (17) and none being zero are in fact linearly independent. Hence in this case the first condition in the definition is unnecessary.

Example 7. Consider the positive definite matrix

$$\boldsymbol{H} = \begin{pmatrix} 4 & \sqrt{6} \\ \sqrt{6} & 4 \end{pmatrix}.$$

Let $\boldsymbol{d}_1 = (1, 0)$ be the first basis vector and let $\boldsymbol{d}_2 = (a, b)$. Condition (17) in this case means

$$\begin{pmatrix} 1 & 0 \end{pmatrix} \begin{pmatrix} 4 & \sqrt{6} \\ \sqrt{6} & 4 \end{pmatrix} \begin{pmatrix} a \\ b \end{pmatrix} = 0 \iff 4a + \sqrt{6}\, b = 0 \iff$$

$$\iff \begin{cases} a = -\sqrt{6}\, t \\ b = 4t. \end{cases}$$

Thus, the vectors $\boldsymbol{d}_1 = (1, 0)$ and $\boldsymbol{d}_2 = (-\sqrt{6}, 4)$ are $\boldsymbol{H}$-conjugate. (They are obviously linearly independent, and, anyway, $\boldsymbol{H}$ is positive definite.) □

For any positive definite $n \times n$ matrix a *basis* of conjugate directions exists and can be constructed by the *Gram-Schmidt orthogonalization process* with respect to the scalar product $\langle \boldsymbol{x}, \boldsymbol{y} \rangle$ defined above. We refer to literature on linear algebra for a proof of this. (See also exercise 3.16, page 90.)

Assume that $\boldsymbol{d}_1, \ldots, \boldsymbol{d}_n$ is a basis of vectors that are conjugate with respect to the positive definite matrix $\boldsymbol{H}$. The coordinates of a point (vector) $\boldsymbol{x}$ in $\mathbf{R}^n$ with respect to the canonical basis are as usual denoted by $x_1, \ldots, x_n$, and for the coordinates of $\boldsymbol{x}$ in the conjugate directions basis we use the notation $\alpha_1, \ldots, \alpha_n$. Thus

$$\boldsymbol{x} = \sum_{j=1}^{n} \alpha_j \boldsymbol{d}_j.$$

In matrix form the formula for the change of coordinates is[9]

$$\boldsymbol{x} = \boldsymbol{S}\boldsymbol{\alpha}, \tag{18}$$

where $\boldsymbol{S}$ is the matrix whose column entries are (the canonical coordinates of) the vectors $\boldsymbol{d}_i$, and $\boldsymbol{\alpha}$ is the column matrix containing the coordinates $\alpha_1, \ldots, \alpha_n$. The definition (17) of $\boldsymbol{H}$-conjugacy implies that

$$\boldsymbol{D} = \boldsymbol{S}^T \boldsymbol{H} \boldsymbol{S}$$

is a diagonal matrix. Consequently

$$\boldsymbol{x}^T \boldsymbol{H} \boldsymbol{x} = \boldsymbol{\alpha}^T \boldsymbol{S}^T \boldsymbol{H} \boldsymbol{S} \boldsymbol{\alpha} = \boldsymbol{\alpha}^T \boldsymbol{D} \boldsymbol{\alpha} = \sum_{i=1}^{n} \xi_i \alpha_i^2, \tag{19}$$

where $\xi_1, \ldots, \xi_n$ are the diagonal entries of $\boldsymbol{D}$. We state this result as a theorem.

THEOREM 3. *Changing the coordinates to a basis of $\boldsymbol{H}$-conjugate directions puts the quadratic form $\boldsymbol{x}^T \boldsymbol{H} \boldsymbol{x}$ on diagonal form.*

Example 8. As an exercise we do some more computing with the above formulas.

Let $q(\boldsymbol{x}) = \frac{1}{2}\boldsymbol{x}^T \boldsymbol{H} \boldsymbol{x} + \boldsymbol{c}^T \boldsymbol{x}$ be a quadratic function of n variables. Assume that $\boldsymbol{H}$ is positive definite and let $\boldsymbol{d}_1, \ldots, \boldsymbol{d}_n$ be a basis of conjugate directions. Denote by $\bar{\boldsymbol{x}}$ the minimum point of q. Since $\boldsymbol{\nabla} q(\boldsymbol{x}) = \boldsymbol{H}\boldsymbol{x} + \boldsymbol{c}$ we have $\boldsymbol{H}\bar{\boldsymbol{x}} = -\boldsymbol{c}$.

[9] As usual the notation $\boldsymbol{x}$ has double roles in a context like this: it denotes a certain point in $\mathbf{R}^n$, but also the column matrix containing the coordinates of this point in the canonical basis.

Let $\bar{\alpha}_1, \ldots, \bar{\alpha}_n$ denote the coordinates of $\bar{\boldsymbol{x}}$ in the conjugate directions basis. By the change of coordinates formula (18) then $\bar{\boldsymbol{x}} = \boldsymbol{S}\bar{\alpha}$. Multiplication by $\boldsymbol{S}^T\boldsymbol{H}$ from the left gives

$$\boldsymbol{S}^T\boldsymbol{H}\bar{\boldsymbol{x}} = \boldsymbol{S}^T\boldsymbol{H}\boldsymbol{S}\bar{\alpha}$$

which simplifies to

$$-\boldsymbol{S}^T\boldsymbol{c} = \boldsymbol{D}\bar{\alpha}.$$

For the individual rows this means that $-\boldsymbol{d}_i^T\boldsymbol{c} - \xi_i\bar{\alpha}_i$. Solving for $\bar{\alpha}_i$ we obtain

$$\bar{\alpha}_i = -\frac{\boldsymbol{d}_i^T\boldsymbol{c}}{\xi_i} = -\frac{\boldsymbol{d}_i^T\boldsymbol{c}}{\boldsymbol{d}_i^T\boldsymbol{H}\boldsymbol{d}_i}.$$

□

3.5.2 Minimization of quadratic functions along conjugate directions

We again consider a quadratic function of n variables

$$q(\boldsymbol{x}) = \frac{1}{2}\boldsymbol{x}^T\boldsymbol{H}\boldsymbol{x} + \boldsymbol{c}^T\boldsymbol{x},$$

with $\boldsymbol{H}$ positive definite. The following lemma will be needed here and elsewhere in the course.

LEMMA 3. *Let $\boldsymbol{x}$ be a point in $\mathbf{R}^n$ and $\boldsymbol{d} \neq \boldsymbol{0}$ a direction. Denote by $F(\lambda)$ the restriction of q to the line through $\boldsymbol{x}$ with direction $\boldsymbol{d}$, i.e.*

$$F(\lambda) = q(\boldsymbol{x} + \lambda\boldsymbol{d}), \qquad \lambda \in \mathbf{R}.$$

Then F has its minimum at $\lambda = -\dfrac{\boldsymbol{d}^T(\boldsymbol{H}\boldsymbol{x} + \boldsymbol{c})}{\boldsymbol{d}^T\boldsymbol{H}\boldsymbol{d}}$.

PROOF. A simple computation gives

$$F(\lambda) = \frac{1}{2}\boldsymbol{x}^T\boldsymbol{H}\boldsymbol{x} + \lambda\boldsymbol{x}^T\boldsymbol{H}\boldsymbol{d} + \frac{1}{2}\lambda^2\boldsymbol{d}^T\boldsymbol{H}\boldsymbol{d} + \boldsymbol{c}^T\boldsymbol{x} + \lambda\boldsymbol{c}^T\boldsymbol{d}.$$

Differentiating we obtain

$$F'(\lambda) = \boldsymbol{x}^T\boldsymbol{H}\boldsymbol{d} + \lambda\boldsymbol{d}^T\boldsymbol{H}\boldsymbol{d} + \boldsymbol{c}^T\boldsymbol{d}.$$

Setting the derivative equal to zero gives the desired result, since $\boldsymbol{c}^T\boldsymbol{d} = \boldsymbol{d}^T\boldsymbol{c}$ and $\boldsymbol{x}^T\boldsymbol{H}\boldsymbol{d} = \boldsymbol{d}^T\boldsymbol{H}\boldsymbol{x}$. □

We are now going to investigate the effect of successively searching for minima along $\boldsymbol{H}$-conjugate directions.

Example 9. Suppose that $q(\boldsymbol{x}) = \frac{1}{2}\boldsymbol{x}^T\boldsymbol{H}\boldsymbol{x}$ with the matrix $\boldsymbol{H}$ taken from example 7, i.e. $q(\boldsymbol{x}) = 2x_1^2 + \sqrt{6}\,x_1x_2 + 2x_2^2$. (Thus $\boldsymbol{c} = \boldsymbol{0}$ here.) We intend to use the $\boldsymbol{H}$-conjugate basis vectors $\boldsymbol{d}_1$ and $\boldsymbol{d}_2$ from that example as search directions in an iterative process to find the minimum. Of course the exact minimum occurs at the origin.

For a start we take as the initial point $\boldsymbol{x}_1 = (4, -2)$ and search for the minimum of q along the line through $\boldsymbol{x}_1$ with direction $\boldsymbol{d}_1 = (1, 0)$. By lemma 3 the minimum occurs at $\boldsymbol{x}_1 + \lambda_1\boldsymbol{d}_1$ where

$$\lambda_1 = -\frac{\boldsymbol{d}_1^T\boldsymbol{H}\boldsymbol{x}_1}{\boldsymbol{d}_1^T\boldsymbol{H}\boldsymbol{d}_1} = -\frac{16 - 2\sqrt{6}}{4} = -4 + \tfrac{1}{2}\sqrt{6}.$$

Put $\boldsymbol{x}_2 = \boldsymbol{x}_1 + \lambda\boldsymbol{d}_1 = (\frac{1}{2}\sqrt{6}, -2)$.

We now search for a minimum along the line through $\boldsymbol{x}_2$ with direction $\boldsymbol{d}_2 = (-\sqrt{6}, 4)$. By the lemma again this minimum occurs at $\boldsymbol{x}_3 = \boldsymbol{x}_2 + \lambda_2\boldsymbol{d}_2$ where

$$\lambda_2 = -\frac{\boldsymbol{d}_2^T\boldsymbol{H}\boldsymbol{x}_2}{\boldsymbol{d}_2^T\boldsymbol{H}\boldsymbol{d}_2} = -\frac{-20}{40} = \frac{1}{2}.$$

The result is $\boldsymbol{x}_3 = (0, 0)$. We have found the *exact* minimum with two iteration steps only. See the figure.

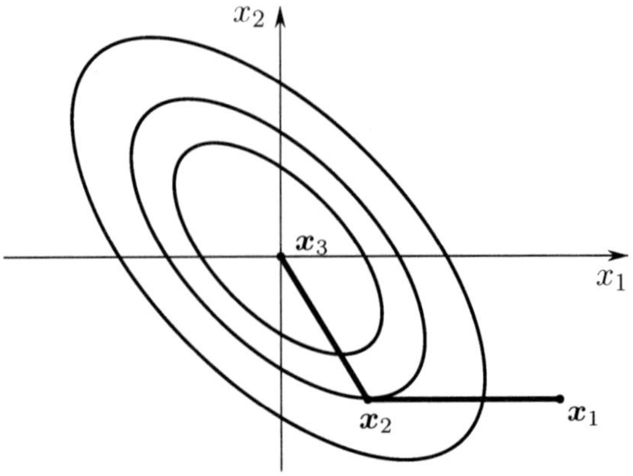

□

We now set out to prove that the phenomenon encountered in the example is no coincidence. In the case of n variables no more than n line searches along conjugate directions are needed to find the exact minimum of a quadratic function.

Consider as before the quadratic function $q(\boldsymbol{x}) = \frac{1}{2}\boldsymbol{x}^T\boldsymbol{H}\boldsymbol{x} + \boldsymbol{c}^T\boldsymbol{x}$. Letting $\bar{\boldsymbol{x}}$ denote its minimum point we rewrite the function as

$$(20) \qquad q(\boldsymbol{x}) = \frac{1}{2}(\boldsymbol{x}-\bar{\boldsymbol{x}})^T\boldsymbol{H}(\boldsymbol{x}-\bar{\boldsymbol{x}}) + a, \qquad \text{where} \quad a = q(\bar{\boldsymbol{x}})$$

(see lemma 1, page 17). Let $\boldsymbol{d}_1, \boldsymbol{d}_2, \ldots, \boldsymbol{d}_n$ be a basis of $\boldsymbol{H}$-conjugate directions. Assume that

$$\boldsymbol{x} = \sum_{j=1}^{n} \alpha_j \boldsymbol{d}_j, \qquad \bar{\boldsymbol{x}} = \sum_{j=1}^{n} \bar{\alpha}_j \boldsymbol{d}_j.$$

By theorem 3 this change of coordinates results in

$$(21) \qquad q(\boldsymbol{x}) = \frac{1}{2}\sum_{i=1}^{n} \xi_i(\alpha_i - \bar{\alpha}_i)^2 + a.$$

Thus we have represented $q(\boldsymbol{x})$ as a sum of uncoupled terms, each dependent on one variable α_i only. For a positive definite $\boldsymbol{H}$ all ξ_i are positive.

Of course a line search along direction $\boldsymbol{d}_i$ is equivalent to minimizing q in (21) with all variables except α_i kept fixed, obviously resulting in $\alpha_i = \bar{\alpha}_i$. Consequently, from (21) we draw the conclusion that if we minimize $q(\boldsymbol{x})$ successively along the directions $\boldsymbol{d}_1, \ldots, \boldsymbol{d}_n$, the order in which these directions are used is of no consequence, and after n steps[10] the exact minimum is found.

Before we state the theorem we introduce the notation $\mathcal{L}(\boldsymbol{d}_1, \ldots, \boldsymbol{d}_k)$ for the set of all linear combinations of the vectors $\boldsymbol{d}_1, \ldots, \boldsymbol{d}_k$ (the linear space *spanned by* $\boldsymbol{d}_1, \ldots, \boldsymbol{d}_k$).

THEOREM 4. *Consider a quadratic function $q(\boldsymbol{x}) = \frac{1}{2}\boldsymbol{x}^T\boldsymbol{H}\boldsymbol{x} + \boldsymbol{c}^T\boldsymbol{x}$, with $\boldsymbol{H}$ positive definite, and a basis $\boldsymbol{d}_1, \ldots, \boldsymbol{d}_n$ of $\boldsymbol{H}$-conjugate vectors.*

a) *Let $\boldsymbol{x}_1$ be any initial point and define successively*

$$\boldsymbol{x}_{k+1} = \boldsymbol{x}_k + \lambda_k \boldsymbol{d}_k, \qquad k = 1, 2, \ldots, n,$$

[10] For certain initial points the minimum may of course be found in fewer steps.

where $\lambda_k = -\dfrac{\boldsymbol{d}_k^T(\boldsymbol{H}\boldsymbol{x}_k + \boldsymbol{c})}{\boldsymbol{d}_k^T\boldsymbol{H}\boldsymbol{d}_k}$. *Then* $\boldsymbol{x}_{n+1}$ *is the exact solution to the problem*

$$\min_{\boldsymbol{x}\in\mathbf{R}^n} q(\boldsymbol{x}).$$

b) Likewise, $\boldsymbol{x}_{k+1}$ *is for any* k *the solution to the problem*

$$\min_{\boldsymbol{x}-\boldsymbol{x}_1\in\mathcal{L}(\boldsymbol{d}_1,\ldots,\boldsymbol{d}_k)} q(\boldsymbol{x}).$$

PROOF. Since the numbers λ_k are defined according to lemma 3 the point $\boldsymbol{x}_{k+1}$ minimizes q in direction $\boldsymbol{d}_k$ from $\boldsymbol{x}_k$. Hence the statements in both a) and b) follow from the representation (21) of q in the conjugate directions basis, as explained above. □

Statement b) in the theorem is sometimes called the *expanding subspace property*. It has an interesting geometrical interpretation. Put

$$E(\boldsymbol{x}) = \frac{1}{2}(\boldsymbol{x}-\bar{\boldsymbol{x}})^T\boldsymbol{H}(\boldsymbol{x}-\bar{\boldsymbol{x}}) = q(\boldsymbol{x}) - q(\bar{\boldsymbol{x}}).$$

Of course E is minimized when q is. (In the metric defined by the scalar product $\langle\boldsymbol{x},\boldsymbol{y}\rangle$ it is connected to the distance between $\boldsymbol{x}$ and $\bar{\boldsymbol{x}}$.) The following picture illustrates statement b) above. Let $\boldsymbol{x}_1 = \boldsymbol{0}$ for simplicity. The point $\boldsymbol{x}_k$ minimizes E over the space $\mathcal{L}(\boldsymbol{d}_1,\ldots,\boldsymbol{d}_{k-1})$, while $\boldsymbol{x}_{k+1}$ minimizes E over $\mathcal{L}(\boldsymbol{d}_1,\ldots,\boldsymbol{d}_{k-1},\boldsymbol{d}_k)$.

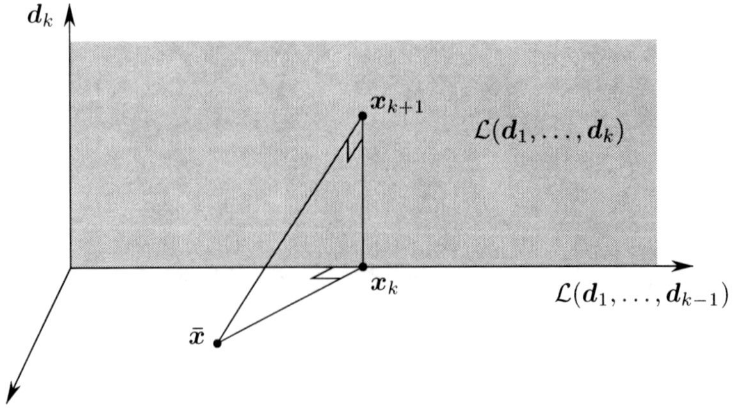

From lemma 1, page 43, we know that exact line search implies that the procured point $\boldsymbol{x}_{k+1}$ has the property that $\nabla q(\boldsymbol{x}_{k+1})$ is orthogonal to the search direction $\boldsymbol{d}_k$. When conjugate directions are used $\nabla q(\boldsymbol{x}_{k+1})$ is in fact orthogonal to *all* earlier search directions, that is, to the whole space $\mathcal{L}(\boldsymbol{d}_1, \ldots, \boldsymbol{d}_k)$. This is the content of the following lemma.

LEMMA 4.*Minimizing (with exact line search) a quadratic function q along a sequence of conjugate directions $\boldsymbol{d}_1, \ldots, \boldsymbol{d}_n$ results in a sequence of points $\boldsymbol{x}_1, \ldots, \boldsymbol{x}_{n+1}$ such that for all k*

$$\nabla q(\boldsymbol{x}_{k+1})^T \boldsymbol{d}_i = 0 \quad \text{for} \quad i = 1, \ldots, k. \tag{22}$$

PROOF. We again use the change of coordinates $\boldsymbol{x} = \boldsymbol{S}\boldsymbol{\alpha}$ from (18). By the chain rule

$$\frac{\partial q}{\partial \alpha_i}(\boldsymbol{\alpha}) = \nabla q(\boldsymbol{x})^T \boldsymbol{d}_i$$

(∇ refers to derivatives with respect to the $\boldsymbol{x}$-variables). In particular,

$$\nabla q(\boldsymbol{x}_{k+1})^T \boldsymbol{d}_i = \frac{\partial q}{\partial \alpha_i}(\boldsymbol{\alpha}_{k+1}).$$

From (21)

$$\frac{\partial q}{\partial \alpha_i}(\boldsymbol{\alpha}) = \xi_i(\alpha_i - \bar{\alpha}_i).$$

The point $\boldsymbol{x}_{k+1}$ has in the conjugate directions basis coordinates

$$\boldsymbol{\alpha}_{k+1} = (\bar{\alpha}_1, \ldots, \bar{\alpha}_k, \alpha_{k+1}, \ldots, \alpha_n).$$

From this the result follows. □

Until now we have worked on the implicit assumption that the conjugate directions are somehow given beforehand. Far more interesting is to construct the directions successively, stepwise in the iterations. It is towards this issue that we now turn our attention.

Suppose that we start by choosing $\boldsymbol{d}_1$ as the direction $-\nabla f(\boldsymbol{x}_1)$ of steepest descent from $\boldsymbol{x}_1$. Then $\boldsymbol{x}_2$ is defined as the solution to a minimization problem on the line through $\boldsymbol{x}_1$ in direction $\boldsymbol{d}_1$. A new direction $\boldsymbol{d}_2$ is constructed based on the results of these computations, and so on. We require that $\boldsymbol{d}_1, \boldsymbol{d}_2, \ldots$ are $\boldsymbol{H}$-conjugate. At a current point $\boldsymbol{x}_{k+1}$ we are supposed to choose a direction $\boldsymbol{d}_{k+1}$ conjugate to all former directions $\boldsymbol{d}_1, \ldots, \boldsymbol{d}_k$. We might try to write this vector as a deflection of the negative gradient $-\nabla q(\boldsymbol{x}_{k+1})$ by a vector in the space $\mathcal{L}(\boldsymbol{d}_1, \ldots, \boldsymbol{d}_k)$, i.e., a linear combination of $\boldsymbol{d}_1, \ldots, \boldsymbol{d}_k$. In fact it can be proved (see theorem 5 below) that we can do better; it is enough to choose $\boldsymbol{d}_{k+1}$ as a linear combination of $-\nabla q(\boldsymbol{x}_{k+1})$ and the last direction $\boldsymbol{d}_k$:

$$\boldsymbol{d}_{k+1} = -\nabla q(\boldsymbol{x}_{k+1}) + \beta_k \boldsymbol{d}_k. \tag{23}$$

(We have normalized so that the coefficient of $-\nabla q(\boldsymbol{x}_{k+1})$ is 1.) We will soon decide the value required for the coefficient β_k, but first we deduce some general consequences of the proposed scheme.

As indicated above the construction is done recursively. To be precise: assume that directions $\boldsymbol{d}_i$ and points $\boldsymbol{x}_{i+1}$, $1 \le i \le k$, have been chosen successively, starting from some initial point $\boldsymbol{x}_1$ and with $\boldsymbol{d}_1 = -\nabla q(\boldsymbol{x}_1)$. The points have been defined via exact line search, and the directions have been defined by the rule

$$\boldsymbol{d}_i = -\nabla q(\boldsymbol{x}_i) + \beta_{i-1} \boldsymbol{d}_{i-1}, \qquad i = 2, \ldots, k, \tag{24}$$

with coefficients β_{i-1} ensuring that $\boldsymbol{d}_1, \ldots, \boldsymbol{d}_k$ are $\boldsymbol{H}$-conjugate.

Multiplying (24) by $\nabla q(\boldsymbol{x}_i)^T$ we obtain

$$\nabla q(\boldsymbol{x}_i)^T \boldsymbol{d}_i = -\nabla q(\boldsymbol{x}_i)^T \nabla q(\boldsymbol{x}_i) + \beta_{i-1} \nabla q(\boldsymbol{x}_i)^T \boldsymbol{d}_{i-1}.$$

By lemma 4 the last term is zero, resulting in

$$\nabla q(\boldsymbol{x}_i)^T \boldsymbol{d}_i = -\|\nabla q(\boldsymbol{x}_i)\|^2, \qquad i = 1, \ldots, k. \tag{25}$$

(This is by definition true also when $i = 1$.) From (25) it follows in particular that $\nabla q(\boldsymbol{x}_i)^T \boldsymbol{d}_i < 0$, unless $\boldsymbol{x}_i$ is a stationary point; thus $\boldsymbol{d}_i$ is a descent direction at $\boldsymbol{x}_i$.

Multiplying (24) by $\nabla q(\boldsymbol{x}_j)^T$ for $i < j\ (\le k+1)$ we likewise obtain

$$\nabla q(\boldsymbol{x}_j)^T \boldsymbol{d}_i = -\nabla q(\boldsymbol{x}_j)^T \nabla q(\boldsymbol{x}_i) + \beta_{i-1} \nabla q(\boldsymbol{x}_j)^T \boldsymbol{d}_{i-1},$$

which by lemma 4 leads to

$$\nabla q(\boldsymbol{x}_j)^T \nabla q(\boldsymbol{x}_i) = 0 \qquad \text{if} \quad j > i. \tag{26}$$

(Also this formula holds when $i = 1$ as well.)

For later reference we make another observation: since $\nabla q(\boldsymbol{x}) = \boldsymbol{H}\boldsymbol{x} + \boldsymbol{c}$ we have

$$\nabla q(\boldsymbol{x}_{i+1}) - \nabla q(\boldsymbol{x}_i) = \boldsymbol{H}(\boldsymbol{x}_{i+1} - \boldsymbol{x}_i) = \lambda_i \boldsymbol{H}\boldsymbol{d}_i. \tag{27}$$

When convenient in the future we will assume that $\lambda_i \neq 0$. If $\lambda_i = 0$ (this may occur for some choices of the initial point) the consequence would be that $\boldsymbol{x}_{i+1} = \boldsymbol{x}_i$. This is of course no complication at all in practice. However, it introduces some exceptional cases in the future discussion, something we prefer to avoid.

We now return to (23) with the intention to find a value of β_k making $\boldsymbol{d}_{k+1}$ $\boldsymbol{H}$-conjugate to $\boldsymbol{d}_k$. Multiplying by $\lambda_k \boldsymbol{H}\boldsymbol{d}_k$ from the right gives, using (27),

$$\begin{aligned}
&\lambda_k \boldsymbol{d}_{k+1}^T \boldsymbol{H}\boldsymbol{d}_k = \\
&\quad = -\nabla q(\boldsymbol{x}_{k+1})^T \lambda_k \boldsymbol{H}\boldsymbol{d}_k + \beta_k \boldsymbol{d}_k^T \lambda_k \boldsymbol{H}\boldsymbol{d}_k = \\
&\quad = -\nabla q(\boldsymbol{x}_{k+1})^T \Big(\nabla q(\boldsymbol{x}_{k+1}) - \nabla q(\boldsymbol{x}_k)\Big) + \beta_k \boldsymbol{d}_k^T \Big(\nabla q(\boldsymbol{x}_{k+1}) - \nabla q(\boldsymbol{x}_k)\Big).
\end{aligned} \tag{28}$$

Since $\boldsymbol{d}_1, \ldots, \boldsymbol{d}_k$ are assumed to be $\boldsymbol{H}$-conjugate, by (26) in particular

$$\nabla q(\boldsymbol{x}_{k+1})^T \nabla q(\boldsymbol{x}_k) = 0$$

and by (22)

$$\boldsymbol{d}_k^T \nabla q(\boldsymbol{x}_{k+1}) = \nabla q(\boldsymbol{x}_{k+1})^T \boldsymbol{d}_k = 0.$$

Hence, using (25) we can simplify (28) to

$$\lambda_k \boldsymbol{d}_{k+1}^T \boldsymbol{H}\boldsymbol{d}_k = -\|\nabla q(\boldsymbol{x}_{k+1})\|^2 + \beta_k \|\nabla q(\boldsymbol{x}_k)\|^2.$$

It follows that $\boldsymbol{d}_{k+1}$ is $\boldsymbol{H}$-conjugate to $\boldsymbol{d}_k$ if and only if

$$-\|\nabla q(\boldsymbol{x}_{k+1})\|^2 + \beta_k \|\nabla q(\boldsymbol{x}_k)\|^2 = 0.$$

Here we can solve for β_k. The result is

$$\beta_k = \frac{\|\nabla q(\boldsymbol{x}_{k+1})\|^2}{\|\nabla q(\boldsymbol{x}_k)\|^2}. \tag{29}$$

We must prove that the directions $\boldsymbol{d}_{k+1}$ defined recursively above with β_k given by (29) are $\boldsymbol{H}$-conjugate to all previous directions, not only to the last one $\boldsymbol{d}_k$. This is the content of the following theorem.

THEOREM 5. *Suppose that we define* $\boldsymbol{d}_1 = -\boldsymbol{\nabla} q(\boldsymbol{x}_1)$ *and recursively*

$$\boldsymbol{x}_{k+1} = \boldsymbol{x}_k + \lambda_k \boldsymbol{d}_k, \qquad k = 1, \ldots, n,$$
$$\boldsymbol{d}_{k+1} = -\boldsymbol{\nabla} q(\boldsymbol{x}_{k+1}) + \beta_k \boldsymbol{d}_k, \qquad k = 1, \ldots, n-1,$$

where

$$\lambda_k = -\frac{\boldsymbol{d}_k^T(\boldsymbol{H}\boldsymbol{x}_k + \boldsymbol{c})}{\boldsymbol{d}_k^T \boldsymbol{H} \boldsymbol{d}_k}, \qquad \beta_k = \frac{\|\boldsymbol{\nabla} q(\boldsymbol{x}_{k+1})\|^2}{\|\boldsymbol{\nabla} q(\boldsymbol{x}_k)\|^2}.$$

Then

$$\boldsymbol{d}_i^T \boldsymbol{H} \boldsymbol{d}_j = 0 \quad \text{if} \quad i \neq j.$$

PROOF. Note that the values of λ_k are those obtained in lemma 3 when minimizing q in direction $\boldsymbol{d}_k$ from $\boldsymbol{x}_k$.

We use induction over k to prove that $\boldsymbol{d}_{k+1}^T \boldsymbol{H} \boldsymbol{d}_l = 0$ for $l = 1, \ldots, k$. We already know that the choice of β_k makes $\boldsymbol{d}_{k+1}^T \boldsymbol{H} \boldsymbol{d}_k = 0$. In particular this covers the case $k = 1$ of the induction proof.

As induction hypothesis we now assume that

$$\boldsymbol{d}_k^T \boldsymbol{H} \boldsymbol{d}_l = 0, \quad l = 1, \ldots, k-1,$$

and consider $\boldsymbol{d}_{k+1}^T \boldsymbol{H} \boldsymbol{d}_l$. For $l = k$ we already know that this is equal to zero. Assuming $l \leq k-1$, by the recursion formula we obtain

$$\boldsymbol{d}_{k+1}^T \boldsymbol{H} \boldsymbol{d}_l = -\boldsymbol{\nabla} q(\boldsymbol{x}_{k+1})^T \boldsymbol{H} \boldsymbol{d}_l + \beta_k \boldsymbol{d}_k^T \boldsymbol{H} \boldsymbol{d}_l.$$

By (27) the first term can be written $-\frac{1}{\lambda_l} \boldsymbol{\nabla} q(\boldsymbol{x}_{k+1})^T (\boldsymbol{\nabla} q(\boldsymbol{x}_{l+1}) - \boldsymbol{\nabla} q(\boldsymbol{x}_l))$, and is thus equal to zero by (26). The second term is equal to zero by the induction hypothesis. This proves the theorem. □

3.5.3 Minimization methods using conjugate directions

The conjugate gradient method

It is not the intention of the foregoing discussion to suggest that conjugate directions should be used to minimize quadratic functions. We know how to find their exact minimum in simpler ways. Rather, the usefulness of conjugate directions comes from the fact that by Taylor's formula any twice differentiable function can be approximated in the vicinity of a local minimum point by a quadratic function. Hence, a method that works well for quadratic functions should be expected to work well for any such function close to the minimum.

The way of choosing the search directions in theorem 5 does not require knowledge of second order derivatives for the computation of β_k. The formula for λ_k however must be abolished in the general case, and a line search method introduced in its place. Of course we can not in the case of a general function f expect to find the exact minimum after n iterations. Instead, after carrying out these iterations we begin another round, starting anew with the negative gradient and successively choosing the other search directions, and so on.

More precisely, the algorithm runs as follows. There is an outer loop, controlled by the variable k, and an inner loop controlled by j. Let an initial point $\boldsymbol{x}_1$ be chosen.

For $k := 1$ *until termination do*

- *Let* $\boldsymbol{y}_1 := \boldsymbol{x}_k$.
- *For* $j := 1$ *to* n *do*
  - *Put* $\boldsymbol{d}_j :=$ *if* $j = 1$ *then* $-\nabla f(\boldsymbol{y}_1)$

    $$\textit{else} \quad -\nabla f(\boldsymbol{y}_j) + \frac{\|\nabla f(\boldsymbol{y}_j)\|^2}{\|\nabla f(\boldsymbol{y}_{j-1})\|^2}\boldsymbol{d}_{j-1}.$$
  - *Let* λ_j *be the solution to the one-dimensional problem*

    $$\min_\lambda f(\boldsymbol{y}_j + \lambda \boldsymbol{d}_j).$$
  - *Put* $\boldsymbol{y}_{j+1} := \boldsymbol{y}_j + \lambda_j \boldsymbol{d}_j$.
- *Let* $\boldsymbol{x}_{k+1} := \boldsymbol{y}_{n+1}$.
- *Decide whether to terminate or continue.*

The formula (29) for β_k can in the case of a quadratic function q be rewritten in several equivalent ways. For a non-quadratic function these expressions can be used to devise other versions of the conjugate gradient algorithm, with slightly different properties. We refer to the literature for a description of these modifications. Our version is known as the *Fletcher-Reeves* method.

k	$\boldsymbol{x}_k = \boldsymbol{y}_1$	$f(\boldsymbol{x}_k)$	j	$\boldsymbol{d}_j$	$\boldsymbol{y}_{j+1}$	$f(\boldsymbol{y}_{j+1})$
1	(0, 2.5)	88.2500	1	(94, −45)	(2.1799, 1.4564)	0.0022
			2	(−0.0083, −0.0574)	(2.1793, 1.4525)	0.0021
2	(2.1793, 1.4525)	0.0021	1	(−0.0510, 0.0074)	(2.1752, 1.4531)	0.0020
			2	(−0.0627, −0.0451)	(2.0673, 1.3756)	0.000103
3	(2.0673, 1.3756)	0.0001	1	(−0.0339, 0.0473)	(2.0660, 1.3774)	0.000038
			2	(−0.0016, −0.0011)	(2.0090, 1.3392)	0.000000
4	(2.0090, 1.3392)	0.000000				

Example 10. As before (example 1, page 45, example 4, page 49, example 6, page 63) we apply the method to the objective function

$$f(x_1, x_2) = 2(x_1 - 2)^4 + (2x_1 - 3x_2)^2.$$

The table on the opposite page gives the results of the computations, which also are illustrated in the figure below. After only three iterations we have arrived at (2.0090, 1.3392). The distance to the exact minimum $(2, \frac{4}{3})$ is approximately 0.01.

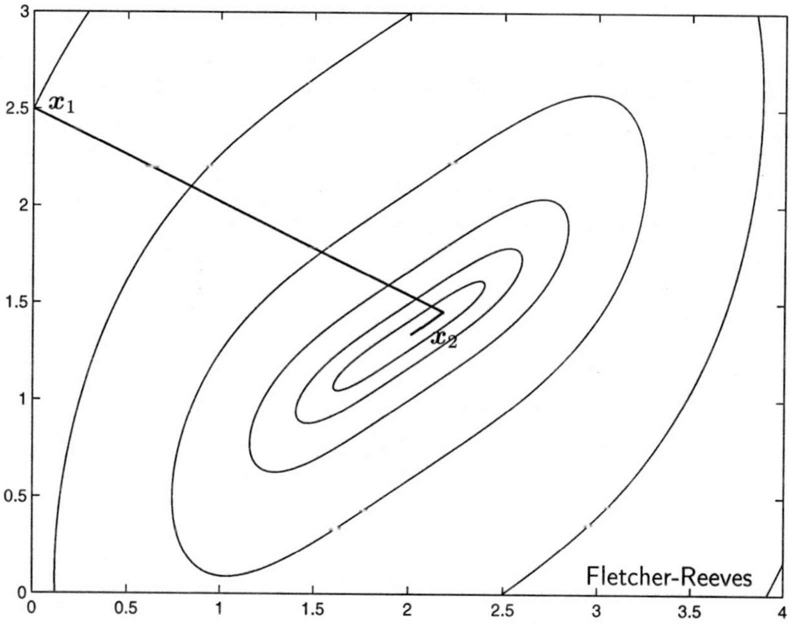

□

An important feature of the conjugate gradient method is that it requires *little computer memory*. The information that must be stored and updated in the iteration steps is contained in a one-dimensional array only. This is in contrast to the following quasi-Newton methods, where *matrices* need to be stored and recomputed at every iteration.

For this reason the conjugate gradient method is suitable for problems containing a large number of variables (meaning about 500 or more). In fact it is almost the only method used for such large problems.

Quasi-Newton methods

The general idea behind all quasi-Newton methods is to replace the inverse Hessian in Newton's method with an approximation, which requires first derivative information only to be computed and no matrix inversion. Hence the general iteration step will look like

$$\boldsymbol{x}_{k+1} = \boldsymbol{x}_k - \lambda_k \boldsymbol{D}_k \nabla f(\boldsymbol{x}_k).$$

The matrix $\boldsymbol{D}_k$, which is updated at every iteration step, should be positive definite to assure that a descent direction is chosen.

The following algorithm, one of many with these properties, is known as the *Davidon-Fletcher-Powell* (DFP) method. As in the Fletcher-Reeves method there is an outer loop (variable k below) and an inner one (variable j). First an initial point $\boldsymbol{x}_1$ is chosen, then:

For $k := 1$ until termination do

- *Let $\boldsymbol{y}_1 := \boldsymbol{x}_k$.*
- *Let $\boldsymbol{D}_1 := \boldsymbol{I}$.*
- *For $j := 1$ to n do*
  - *Put $\boldsymbol{d}_j := -\boldsymbol{D}_j \nabla f(\boldsymbol{y}_j)$.*
  - *Let λ_j be the solution to the one-dimensional problem*
  $$\min_{\lambda} f(\boldsymbol{y}_j + \lambda \boldsymbol{d}_j).$$
  - *Put $\boldsymbol{y}_{j+1} := \boldsymbol{y}_j + \lambda_j \boldsymbol{d}_j$.*
  - *Update $\boldsymbol{D}_j$ by formulas (30) below.*
- *Let $\boldsymbol{x}_{k+1} := \boldsymbol{y}_{n+1}$.*
- *Decide whether to terminate or continue.*

The formulas for updating the matrix $\boldsymbol{D}_j$ are as follows:

$$\begin{aligned} \boldsymbol{p}_j &:= \lambda_j \boldsymbol{d}_j \equiv \boldsymbol{y}_{j+1} - \boldsymbol{y}_j, \\ \boldsymbol{q}_j &:= \nabla f(\boldsymbol{y}_{j+1}) - \nabla f(\boldsymbol{y}_j), \\ \boldsymbol{D}_{j+1} &:= \boldsymbol{D}_j + \frac{1}{\boldsymbol{p}_j^T \boldsymbol{q}_j} \boldsymbol{p}_j \boldsymbol{p}_j^T - \frac{1}{\boldsymbol{q}_j^T \boldsymbol{D}_j \boldsymbol{q}_j} \boldsymbol{D}_j \boldsymbol{q}_j \boldsymbol{q}_j^T \boldsymbol{D}_j. \end{aligned} \tag{30}$$

k	$\boldsymbol{x}_k = \boldsymbol{y}_1$	$f(\boldsymbol{x}_k)$	j	$\boldsymbol{D}_j$	$\boldsymbol{d}_j$	$\boldsymbol{y}_{j+1}$	$f(\boldsymbol{y}_{j+1})$
1	(0 , 2.5)	88.2500	1	$\begin{pmatrix} 1 & 0 \\ 0 & 1 \end{pmatrix}$	(94 , −45)	(2.1799, 1.4564)	0.0022
			2	$\begin{pmatrix} 0.2049 & 0.3801 \\ 0.3801 & 0.8183 \end{pmatrix}$	(−0.0235, −0.0501)	(2.1766, 1.4494)	0.0020
2	(2.1766, 1.4494)	0.0020	1	$\begin{pmatrix} 1 & 0 \\ 0 & 1 \end{pmatrix}$	(−0.0641, 0.0301)	(2.1733, 1.4509)	0.0018
			2	$\begin{pmatrix} 0.7112 & 0.4508 \\ 0.4508 & 0.3397 \end{pmatrix}$	(−0.0288, −0.0202)	(2.0560, 1.3685)	0.000062
3	(2.0560, 1.3685)	0.000062	1	$\begin{pmatrix} 1 & 0 \\ 0 & 1 \end{pmatrix}$	(−0.0275, 0.0392)	(2.0550, 1.3700)	0.000018
			2	$\begin{pmatrix} 0.7038 & 0.4440 \\ 0.4440 & 0.3347 \end{pmatrix}$	(−0.0009, −0.0006)	(2.0064, 1.3376)	0.000000
4	(2.0064, 1.3376)	0.000000					

Example 11. We study the performance of the algorithm on our old objective function,

$$f(x_1, x_2) = 2(x_1 - 2)^4 + (2x_1 - 3x_2)^2.$$

As in the case of the Fletcher-Reeves method (example 10) we obtain convergence to the minimum $(2, \frac{4}{3})$ in 3 iterations. The table (previous page) and the figure show the results of the computations.

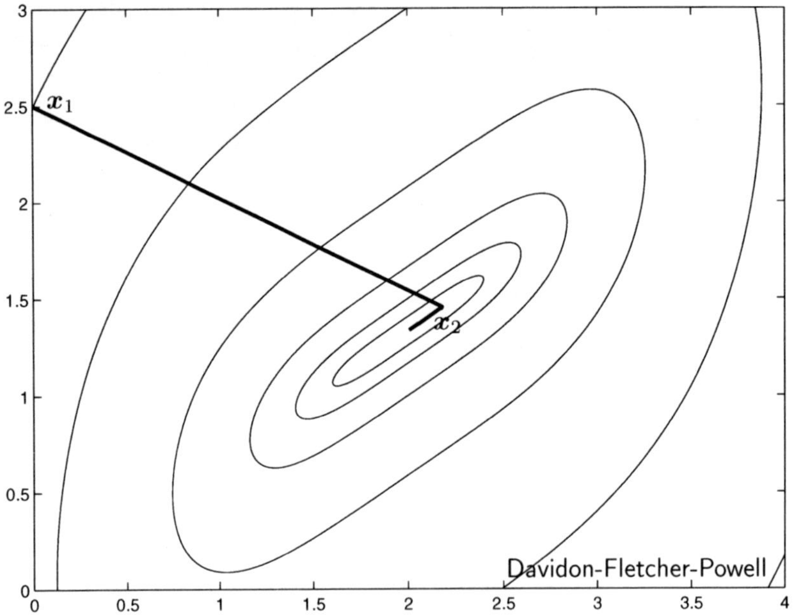

□

It must of course be proved that the matrices $\boldsymbol{D}_j$ defined in (30) are positive definite. We will do that in a moment, but first we give some general comments on the algorithm.

The first step is taken in the direction of the negative gradient, meaning that the algorithm starts out as the steepest descent method. In the case of a quadratic objective function we later prove that the last matrix $\boldsymbol{D}_{n+1}$ is equal to the inverse Hessian $\boldsymbol{H}^{-1}$. The algorithm thus ends up as the Newton method, with its fast convergence near a minimum. One might say that the algorithm is able to use the advantages of both methods.

Furthermore, it will be proved soon that in the case of a quadratic function the directions generated are $\boldsymbol{H}$-conjugate. By theorem 4 this implies that the exact minimum is found after n iterations (only one outer loop in the algorithm). — In the case of a non-quadratic objective function we can of course not expect this. Then we start another sequence of n (inner) iterations, beginning as before with the steepest descent direction, etc.

The DFP method is considered more effective than the Fletcher-Reeves method discussed before. However, it requires more computer storage, and hence can not be used on too large problems.[11]

In both methods it is sometimes advantageous to stop the inner iterations prematurely, at $j = m$ for some $m < n$, and resetting the search direction to steepest descent. The reason for this has to do with the eigenvalue distribution of the Hessian. We refer to other literature.

We now turn to proofs of some of the properties of the DFP algorithm stated above.[12]

LEMMA 5. *The matrices $\boldsymbol{D}_j$ appearing in the DFP algorithm above are all symmetric and positive definite.*

PROOF. The proof is done by induction. The initial matrix is $\boldsymbol{D}_1 = \boldsymbol{I}$, which is obviously positive definite. Assume that $\boldsymbol{D}_j$ is symmetric and positive definite. Then $\boldsymbol{D}_{j+1}$ is easily seen to be symmetric; we now prove that it is positive definite.

Let $\boldsymbol{x} \neq \boldsymbol{0}$ and consider $\boldsymbol{x}^T \boldsymbol{D}_{j+1} \boldsymbol{x}$. By the recursive definition in (30)

$$\boldsymbol{x}^T \boldsymbol{D}_{j+1} \boldsymbol{x} = \boldsymbol{x}^T \boldsymbol{D}_j \boldsymbol{x} + \frac{\boldsymbol{x}^T \boldsymbol{p}_j \boldsymbol{x}^T \boldsymbol{p}_j}{\boldsymbol{p}_j^T \boldsymbol{q}_j} - \frac{\boldsymbol{x}^T \boldsymbol{D}_j \boldsymbol{q}_j \boldsymbol{x}^T \boldsymbol{D}_j \boldsymbol{q}_j}{\boldsymbol{q}_j^T \boldsymbol{D}_j \boldsymbol{q}_j}. \tag{31}$$

Since $\boldsymbol{D}_j$ is assumed positive definite it has a Cholesky factorization $\boldsymbol{D}_j = \boldsymbol{L}_j \boldsymbol{L}_j^T$. Put

$$\boldsymbol{a}_j = \boldsymbol{L}_j^T \boldsymbol{x}, \qquad \boldsymbol{b}_j = \boldsymbol{L}_j^T \boldsymbol{q}_j.$$

[11] It can be shown that for a quadratic objective function these two methods are equivalent, yielding the same result at every iteration step.

[12] We ignore the possibility of $\nabla f(\boldsymbol{y}_j)$ being zero for $j = 1, \ldots, n$, that is, that we by an accidental lucky choice of initial point find the exact minimum prematurely.

Now the first and last term in the right hand side of (31) can be written

$$\boldsymbol{a}_j^T \boldsymbol{a}_j - \frac{\boldsymbol{a}_j^T \boldsymbol{b}_j \boldsymbol{a}_j^T \boldsymbol{b}_j}{\boldsymbol{b}_j^T \boldsymbol{b}_j} = \frac{\boldsymbol{a}_j^T \boldsymbol{a}_j \boldsymbol{b}_j^T \boldsymbol{b}_j - (\boldsymbol{a}_j^T \boldsymbol{b}_j)^2}{\boldsymbol{b}_j^T \boldsymbol{b}_j}.$$

By the Cauchy-Schwarz inequality this expression is non-negative. Consider now the middle term in (31). The denominator can be written

$$\boldsymbol{p}_j^T \boldsymbol{q}_j = \lambda_j \boldsymbol{d}_j^T (\boldsymbol{\nabla} f(\boldsymbol{y}_{j+1}) - \boldsymbol{\nabla} f(\boldsymbol{y}_j)) = -\lambda_j \boldsymbol{d}_j^T \boldsymbol{\nabla} f(\boldsymbol{y}_j)$$

by lemma 1, page 43. Using the definition of $\boldsymbol{d}_j$ this can be rewritten

$$\boldsymbol{p}_j^T \boldsymbol{q}_j = \lambda_j \boldsymbol{\nabla} f(\boldsymbol{y}_j)^T \boldsymbol{D}_j \boldsymbol{\nabla} f(\boldsymbol{y}_j). \tag{32}$$

By the induction hypothesis this number is strictly positive if $\boldsymbol{\nabla} f(\boldsymbol{y}_j) \neq \boldsymbol{0}$. Thus the middle term in (31) is non-negative, since the nominator is $(\boldsymbol{x}^T \boldsymbol{p}_j)^2$, which is ≥ 0.

We must prove that the two expressions into which we have separated (31) are not both zero. But equality in the Cauchy-Schwarz inequality occurs only when $\boldsymbol{a}_j$ and $\boldsymbol{b}_j$ are proportional, $\boldsymbol{a}_j = \gamma \boldsymbol{b}_j$ say. Then $\boldsymbol{x} = \gamma \boldsymbol{q}_j$ and consequently by (32)

$$\boldsymbol{x}^T \boldsymbol{p}_j = \gamma \boldsymbol{q}_j^T \boldsymbol{p}_j = \gamma \lambda_j \boldsymbol{\nabla} f(\boldsymbol{y}_j)^T \boldsymbol{D}_j \boldsymbol{\nabla} f(\boldsymbol{y}_j) \neq 0.$$

Thus the proof is finished. □

Consider now the special case when the objective function is quadratic,

$$q(\boldsymbol{x}) = \frac{1}{2} \boldsymbol{x}^T \boldsymbol{H} \boldsymbol{x} + \boldsymbol{c}^T \boldsymbol{x},$$

with $\boldsymbol{H}$ positive definite. Then $\boldsymbol{\nabla} q(\boldsymbol{x}) = \boldsymbol{H}\boldsymbol{x} + \boldsymbol{c}$, implying that, with the notation in (30),

$$\boldsymbol{q}_j = \boldsymbol{H} \boldsymbol{y}_{j+1} - \boldsymbol{H} \boldsymbol{y}_j = \boldsymbol{H} \boldsymbol{p}_j. \tag{33}$$

THEOREM 6. *Assume that a sequence of points* $\{\boldsymbol{y}_j\}_1^n$ *and a sequence of directions* $\{\boldsymbol{d}_j\}_1^n$ *are generated by one (outer) step in the DFP algorithm for the quadratic function* q. *Then*

a) $\boldsymbol{d}_j^T \boldsymbol{H} \boldsymbol{d}_i = 0$ *if* $i \neq j$.

b) $\boldsymbol{D}_{j+1} \boldsymbol{H} \boldsymbol{d}_i = \boldsymbol{d}_i$ *for* $i = 1, \ldots, j$.

PROOF. Since $\boldsymbol{p}_i = \lambda_i \boldsymbol{d}_i$ we can substitute $\boldsymbol{p}_i$ for $\boldsymbol{d}_i$ in the statement. We are going to use induction on j to prove both statements simultaneously. To be precise, the statement we intend to work with is

$$\left.\begin{aligned} &\text{(a)}\quad \boldsymbol{p}_{j+1}^T \boldsymbol{H} \boldsymbol{p}_i = 0, \\ &\text{(b)}\quad \boldsymbol{D}_{j+1} \boldsymbol{H} \boldsymbol{p}_i = \boldsymbol{p}_i, \end{aligned}\right\} \quad \text{for} \quad i \le j.$$

By (33) and the recursive definition of $\boldsymbol{D}_j$

$$\begin{aligned} \boldsymbol{D}_{j+1}\boldsymbol{H}\boldsymbol{p}_j &= \boldsymbol{D}_{j+1}\boldsymbol{q}_j = \boldsymbol{D}_j\boldsymbol{q}_j + \frac{\boldsymbol{p}_j\boldsymbol{p}_j^T\boldsymbol{q}_j}{\boldsymbol{p}_j^T\boldsymbol{q}_j} - \frac{\boldsymbol{D}_j\boldsymbol{q}_j\boldsymbol{q}_j^T\boldsymbol{D}_j\boldsymbol{q}_j}{\boldsymbol{q}_j^T\boldsymbol{D}_j\boldsymbol{q}_j} = \\ &= \boldsymbol{D}_j\boldsymbol{q}_j + \boldsymbol{p}_j - \boldsymbol{D}_j\boldsymbol{q}_j = \boldsymbol{p}_j. \end{aligned} \tag{34}$$

In particular this proves the case $j = 1$ of part (b). From (34) it also follows that

$$\begin{aligned} \boldsymbol{p}_{j+1}^T\boldsymbol{H}\boldsymbol{p}_j &= -\lambda_{j+1}(\boldsymbol{D}_{j+1}\nabla f(\boldsymbol{y}_{j+1}))^T\boldsymbol{H}\boldsymbol{p}_j = \\ &= -\lambda_{j+1}\nabla f(\boldsymbol{y}_{j+1})^T\boldsymbol{D}_{j+1}\boldsymbol{H}\boldsymbol{p}_j = -\lambda_{j+1}\nabla f(\boldsymbol{y}_{j+1})^T\boldsymbol{p}_j. \end{aligned} \tag{35}$$

Using lemma 1, page 43, (35) provides a proof of the case $j = 1$ of part (a).

For the induction step (from $j-1$ to j, $j \ge 2$) assume that $\boldsymbol{p}_j^T\boldsymbol{H}\boldsymbol{p}_i = 0$ and $\boldsymbol{D}_j\boldsymbol{H}\boldsymbol{p}_i = \boldsymbol{p}_i$ for all $i < j$. Consider first $\boldsymbol{D}_{j+1}\boldsymbol{H}\boldsymbol{p}_i$. If $i = j$ we already know by (34) that this equals $\boldsymbol{p}_i$. If $i < j$ then

$$\boldsymbol{D}_{j+1}\boldsymbol{H}\boldsymbol{p}_i = \boldsymbol{D}_j\boldsymbol{H}\boldsymbol{p}_i + \frac{\boldsymbol{p}_j\boldsymbol{p}_j^T\boldsymbol{H}\boldsymbol{p}_i}{\boldsymbol{p}_j^T\boldsymbol{q}_j} - \frac{\boldsymbol{D}_j\boldsymbol{q}_j\boldsymbol{q}_j^T\boldsymbol{D}_j\boldsymbol{H}\boldsymbol{p}_i}{\boldsymbol{q}_j^T\boldsymbol{D}_j\boldsymbol{q}_j}.$$

Here

$$\boldsymbol{p}_j^T\boldsymbol{H}\boldsymbol{p}_i = 0$$

by the induction hypothesis on (a), and

$$\boldsymbol{D}_j\boldsymbol{q}_j\boldsymbol{q}_j^T\boldsymbol{D}_j\boldsymbol{H}\boldsymbol{p}_i = \boldsymbol{D}_j\boldsymbol{q}_j\boldsymbol{q}_j^T\boldsymbol{p}_i - \boldsymbol{D}_j\boldsymbol{q}_j\boldsymbol{p}_j^T\boldsymbol{H}\boldsymbol{p}_i = 0$$

by (33) and by the induction hypothesis again. Hence

$$\boldsymbol{D}_{j+1}\boldsymbol{H}\boldsymbol{p}_i = \boldsymbol{D}_j\boldsymbol{H}\boldsymbol{p}_i = \boldsymbol{p}_i,$$

by the induction hypothesis once more.

It remains to prove that $\boldsymbol{p}_{j+1}^T\boldsymbol{H}\boldsymbol{p}_i = 0$. As in (35) we obtain

$$\boldsymbol{p}_{j+1}^T\boldsymbol{H}\boldsymbol{p}_i = -\lambda_{j+1}\nabla f(\boldsymbol{y}_{j+1})^T\boldsymbol{p}_i.$$

By the definition of $\boldsymbol{q}_j$ this can for $i < j$ be rewritten as

$$\boldsymbol{p}_{j+1}^T \boldsymbol{H} \boldsymbol{p}_i = -\lambda_{j+1} \Big(\sum_{l=i+1}^{j} \boldsymbol{q}_l \Big)^T \boldsymbol{p}_i - \lambda_{j+1} \nabla f(\boldsymbol{y}_{i+1})^T \boldsymbol{p}_i =$$
$$= -\lambda_{j+1} \textstyle\sum_{l=i+1}^{j} \boldsymbol{q}_l^T \boldsymbol{p}_i - \lambda_{j+1} \nabla f(\boldsymbol{y}_{i+1})^T \boldsymbol{p}_i.$$

Here all terms are zero by the induction hypothesis (since $\boldsymbol{q}_l = \boldsymbol{H}\boldsymbol{p}_l$) and lemma 1, page 43. When $i = j$ no rewriting is necessary to arrive at the same result.

The proof is finished. □

COROLLARY 1. *For a quadratic function* $q(\boldsymbol{x}) = \frac{1}{2}\boldsymbol{x}^T \boldsymbol{H} \boldsymbol{x} + \boldsymbol{c}^T \boldsymbol{x}$ *with* $\boldsymbol{H}$ *positive definite, the generated directions are* $\boldsymbol{H}$*-conjugate. Consequently, after n iterations we have reached the exact minimum. Furthermore, the last matrix* $\boldsymbol{D}_{n+1}$ *is equal to* $\boldsymbol{H}^{-1}$.

PROOF. The first statement is a) in the theorem. The second statement follows from theorem 4, page 73. The third statement follows from b), which in particular states that

$$\boldsymbol{D}_{n+1} \boldsymbol{H} \boldsymbol{d}_i = \boldsymbol{d}_i, \qquad i = 1, \ldots, n.$$

Since the $\boldsymbol{d}_i$ are $\boldsymbol{H}$-conjugate they are linearly independent. Thus

$$\boldsymbol{D}_{n+1} \boldsymbol{H} = \boldsymbol{I},$$

which shows that $\boldsymbol{D}_{n+1} = \boldsymbol{H}^{-1}$. □

In statement b) of the theorem we see that the vectors $\boldsymbol{d}_i$, $i \leq j$, are eigenvectors of $\boldsymbol{D}_{j+1}\boldsymbol{H}$ with eigenvalue 1. This shows how we successively build up the matrices $\boldsymbol{D}_j$ to become better and better approximations to $\boldsymbol{H}^{-1}$. We use knowledge from the computation of successive gradients to obtain second order information.

The eigenvector property can also be stated as

$$\boldsymbol{D}_{j+1} \boldsymbol{q}_i = \boldsymbol{p}_i, \qquad i = 1, \ldots, j.$$

This is sometimes called the *quasi-Newton condition*. It can be used to explain the ideas behind the updating formula (30). Suppose that we want to obtain $\boldsymbol{D}_{j+1}$ by correcting $\boldsymbol{D}_j$ with two matrices of rank one.[13] That means putting

$$\boldsymbol{D}_{j+1} = \boldsymbol{D}_j + a\boldsymbol{u}\boldsymbol{u}^T + b\boldsymbol{v}\boldsymbol{v}^T$$

[13]There is also a method which uses only one matrix of rank one to correct $\boldsymbol{D}_j$, but it does not behave satisfactorily.

for some constants a and b and some column matrices $\boldsymbol{u}$ and $\boldsymbol{v}$. Then the quasi-Newton condition for $i = j$ is

$$\boldsymbol{p}_j = \boldsymbol{D}_{j+1}\boldsymbol{q}_j = \boldsymbol{D}_j\boldsymbol{q}_j + a\boldsymbol{u}\boldsymbol{u}^T\boldsymbol{q}_j + b\boldsymbol{v}\boldsymbol{v}^T\boldsymbol{q}_j.$$

This can be achieved by putting $\boldsymbol{u} = \boldsymbol{p}_j$, $a = 1/\boldsymbol{p}_j^T\boldsymbol{q}_j$ and $\boldsymbol{v} = \boldsymbol{D}_j\boldsymbol{q}_j$, $b = -1/\boldsymbol{q}_j^T\boldsymbol{D}_j\boldsymbol{q}_j$. This is precisely the updating formula (30) of the DFP method.

The discussion here of the quasi-Newton condition shows that there is considerable latitude in the choice of corrections to $\boldsymbol{D}_j$. In fact there is a whole one-parameter family of choices (the *Broyden family*). We will not go into this, but mention only the following update formula, known as the BFGS (*Broyden-Fletcher-Goldfarb-Shanno*) method:

$$\boldsymbol{D}_{j+1} = \boldsymbol{D}_j + \left(1 + \frac{\boldsymbol{q}_j^T\boldsymbol{D}_j\boldsymbol{q}_j}{\boldsymbol{p}_j^T\boldsymbol{q}_j}\right)\frac{1}{\boldsymbol{p}_j^T\boldsymbol{q}_j}\boldsymbol{p}_j\boldsymbol{p}_j^T - \frac{1}{\boldsymbol{p}_j^T\boldsymbol{q}_j}\left(\boldsymbol{p}_j\boldsymbol{q}_j^T\boldsymbol{D}_j + \boldsymbol{D}_j\boldsymbol{q}_j\boldsymbol{p}_j^T\right).$$

The BFGS method is known to perform even better than the DFP method in practice.

Exercises

3.13 Let $\boldsymbol{H}$ be a symmetric matrix. Show that if $\boldsymbol{u}$ and $\boldsymbol{v}$ are eigenvectors of $\boldsymbol{H}$ corresponding to distinct eigenvalues, then $\boldsymbol{u}$ and $\boldsymbol{v}$ are $\boldsymbol{H}$-conjugate.

S 3.14 Show that if the non-zero vectors $\boldsymbol{d}_1, \boldsymbol{d}_2, \ldots, \boldsymbol{d}_n$ satisfiy $\boldsymbol{d}_i^T\boldsymbol{H}\boldsymbol{d}_j = 0$ when $i \neq j$, for a positive definite matrix $\boldsymbol{H}$, then these vectors are linearly independent, and hence $\boldsymbol{H}$-conjugate.

3.15 Put $\boldsymbol{H} = \begin{pmatrix} 1 & 1 & -2 \\ 1 & 2 & -3 \\ -2 & -3 & 10 \end{pmatrix}$. $\boldsymbol{H}$ is positive definite. (You need not verify this.)

a) Verify that $\boldsymbol{d}_1 = (2,1,0)$ and $\boldsymbol{d}_2 = (1,1,1)$ are $\boldsymbol{H}$-conjugate.

b) Find $\boldsymbol{d}_3$ so that $\boldsymbol{d}_1$, $\boldsymbol{d}_2$, $\boldsymbol{d}_3$ are $\boldsymbol{H}$-conjugate.

c) Put

$$q(\boldsymbol{x}) = \frac{1}{2}\boldsymbol{x}^T\boldsymbol{H}\boldsymbol{x} + \boldsymbol{c}^T\boldsymbol{x}, \qquad \text{where} \quad \boldsymbol{c} = \begin{pmatrix} 2 \\ -3 \\ 1 \end{pmatrix}.$$

Starting at $\boldsymbol{x}^{(1)} = (0,1,1)$, the function q is minimized in direction $\boldsymbol{d}_1$ (exact line search). Find the resulting point $\boldsymbol{x}^{(2)}$.

d) Starting from $\boldsymbol{x}^{(2)}$ line search in direction $\boldsymbol{d}_2$ can be carried out in the same way. Denote the resulting point by $\boldsymbol{x}^{(3)}$. Doing line search from $\boldsymbol{x}^{(3)}$ in direction $\boldsymbol{d}_3$ gives $\boldsymbol{x}^{(4)}$. Find $\boldsymbol{x}^{(4)}$.

(1996–12–16, 4)

3.16 Let

$$\boldsymbol{H} = \begin{pmatrix} 1 & 1 & 0 \\ 1 & 2 & 1 \\ 0 & 1 & 2 \end{pmatrix}, \quad \boldsymbol{e}_1 = \begin{pmatrix} 1 \\ 0 \\ 0 \end{pmatrix}, \quad \boldsymbol{e}_2 = \begin{pmatrix} 0 \\ 1 \\ 0 \end{pmatrix}, \quad \boldsymbol{e}_3 = \begin{pmatrix} 0 \\ 0 \\ 1 \end{pmatrix}.$$

a) Verify that $\boldsymbol{H}$ is positive definite.

b) Determine three $\boldsymbol{H}$-conjugate directions $\boldsymbol{d}_1, \boldsymbol{d}_2, \boldsymbol{d}_3$ by means of the Gram-Schmidt orthogonalization process, i.e.,

i. by taking $\boldsymbol{d}_1 = \boldsymbol{e}_1$,

ii. by determining α_1 such that $\boldsymbol{d}_2 = \boldsymbol{e}_2 - \alpha_1 \boldsymbol{d}_1$ is $\boldsymbol{H}$-conjugate to $\boldsymbol{d}_1$,

iii. by determining β_1, β_2, such that $\boldsymbol{d}_3 = \boldsymbol{e}_3 - \beta_1 \boldsymbol{d}_1 - \beta_2 \boldsymbol{d}_2$ is $\boldsymbol{H}$-conjugate to $\boldsymbol{d}_1$ and $\boldsymbol{d}_2$.

c) Find the minimum point of

$$f(\boldsymbol{x}) = \frac{1}{2}\boldsymbol{x}^T \boldsymbol{H} \boldsymbol{x} + \boldsymbol{c}^T \boldsymbol{x}, \qquad \boldsymbol{c} = \begin{pmatrix} -1 \\ -2 \\ -3 \end{pmatrix},$$

obtained by successively using the conjugate directions $\boldsymbol{d}_1, \boldsymbol{d}_2, \boldsymbol{d}_3$ and starting from $\boldsymbol{x}_1 = \boldsymbol{0}$. (Use exact line search, exercise 3.1.)

d) Verify that your minimizer $\bar{\boldsymbol{x}}$ satisfies $\boldsymbol{H}\bar{\boldsymbol{x}} + \boldsymbol{c} = \boldsymbol{0}$.

S 3.17 a) Minimize the quadratic function

$$f(\boldsymbol{x}) = x_1^2 + 4x_2^2 - 4x_1 - 8x_2$$

by the Fletcher-Reeves method starting from $\boldsymbol{x}_1 = \boldsymbol{0}$.

b) Verify that the chosen directions are $\boldsymbol{H}$-conjugate and that they are descent directions.

S 3.18 a) Minimize the quadratic function

$$f(\boldsymbol{x}) = x_1^2 + 4x_2^2 - 4x_1 - 8x_2$$

by the DFP method starting from $\boldsymbol{x}_1 = \boldsymbol{0}$.

b) Verify that the chosen directions are $\boldsymbol{H}$-conjugate and that they are descent directions.

S 3.19 Consider the quadratic function $\frac{1}{2}\boldsymbol{x}^T\boldsymbol{H}\boldsymbol{x} + \boldsymbol{c}^T\boldsymbol{x}$ where

$$\boldsymbol{H} = \begin{pmatrix} 3 & -1 & 0 & 0 \\ -1 & 3 & -1 & 0 \\ 0 & -1 & 3 & -1 \\ 0 & 0 & -1 & 3 \end{pmatrix}, \qquad \boldsymbol{c} = \begin{pmatrix} 2 \\ 1 \\ 1 \\ 2 \end{pmatrix}.$$

Apply the conjugate gradient method with exact line search to this problem starting from $\boldsymbol{x}_1 = \boldsymbol{0}$. Show that it converges in two iterations.

3.6 The least squares problem

3.6.1 Introduction

One particular class of minimization problems which appears frequently in practice is when the objective function is a sum of squares,

$$f(\boldsymbol{x}) = \sum_{i=1}^{m} (r_i(\boldsymbol{x}))^2, \qquad \boldsymbol{x} \in \mathbf{R}^n. \tag{36}$$

Example 12. *(Curve fitting.)* Suppose that a number m of points in the plane

$$(t_i, b_i), \qquad i = 1, \dots, m,$$

are given, obtained e.g. by experiment. We wish to fit these data to some particular model $b = \varphi(\boldsymbol{x}, t)$, where $\boldsymbol{x} \in \mathbf{R}^n$ is a set of n *parameters*. These should ideally be chosen so that

$$\varphi(\boldsymbol{x}, t_i) = b_i, \qquad i = 1, \dots, m.$$

Normally $m \gg n$. Due to experimental and/or numerical errors or other reasons such conditions can generally not be satisfied exactly. We have to be content with finding values $\boldsymbol{x}^*$ of the parameters which in some sense to be specified minimizes the *residuals*

$$r_i(\boldsymbol{x}) = \varphi(\boldsymbol{x}, t_i) - b_i.$$

This is often (but now always) to be understood as minimizing the sum of squares

$$r_1(\boldsymbol{x})^2 + \ldots + r_m(\boldsymbol{x})^2 = \boldsymbol{r}(\boldsymbol{x})^T\boldsymbol{r}(\boldsymbol{x}) = \|\boldsymbol{r}(\boldsymbol{x})\|^2,$$

where we have put $\boldsymbol{r}(\boldsymbol{x}) = (r_1(\boldsymbol{x}), \ldots, r_m(\boldsymbol{x}))^T$. In this way we obtain the objective function f in (36). □

Methods exist which take advantage of the particular structure of this minimization problem. We begin with a study of the case when all functions r_i are affine. Then the problem can be interpreted and solved using concepts from linear algebra.

3.6.2 The linear case

Assume that $\boldsymbol{r}(\boldsymbol{x})$ is affine, i.e., of the form

$$\boldsymbol{r}(\boldsymbol{x}) = \boldsymbol{A}\boldsymbol{x} - \boldsymbol{b},$$

where $\boldsymbol{A}$ is an $m \times n$-matrix and $\boldsymbol{b}$ an m-vector, and $m \geq n$. The function to be minimized is

$$f(\boldsymbol{x}) = \|\boldsymbol{A}\boldsymbol{x} - \boldsymbol{b}\|^2 = \sum_{i=1}^{m} (r_i(\boldsymbol{x}))^2.$$

Let $\mathcal{L}$ denote the set $\{\, \boldsymbol{A}\boldsymbol{x}\,;\, \boldsymbol{x} \in \mathbf{R}^n \,\}$ in $\mathbf{R}^m$, i.e., the set of linear combinations of the columns of $\boldsymbol{A}$ (the *column space* of $\boldsymbol{A}$). Geometrically, our problem is to find that point $\boldsymbol{A}\boldsymbol{x}^* \in \mathcal{L}$ which minimizes the distance to the point $\boldsymbol{b}$ in $\mathbf{R}^m$. If $\boldsymbol{y}$ denotes the orthogonal projection of $\boldsymbol{b}$ onto $\mathcal{L}$ then $\boldsymbol{x}^*$ solves the equation $\boldsymbol{A}\boldsymbol{x} = \boldsymbol{y}$.

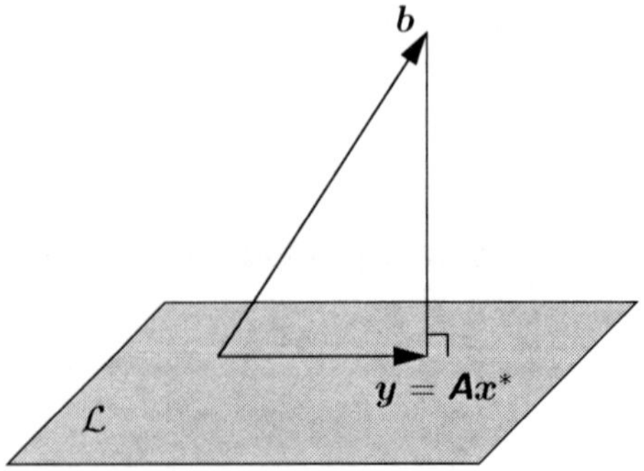

The condition on $\boldsymbol{x}^*$ is that the vector $\boldsymbol{b}-\boldsymbol{A}\boldsymbol{x}^*$ should be orthogonal to $\mathcal{L}$, that is, to the columns of $\boldsymbol{A}$. This implies that

$$\boldsymbol{A}^T(\boldsymbol{b}-\boldsymbol{A}\boldsymbol{x}^*) = \boldsymbol{0},$$

which can be rewritten as

$$\boldsymbol{A}^T\boldsymbol{A}\boldsymbol{x}^* = \boldsymbol{A}^T\boldsymbol{b}. \tag{37}$$

We can easily prove that any $\boldsymbol{x}^*$ satisfying (37) minimizes f. For any $\boldsymbol{x}$ in $\mathbf{R}^n$ we have

$$\begin{aligned} f(\boldsymbol{x}) &= \|\boldsymbol{A}\boldsymbol{x}-\boldsymbol{A}\boldsymbol{x}^* + \boldsymbol{A}\boldsymbol{x}^* - \boldsymbol{b}\|^2 = \\ &= \|\boldsymbol{A}\boldsymbol{x}-\boldsymbol{A}\boldsymbol{x}^*\|^2 + 2(\boldsymbol{A}\boldsymbol{x}-\boldsymbol{A}\boldsymbol{x}^*)^T(\boldsymbol{A}\boldsymbol{x}^*-\boldsymbol{b}) + \|\boldsymbol{A}\boldsymbol{x}^*-\boldsymbol{b}\|^2. \end{aligned}$$

Here the middle term is $2(\boldsymbol{x}-\boldsymbol{x}^*)^T\boldsymbol{A}^T(\boldsymbol{A}\boldsymbol{x}^*-\boldsymbol{b}) = 0$ by (37). Hence

$$f(\boldsymbol{x}) \geq \|\boldsymbol{A}\boldsymbol{x}^*-\boldsymbol{b}\|^2 \quad \text{for all } \boldsymbol{x},$$

obviously with equality when $\boldsymbol{x} = \boldsymbol{x}^*$.

Equations (37) are called the *normal equations*.

The matrix $\boldsymbol{A}^T\boldsymbol{A}$ is invertible if and only if $\boldsymbol{A}$ has rank n.[14] In this case the solution to the normal equations is unique:

$$\boldsymbol{x}^* = (\boldsymbol{A}^T\boldsymbol{A})^{-1}\boldsymbol{A}^T\boldsymbol{b}.$$

The matrix $(\boldsymbol{A}^T\boldsymbol{A})^{-1}\boldsymbol{A}^T$ is sometimes called the *pseudoinverse* of $\boldsymbol{A}$.

In practice the normal equations are never solved by computation of an inverse matrix. The method is to find and use a Cholesky factorization $\boldsymbol{A}^T\boldsymbol{A} = \boldsymbol{L}^T\boldsymbol{L}$ with $\boldsymbol{L}$ left triangular. Care should be taken here not to worsen a usually ill-conditioned problem. It is preferable to avoid first computing the product $\boldsymbol{A}^T\boldsymbol{A}$; a so called QR-factorization of $\boldsymbol{A}$ should be used. We refer to literature on numerical linear algebra for a discussion of these problems.

---

[14] PROOF: We have $\boldsymbol{x}^T\boldsymbol{A}^T\boldsymbol{A}\boldsymbol{x} = (\boldsymbol{A}\boldsymbol{x})^T(\boldsymbol{A}\boldsymbol{x}) = \|\boldsymbol{A}\boldsymbol{x}\|^2$. Hence $\boldsymbol{A}^T\boldsymbol{A}$ is positive definite if and only if $\boldsymbol{A}\boldsymbol{x} = \boldsymbol{0}$ for $\boldsymbol{x} = \boldsymbol{0}$ only, which is equivalent to $\boldsymbol{A}$ being of rank n.

3.6.3 The Gauss-Newton method

We now turn to the general problem of minimizing the sum of squares $f(\boldsymbol{x})$ in (36). Differentiating we obtain

$$\nabla f(\boldsymbol{x}) = \sum_{i=1}^{m} 2r_i(\boldsymbol{x})\nabla r_i(\boldsymbol{x}). \tag{38}$$

Introducing the Jacobian of $\boldsymbol{r} = (r_1, \ldots, r_m)^T$, i.e., the $m \times n$-matrix $\boldsymbol{J}(\boldsymbol{x}) = \left(\frac{\partial r_i}{\partial x_k}\right)$ whose rows are the (transposed) gradients

$$\nabla r_1(\boldsymbol{x})^T, \quad \ldots, \quad \nabla r_m(\boldsymbol{x})^T,$$

(38) can be written

$$\nabla f(\boldsymbol{x}) = 2\boldsymbol{J}(\boldsymbol{x})^T \boldsymbol{r}(\boldsymbol{x}).$$

For the Hessian $\boldsymbol{H}(\boldsymbol{x})$ of f we have the formula

$$\boldsymbol{H}(\boldsymbol{x}) = 2\boldsymbol{J}(\boldsymbol{x})^T \boldsymbol{J}(\boldsymbol{x}) + 2\sum_{i=1}^{m} r_i(\boldsymbol{x})\nabla^2 r_i(\boldsymbol{x}). \tag{39}$$

To minimize f we can of course use Newton's method or some quasi-Newton method in the conventional way. However, in the former case it is required to compute $\boldsymbol{H}(\boldsymbol{x})$ at every iteration step, which implies computing the Hessians $\nabla^2 r_i$ of all r_i.

As an alternative we take the following heuristic attitude. By the nature of the problem all $r_i(\boldsymbol{x})$ should be expected to be if not close to zero at least not too large near the minimum. Assuming them to be small (the *small residuals case*) we decide to neglect the second term in (39). This leads to

$$\boldsymbol{H}(\boldsymbol{x}) \approx 2\boldsymbol{J}(\boldsymbol{x})^T \boldsymbol{J}(\boldsymbol{x}).$$

The computation now requires only first derivatives of the r_i.

Application of Newton's method to the least squares problem with this approximation is called the *Gauss-Newton method*. The iteration scheme becomes

$$\boldsymbol{x}_{k+1} = \boldsymbol{x}_k - \left(\boldsymbol{J}(\boldsymbol{x}_k)^T \boldsymbol{J}(\boldsymbol{x}_k)\right)^{-1} \boldsymbol{J}(\boldsymbol{x}_k)^T \boldsymbol{r}(\boldsymbol{x}_k)$$

(the factor 2 cancels out). Compare with the linear case above, where $\boldsymbol{J} = \boldsymbol{A}$.

The comments on numerical methods in the linear case above are valid also in the nonlinear one.

For the Gauss-Newton method to converge it is required not only that the initial point is sufficiently near the minimum, but also that the neglected terms in the Hessian (39) are small enough. It can be shown that the convergence is linear.

To improve convergence it is desirable to include line search (exact or inexact) in the algorithm. Then we have the following iteration scheme:

- Solve the system $\boldsymbol{J}(\boldsymbol{x}_k)^T\boldsymbol{J}(\boldsymbol{x}_k)\boldsymbol{d}_k = -\boldsymbol{J}(\boldsymbol{x}_k)^T\boldsymbol{r}(\boldsymbol{x}_k)$ for the direction $\boldsymbol{d}_k$.
- Find some λ_k such that $f(\boldsymbol{x}_k+\lambda_k\boldsymbol{d}_k) \leq f(\boldsymbol{x}_k)$. (This can be done, since it can be shown that $\boldsymbol{d}_k$ is a descent direction.)
- Let $\boldsymbol{x}_{k+1} := \boldsymbol{x}_k + \lambda_k\boldsymbol{d}_k$.

This is sometimes called the *damped* Gauss-Newton method.

Example 13. Our all-pervading objective function in this chapter is in fact a sum of squares:

$$f(x_1,x_2) = 2(x_1-2)^4 + (2x_1-3x_2)^2 = r_1(x_1,x_2)^2 + r_2(x_1,x_2)^2$$

where

$$r_1(x_1,x_2) = \sqrt{2}\,(x_1-2)^2, \qquad r_2(x_1,x_2) = 2x_1 - 3x_2.$$

Thus it should be interesting to see how the Gauss-Newton method behaves on this function.

The Jacobian $\boldsymbol{J}(x_1,x_2)$ is in this case the matrix

$$\boldsymbol{J}(x_1,x_2) = \begin{pmatrix} 2\sqrt{2}\,(x_1-2) & 0 \\ 2 & -3 \end{pmatrix}.$$

The first picture on page 97 shows the result of 5 iterations when no line search is incorporated. The final point is $\boldsymbol{x}_6 = (1.9375,\ 1.2917)$, which is within a circle of radius 0.1 centered at the exact minimum. The table on page 96 gives the details of the computations.

k	$\boldsymbol{x}_k$	$f(\boldsymbol{x}_k)$	$\boldsymbol{r}(\boldsymbol{x}_k)$	$\boldsymbol{J}(\boldsymbol{x}_k)$	$\boldsymbol{d}_k$
1	(0 , 2.5)	88.2500	(5.6569, −7.5)	$\begin{pmatrix} -5.6569 & 0 \\ 2 & -3 \end{pmatrix}$	(1.0000, −1.8333)
2	(1.0000, 0.6667)	2.0000	(1.4142, 0.0)	$\begin{pmatrix} -2.8284 & 0 \\ 2 & -3 \end{pmatrix}$	(0.5000, 0.3333)
3	(1.5000, 1.0000)	0.1250	(0.3536, 0)	$\begin{pmatrix} -1.4142 & 0 \\ 2 & -3 \end{pmatrix}$	(0.2500, 0.1667)
4	(1.7500, 1.1667)	0.0078	(0.0884, 0)	$\begin{pmatrix} -0.7071 & 0 \\ 2 & -3 \end{pmatrix}$	(0.1250, 0.0833)
5	(1.8750, 1.2500)	0.0005	(0.0221, 0)	$\begin{pmatrix} -0.3536 & 0 \\ 2 & -3 \end{pmatrix}$	(0.0625, 0.0417)
6	(1.9375, 1.2917)	0.00003			

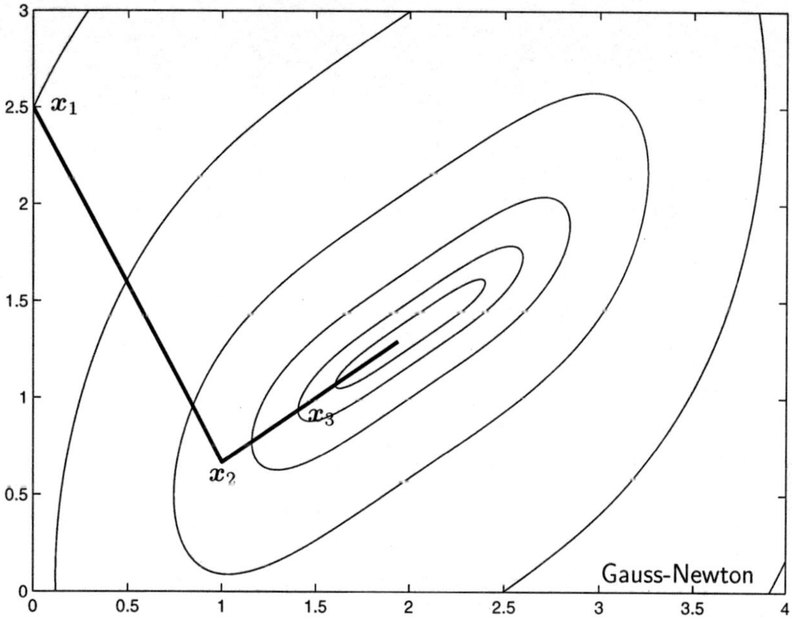

3
2.5
2
1.5
1
0.5
0
0
0.5
1
1.5
2
2.5
3
3.5
4
x_1
x_2
x_3
Gauss-Newton

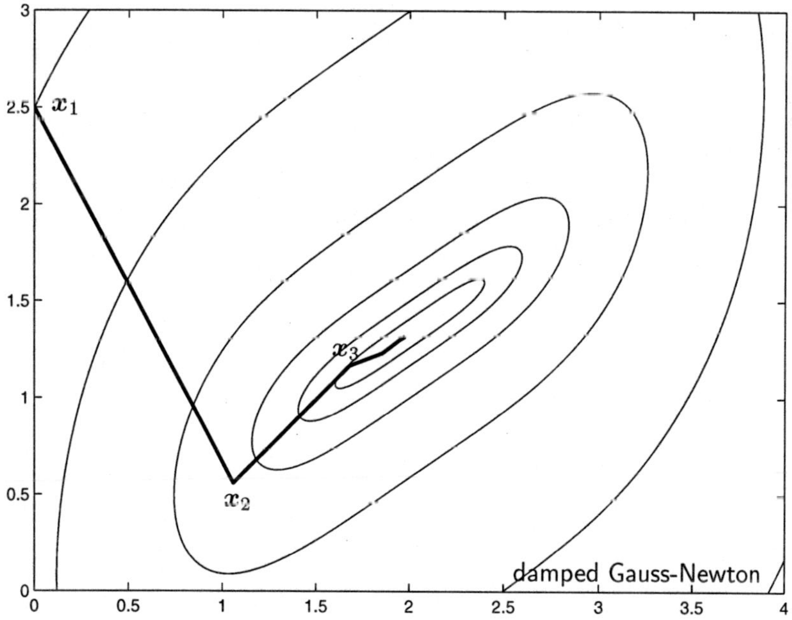

3
2.5
2
1.5
1
0.5
0
0
0.5
1
1.5
2
2.5
3
3.5
4
x_1
x_2
x_3
damped Gauss-Newton

k	$\boldsymbol{x}_k$	$f(\boldsymbol{x}_k)$	$\boldsymbol{r}(\boldsymbol{x}_k)$	$\boldsymbol{J}(\boldsymbol{x}_k)$	$\boldsymbol{d}_k$
1	(0 , 2.5)	88.2500	(5.6569, −7.5)	$\begin{pmatrix} -5.6569 & 0 \\ 2 & -3 \end{pmatrix}$	(1.0000, −1.8333)
2	(1.0592, 0.5581)	1.7640	(1.2517, 0.4441)	$\begin{pmatrix} -2.6609 & 0 \\ 2 & -3 \end{pmatrix}$	(0.4704, 0.4616)
3	(1.6786, 1.1660)	0.0411	(0.1461, −0.1407)	$\begin{pmatrix} -0.9091 & 0 \\ 2 & -3 \end{pmatrix}$	(0.1607, 0.0602)
4	(1.8552, 1.2321)	0.0011	(0.0297, 0.0139)	$\begin{pmatrix} -0.4097 & 0 \\ 2 & -3 \end{pmatrix}$	(0.0724, 0.0529)
5	(1.9453, 1.2980)	0.000029	(0.0042, −0.0034)	$\begin{pmatrix} -0.1546 & 0 \\ 2 & -3 \end{pmatrix}$	(0.0273, 0.0171)
6	(1.9762, 1.3173)	0.000001			

As should be expected the result is even better when line search is used (the second figure on page 97). After 5 iterations we arrive at $\boldsymbol{x}_6 = (1.9762,\ 1.3173)$. Four iterations suffice to reach a distance of 0.1 to the true minimum. The table on page 98 gives the intermediate results.

□

Exercises

3.20 Consider the problem to minimize $\|\boldsymbol{A}\boldsymbol{x} - \boldsymbol{b}\|^2$, where $\boldsymbol{A}$ is an $m \times n$ matrix and $\boldsymbol{b}$ is an m-vector. Solve it using the normal equations when

$$\boldsymbol{A} = \begin{pmatrix} 1 & -1 & 0 \\ 0 & 2 & 1 \\ 0 & 1 & 0 \\ 1 & 0 & 1 \end{pmatrix}, \qquad \boldsymbol{b} = \begin{pmatrix} 2 \\ 1 \\ 1 \\ 0 \end{pmatrix}.$$

3.21 Try to fit an exponential function $y = Ce^{-kx}$ to the following data:

x	0	1	2
y	4	3	2

The assignment is to determine the constants C and k by least squares approximation. Use the undamped Gauss-Newton method and carry out one iteration step. As initial values take $C = 4$, $k = \ln \frac{4}{3} \approx 0.28768$ (which fit the first two points exactly).

3.7 The Nelder-Mead (simplex) method

We finally describe a minimization method which works from different principles from those we have seen so far. Instead of generating a sequence of points converging towards the minimum we consider a set of simplices (to be explained shortly) which is updated in every iteration step and which in the limit converges to a single point.

In spite of it being called the *simplex method* the scheme has nothing to do with the simplex algorithm of linear programming. Other names used are the *polytope algorithm* or the *Nelder-Mead method*. The method is used in one of Matlab's optimization toolboxes.

A *simplex* in $\mathbf{R}^n$ is a set of $n+1$ *corner* points $\boldsymbol{x}_1, \ldots, \boldsymbol{x}_{n+1}$ such that the vectors $\boldsymbol{x}_i - \boldsymbol{x}_1$, $i = 2, \ldots, n+1$, are linearly independent. In particular, in two dimensions a simplex is a (non-degenerate) triangle, in three dimensions it is a tetrahedron.

Denote by f the function to be minimized. For a given simplex we define $\boldsymbol{x}_m$ and $\boldsymbol{x}_M$ as those corner points which have the smallest and the largest function value respectively, i.e.,

$$f(\boldsymbol{x}_m) = \min_{1 \le i \le n+1} f(\boldsymbol{x}_i), \qquad f(\boldsymbol{x}_M) = \max_{1 \le i \le n+1} f(\boldsymbol{x}_i).$$

The goal of the iteration step is to replace the simplex by another one, by removing $\boldsymbol{x}_M$ from the set of corner points and introducing a new point $\boldsymbol{x}_{\text{new}}$ to replace it. We shall give an outline of how this is done, without going into too much detail or discussing some speedup modifications used in practical implementations.

Three parameters α, β, γ are used in the process, chosen in advance so that $\alpha > 0$, $\beta > 1$, $0 < \gamma < 1$. Denote by $\boldsymbol{x}_g$ the center of gravity of the face of the simplex opposite to $\boldsymbol{x}_M$. This is defined by

$$\boldsymbol{x}_g = \frac{1}{n} \sum_{\boldsymbol{x}_i \neq \boldsymbol{x}_M} \boldsymbol{x}_i.$$

The first measure taken in the iteration step is the introduction of the point

$$\boldsymbol{x}_r = \boldsymbol{x}_g + \alpha(\boldsymbol{x}_g - \boldsymbol{x}_M).$$

This point is often described as a *reflection* of $\boldsymbol{x}_M$ in the face of the simplex defined by the other corner points.[15] We now split into three cases.

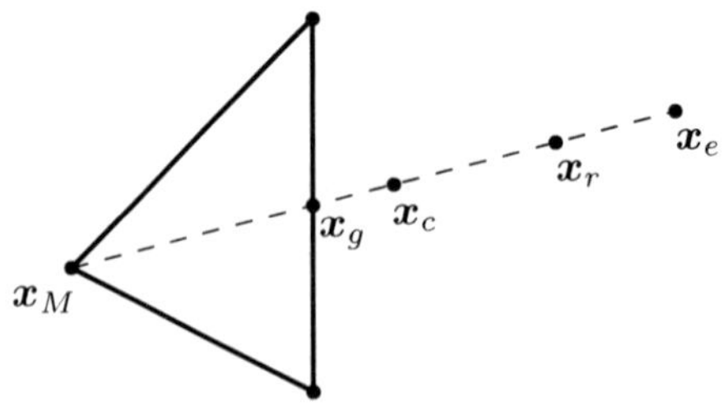

[15] The terminology derives its origin from earlier versions of the method, when $\boldsymbol{x}_r$ was truly a reflection point.

Case 1. $f(\boldsymbol{x}_m) < f(\boldsymbol{x}_r) < f(\boldsymbol{x}_M)$. Then we put

$$\boldsymbol{x}_{\text{new}} := \boldsymbol{x}_r.$$

The new simplex is thereby defined and the iteration step completed.

Case 2. $f(\boldsymbol{x}_r) \leq f(\boldsymbol{x}_m)$. Then we seem to have found a promising direction in which to move, and we try to *expand* the simplex in that direction. We let

$$\boldsymbol{x}_e = \boldsymbol{x}_g + \beta(\boldsymbol{x}_r - \boldsymbol{x}_g)$$

and put

$$\boldsymbol{x}_{\text{new}} := \begin{cases} \boldsymbol{x}_e & \text{if } f(\boldsymbol{x}_e) < f(\boldsymbol{x}_r) \\ \boldsymbol{x}_r & \text{otherwise.} \end{cases}$$

This completes the iteration step in case 2.

Case 3. $f(\boldsymbol{x}_r) \geq f(\boldsymbol{x}_M)$. In this case we do not seem to be moving in a good direction, and we *contract* the simplex by letting

$$\boldsymbol{x}_c = \boldsymbol{x}_g + \gamma(\boldsymbol{x}_r - \boldsymbol{x}_g).$$

If $f(\boldsymbol{x}_c) < f(\boldsymbol{x}_M)$ things have improved and we put

$$\boldsymbol{x}_{\text{new}} := \boldsymbol{x}_c;$$

else we *shrink* the simplex by putting

$$\boldsymbol{x}_i := \frac{1}{2}(\boldsymbol{x}_i + \boldsymbol{x}_m) \quad \text{for all } i$$

and then restart the iteration step.

As said above, in practice there are several modifications of this scheme. We refer to other literature for a fuller description of the method.

Appendix 1 to chapter 3: Convergence rate

Here we discuss some concepts used to describe the speed with which certain sequences converge towards their limit point. These concepts are useful when judging the merits of an optimization algorithm. However, one should be aware that they do not tell the whole truth. They are theoretical tools, and should be handled with caution. They describe the *asymptotic behaviour* of a generated sequence; if you in a certain algorithm terminate the computations prematurely they are of little relevance.

Sequences of real numbers

DEFINITION 2. *Let $\{x_k\}_1^\infty$ be a convergent sequence of real numbers, with limit point x. The* **order of convergence** *of this sequence is the largest number p such that*[16]

$$\lim_{k\to\infty} \frac{|x_{k+1}-\bar{x}|}{|x_k-\bar{x}|^p} \tag{40}$$

exists.

Example 14. Let $x_k = \dfrac{1}{\sqrt{k}}$, with the limit $\bar{x}=0$ as $k\to\infty$. Then

$$\frac{|x_{k+1}-\bar{x}|}{|x_k-\bar{x}|^p} = \frac{\dfrac{1}{\sqrt{k+1}}}{\dfrac{1}{(\sqrt{k})^p}} = \frac{k^{p/2}}{(k+1)^{1/2}} = \frac{k^{(p-1)/2}}{\left(1+\dfrac{1}{k}\right)^{1/2}}.$$

Clearly this quotient has a limit exactly when $p \le 1$. Hence the order of convergence is 1. □

Example 15. For the sequence defined by $x_k = \dfrac{1}{2^k}$ we similarly obtain

$$\frac{|x_{k+1}-0|}{|x_k-0|^p} = \frac{\dfrac{1}{2^{k+1}}}{\dfrac{1}{2^{kp}}} = 2^{k(p-1)-1}.$$

The limit exists if and only if $p \le 1$. The order of convergence is thus 1. □

[16]This should really be the limit superior, but we ignore that here.

Example 16. Let $x_k = \dfrac{1}{k!}$. Then

$$\frac{|x_{k+1} - 0|}{|x_k - 0|^p} = \frac{(k!)^p}{(k+1)!} = \frac{1}{k+1}(k!)^{p-1}.$$

Again the limit exists if and only if $p \le 1$, making the order of convergence equal to 1. □

Now let p denote the order of convergence and C the limit in (40). When $p = 1$ and $C < 1$ we talk about **linear convergence**. The constant C is then called the **convergence ratio**. If $p = 1$ and $C = 0$, or if $p > 1$, the sequence is said to converge **superlinearly**.

Examples 15–16 above exhibit linear convergence. The convergence ratio is $\frac{1}{2}$ in example 15. In the last example the convergence is superlinear.

Example 17. Let a sequence $\{x_k\}_1^\infty$ be defined recursively by

$$x_{k+1} = 1 + \frac{x_k - 1}{2^k}.$$

It can be shown that the limit is equal to 1, whatever the initial point. By the recursion formula

$$\frac{|x_{k+1} - 1|}{|x_k - 1|} = \frac{1}{2^k} \to 0 \quad \text{as} \quad k \to \infty.$$

Hence the convergence is superlinear. □

For a linearly convergent sequence definition 2 implies the estimate

$$|x_{k+1} - \bar{x}| \le (C+\varepsilon)|x_k - \bar{x}|,$$

true for any positive number ε if k is chosen large enough. By iteration, for some constant K,

$$|x_{k+1} - \bar{x}| \le K(C+\varepsilon)^k \quad \text{for all } k \ge 1.$$

For this reason linear convergence is sometimes called *geometric* convergence. Note that ε can be chosen so that $C+\varepsilon < 1$.

Convergence of order 2 is called **quadratic convergence**. In section 2.2 it is proved that sequences generated by Newton's method of optimization converge quadratically.

One rule of thumb says that with quadratic convergence the number of significant figures in x_k is doubled at every iteration step. This fast convergence in the quadratic case is indicated by the examples above, where sequences which one intuitively feels should have a high convergence rate in fact exhibit convergence of order 1 only.

Example 18. The sequence definied by $x_k = e^{-2^k}$ converges quadratically. This follows from the definition:

$$\frac{|x_{k+1} - 0|}{|x_k - 0|^2} = \frac{e^{-2^{k+1}}}{e^{-2^k \cdot 2}} = 1 \to 1 \quad \text{as} \quad k \to \infty.$$

□

Example 19. A well known method for the computation of $\sqrt{2}$ is to use the recursion formula

$$x_{k+1} = \frac{1}{2}\left(x_k + \frac{2}{x_k}\right).$$

For any positive initial point it can be shown that the generated sequence tends to $\sqrt{2}$. Furthermore

$$x_{k+1} - \sqrt{2} = \frac{1}{2}\left(x_k - 2\sqrt{2} + \frac{2}{x_k}\right) = \frac{x_k^2 - 2\sqrt{2}x_k + 2}{2x_k} = \frac{(x_k - \sqrt{2})^2}{2x_k}.$$

Hence

$$\frac{|x_{k+1} - \sqrt{2}|}{|x_k - \sqrt{2}|^2} = \frac{1}{2x_k} \to \frac{1}{2\sqrt{2}} \quad \text{as} \quad k \to \infty,$$

proving that the convergence is quadratic. □

Sequences of points in $\mathbf{R}^n$

For sequences in $\mathbf{R}^n$ we only have to repeat definition 2 with a suitable norm to represent the distances.

DEFINITION 3. *Let* $\{\boldsymbol{x}_k\}_1^\infty$ *be a convergent sequence in* $\mathbf{R}^n$, *with limit point* $\bar{\boldsymbol{x}}$. *The* **order of convergence** *of this sequence is the largest number* p *such that*

$$\lim_{k\to\infty} \frac{\|\boldsymbol{x}_{k+1} - \bar{\boldsymbol{x}}\|}{\|\boldsymbol{x}_k - \bar{\boldsymbol{x}}\|^p}$$

exists.

Linear and quadratic convergence are defined as in the one-dimensional case.

Exercises

3.22 The sequences below all tend to zero as $k \to \infty$. Characterize the convergence on the scale

$$\text{order } p \text{ with } p > 1 \longrightarrow \text{superlinear} \longrightarrow \text{linear} \longrightarrow \text{order } 1.$$

a) $1/\sqrt{k}$, b) $1/k^{100}$, c) $1/2^k$, d) $1/k^k$,

e) $1/k!$, f) $2^k/k!$, g) $e^{-(e^k)}$.

3.23 Let $a > 0$. The sequence $\{x_k\}_1^\infty$ defined by

$$x^{(k+1)} = \frac{1}{2}\left(x^{(k)} + \frac{a}{x^{(k)}}\right), \qquad k \geq 1,$$

and $x^{(1)} = 1$ has the limit $\sqrt{a}$ as $k \to \infty$. Verify that the convergence is of order 2.

S 3.24 Let $f \in \mathcal{C}^1(\mathbf{R})$ and suppose that $f(x^*) = x^*$ and $|f'(x^*)| < 1$. Prove that the sequence $\{x^{(k)}\}_1^\infty$ defined by

$$x^{(k+1)} = f(x^{(k)}), \qquad k \geq 1,$$

converges linearly to x^* as $k \to \infty$. Give an estimate of the convergence ratio. You may assume that $x^{(1)}$ and $x^{(2)}$ are sufficiently close to x^*.

Appendix 2 to chapter 3: Termination criteria

In most cases the sequence generated by an optimization algorithm needs an infinite number of steps to reach its limit (assuming convergence). Hence some criterion is required on how to decide when to terminate the procedure. In order to be able to interpret the computed results correctly a user must of course be aware of and understand the criteron (criteria) used.

The design of a termination criterion is a very delicate matter. There are several sometimes conflicting goals that it should fulfil. Of course there is the issue of ensuring a certain accuracy of the computed minimum point (or value). But a termination criterion should also have the ability to recognize, preferably as early as possible, that a certain generated sequence is in fact *not* converging, and then stop the computations. Furthermore, efficiency should be promoted; waste of computer time should be avoided by not carrying out unnecessary iterations.

The decision to use a certain termination criterion must be based on the actual application and on the optimization method used. Assumptions have to be made on what kind of problems to be solved, and what the requirements on the answers are. You may e.g. wish to ensure an accurate value for the limit of the generated sequence. This will probably require carrying out many iteration steps. Whether the cost of this is worth while depends on the price you have to pay for making a wrong decision based on an inaccurate estimate of the minimum point or minimum value.

Here we content ourselves with giving some hints. First of all, you must be clear about you own objective. Which is most important, the minimum point or the minimum value? Perhaps you are satisfied with coming across some point x satisfying $f(x) \leq b$, where b is some number given in advance, and don't care about the existence of other points with much lesser values. Or are you much concerned about finding a really good point, but the minimum value is of no interest?

It must, as always with optimization methods, be stressed that the choice of termination criterion (not only the parameter values used in it) should be confronted with some test examples. The more important the issues involved are, the more experimenting is needed. Often one may want to use some combination of several stopping criteria.

We now list some suggestions for termination criteria. Others may be designed using the concept of *descent function* introduced in chapter 10. The numbers ε, N and b below are parameters to be decided in advance.

- $\|\nabla f(\boldsymbol{x}_k)\| \leq \varepsilon$. You are close to a stationary point (which might of course not be a minimum).

- $f(\boldsymbol{x}_k) \leq b$. This requires some a priori estimate of the minimum value to make a good choice for b. When you expect the minimum to be close to 0, this criterion might be a good alternative.

- $\|\boldsymbol{x}_{k+N} - \boldsymbol{x}_k\| \leq \varepsilon$. We have not moved very far in the last N iterations. It is time to stop.

- $\dfrac{\|\boldsymbol{x}_{k+1} - \boldsymbol{x}_k\|}{\|\boldsymbol{x}_k\|} \leq \varepsilon$. The relative distance moved in the last step is small.

- $|f(\boldsymbol{x}_{k+N}) - f(\boldsymbol{x}_k)| < \varepsilon$. Not much change in the function value lately.

- $\dfrac{|f(\boldsymbol{x}_{k+1}) - f(\boldsymbol{x}_k)|}{|f(\boldsymbol{x}_k)|} < \varepsilon$. The relative change in the function value is small.

Sometimes *duality theory* (see section 5.3 and chapter 8) can be used to suggest good termination criteria.

Finally, it should be pointed out that although you feel confident that the generated sequence is convergent and that you have a good estimate of its limit, this may in fact not be the result you are looking for. You may have found a local minimum, and the required global minimum is somewhere else. A global minimum might not exist. These kinds of problems are the most difficult ones to handle. You will certainly require some help from theory.

Miscellanous exercises

3.25 Let $f(\boldsymbol{x}) = \frac{1}{2}\boldsymbol{x}^T \boldsymbol{H}\boldsymbol{x} + \boldsymbol{c}^T\boldsymbol{x} + d$, with $\boldsymbol{H}$ positive definite, and denote its minimum point by $\boldsymbol{x}^*$. Let $\boldsymbol{x}_1, \boldsymbol{x}_2, \ldots$ be a sequence generated by the steepest descent method (with exact line search) starting from $\boldsymbol{x}_1 \neq \boldsymbol{x}^*$. Prove that one reaches the exact minimum $\boldsymbol{x}^*$ in a finite number of steps if and only if $\boldsymbol{x}_1 - \boldsymbol{x}^*$ is an eigenvector of $\boldsymbol{H}$.

(*Hint:* It is no restriction to assume that $\boldsymbol{c} = \boldsymbol{0}$ and $d = 0$. Why?)

(1999–10–21, 6)

3.26 As an alternative to minimizing a function $f(\boldsymbol{x})$ using the steepest descent method one might consider minimizing $F(\boldsymbol{x}) = \|\nabla f(\boldsymbol{x})\|^2$ using the same method. Assume the case that f is a quadratic function with a positive definite Hessian. Compare the rates of convergence of the two schemes and hence justify why the equivalent minimization of F is not an attractive strategy.

3.27 Consider the following problem: m points (x_j, y_j), $j = 1, 2, \ldots, m$, in the plane are given. Find the point (x, y) which minimizes the sum of the distances to these points, i.e., minimize

$$f(x,y) = \sum_{j=1}^{m} \sqrt{(x_j - x)^2 + (y_j - y)^2}.$$

Assume here that $m = 3$ and that the three points are $(0, 2)$, $(0, 0)$ and $(6, -2)$.

a) Let $(0, 1)$ be the initial point. What search direction do you obtain in the steepest descent method?

b) With the same starting point, which is the search direction i Newton's method?

c) Carry out one iteration step with Newton's method.

d) The result does not seem satisfactory. Can you make any comment on or explanation of this fact?

(1999–04–09, 5)

3.28 The figure on the next page shows some (not equidistant) level curves of the function

$$f(x_1, x_2) = 10x_1^2 - 5x_1x_2 - 2x_1 + 3x_2 - \frac{1}{6}x_2^3 + 17.$$

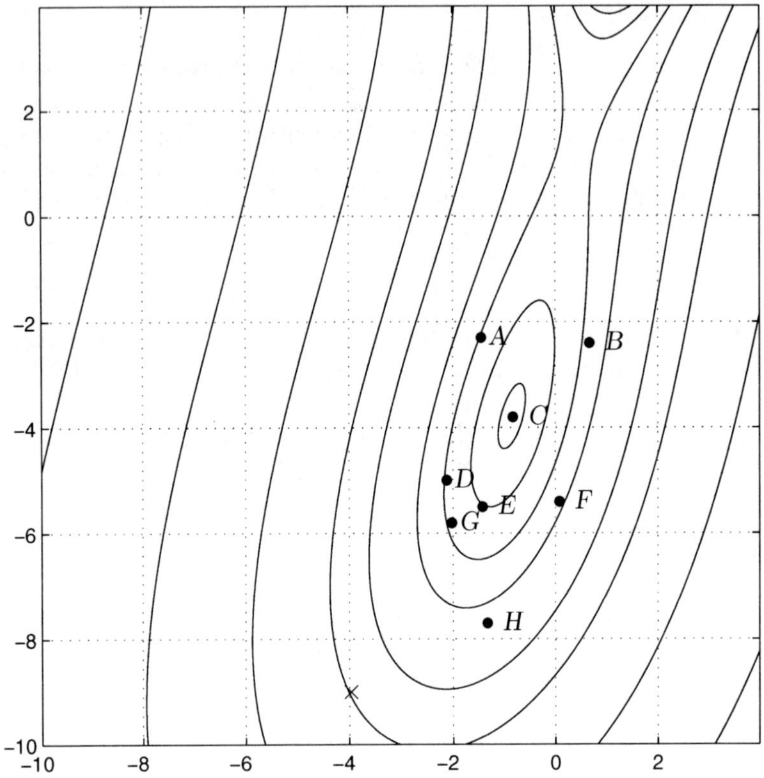

Points A, B, C, D, E, F, G, H are marked. Assuming that you start at $\boldsymbol{x}_1 = (-4, -9)$, one of these will (approximately) be the result of the following computations a) – d). Find out which one in each case.

a) Cyclic coordinates, two iterations.

b) Steepest descent method, two iterations.

c) Newton's method, one iteration.

d) Fletcher-Reeves' method, one iteration.

Don't forget to justify your answers. In some cases it might be unnecessary to carry out the actual computations. (1998–04–16, 4)

3.29 Put

$$f(x_1 x_2) = 2x_1^2 - x_1 x_2 - x_2^2 - 2x_1 + 5x_2 - \frac{1}{6}(x_2 - 2)^3,$$

and let $\boldsymbol{H}(x_1, x_2)$ denote its Hessian.

a) For what positive values of ε is $\boldsymbol{H}(0,0)+\varepsilon\boldsymbol{I}$ positive definite?

b) Put $\boldsymbol{x}^{(0)}=\begin{pmatrix}0\\0\end{pmatrix}$. Let $\varepsilon=1$ and use the modified Newton method to compute a new approximation $\boldsymbol{x}^{(1)}$ to the minimum point of f. (No line search; put $\lambda=1$.) Compute $f(\boldsymbol{x}^{(0)})$ and $f(\boldsymbol{x}^{(1)})$.

c) Verify that $\boldsymbol{H}(\boldsymbol{x}^{(1)})$ is positive definite. Carry out one more iteration step from $\boldsymbol{x}^{(1)}$ using the unmodified Newton method. What is $f(\boldsymbol{x}^{(2)})$?

(1997–12–17, 2)

3.30 Let $\boldsymbol{H}$ be a positive definite matrix of type $n\times n$, and $\boldsymbol{d}\neq\boldsymbol{0}$ an n-vector. The line through the origin in direction $\boldsymbol{d}$ intersects the surface $\boldsymbol{x}^T\boldsymbol{H}\boldsymbol{x}=1$ in two points $\pm\boldsymbol{x}_0$. Show that any vector parallel to the tangent plane at $\pm\boldsymbol{x}_0$ is $\boldsymbol{H}$-conjugate to $\boldsymbol{d}$. (1998–12–14, 6)

3.31 Consider the function $f(\boldsymbol{x})=18(x_1^2-x_2^2)+x_2^4$.

a) Show that the global minimum exists and calculate it.

b) Do 3 steps of Newton's algorithm starting with the initial point $(1,1)$. Does the algorithm seem to converge? Will the answer be satisfactory if you continue the iterations further?

c) Make all necessary changes and describe an improvement of Newton's algorithm that converges to a local minimum.

(2006 01 12, 1)

3.32 a) Consider the function $f(\boldsymbol{x})=2x_1^2+2x_1-2x_1x_2+x_2^2-2x_2$. Calculate all stationary points and extreme points (including their type).

b) The steepest descent method applied to the function f above and the initial point $\boldsymbol{x}_1=(0\ 0)^T$ generates a sequence $\{\boldsymbol{x}_n\}$, $n\geq 1$. It turns out that

$$\boldsymbol{x}_{2k+1}=\begin{pmatrix}0\\1-5^{-k}\end{pmatrix}\qquad\text{for all } k\geq 0.$$

Calculate $\boldsymbol{x}_{2k+2}$ for all $k\geq 0$ and sketch the sequence $\{\boldsymbol{x}_n\}$ in the (x_1,x_2)-plane.

(2009–01–09, 1ab)

Chapter 4

CONVEX SETS. SEPARATION

With this chapter we begin the study of the theory of optimization. An important concept in this context is *convexity*. We start here with convex sets, while convex functions will appear in a later chapter. The theory of convex sets reaches its climax with the so called Farkas' theorem, which will be important in connection with constrained optimization in chapter 7. Convex sets also serve as a background to the chapter on linear optimization.

However, first we review some concepts from topology, notably open and closed sets and compactness. These topics will be important on many occasions in the future. The first section also introduces some examples which will appear extensively later on.

4.1 Review of some concepts from topology

4.1.1 Open and closed sets. Compact sets

DEFINITION 1. *Let S be a subset of $\mathbf{R}^n$.*
$1°$. *A point $\boldsymbol{a} \in \mathbf{R}^n$ is called an* ***interior point*** *of S if there is an open ball $B_\varepsilon(\boldsymbol{a})$ which is contained in S.*
$2°$. *A point $\boldsymbol{a} \in \mathbf{R}^n$ is called an* ***outer point*** *of S if there is an open ball $B_\varepsilon(\boldsymbol{a})$ of which no point belongs to S.*
$3°$. *A point $\boldsymbol{a} \in \mathbf{R}^n$ is called a* ***boundary point*** *of S if any ball $B_\varepsilon(\boldsymbol{a})$ contains as well a point in S as a point outside S.*

See the picture on the next page.

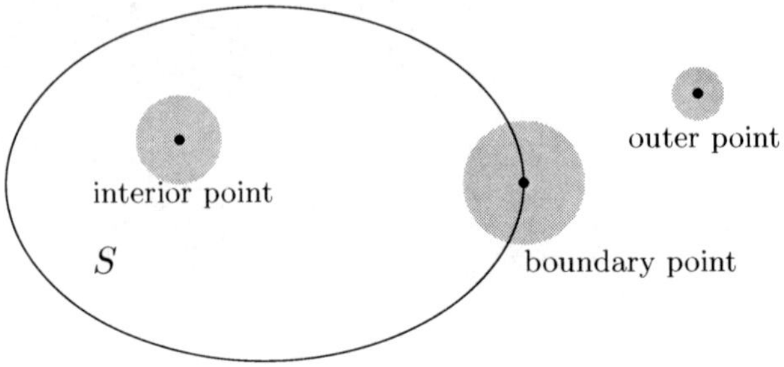

Obviously an interior point of S is an outer point of the complement of S and vice versa. The definitions cover all possibilities and are mutually exclusive; relative to a given set S any point $\boldsymbol{a}$ is either an interior point, an outer point or a boundary point.

The set of interior points of S is denoted $\text{int}(S)$ and is called the *interior* of S. The set of boundary points (the *boundary* of S) is denoted by ∂S.

DEFINITION 2. *The* ***closure*** *of a set S in $\mathbf{R}^n$ is the set*

$$\text{cl}(S) = \text{int}(S) \cup \partial S.$$

DEFINITION 3. *A subset S of $\mathbf{R}^n$ is called* ***closed*** *if $S = \text{cl}(S)$. A subset S of $\mathbf{R}^n$ is called* ***open*** *if $S = \text{int}(S)$.*

Thus S is closed if it contains all its boundary points, and is open if it contains none of its boundary points.

The reader should provide an example of a set which is neither closed nor open. Another exercise: show that the closure of S is a closed set.

Example 1. A *(hyper)plane* in $\mathbf{R}^n$, with normal direction $\boldsymbol{p}$ and containing the point $\bar{\boldsymbol{x}}$, has equation

$$\boldsymbol{p}^T(\boldsymbol{x} - \bar{\boldsymbol{x}}) = 0.$$

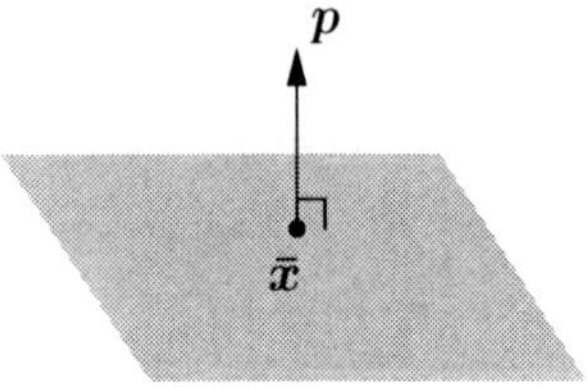

Putting $\boldsymbol{p}^T\bar{\boldsymbol{x}} = \alpha$ the equation can be written

$$\boldsymbol{p}^T\boldsymbol{x} = \alpha.$$

Every point in the hyperplane, and no other, is a boundary point. Thus it is a closed set. It has no interior points. □

Example 2. A *closed half-space* in $\mathbf{R}^n$ is defined by an inequality

$$\boldsymbol{p}^T\boldsymbol{x} \le \alpha \quad \text{or} \quad \boldsymbol{p}^T\boldsymbol{x} \ge \alpha.$$

Its boundary consists of the hyperplane in example 1. It is a closed set.

With strict inequalities we obtain an *open half-space*, which is an open set. □

Example 3. A *sphere* $\|\boldsymbol{x} - \boldsymbol{a}\| = r$ is a closed set. A closed ball, $\|\boldsymbol{x} - \boldsymbol{a}\| \le r$, is a closed set, and an open ball is an open set. □

The closure of a set S is the least closed set containing S. This can also be phrased as

$$\mathrm{cl}(S) = \bigcap_{\substack{T \text{ closed} \\ T \supseteq S}} T.$$

It is left to the reader to prove these statements.

Closedness can often be proved by use of the following lemma.

LEMMA 1. *A set S is closed if and only if, for any convergent sequence $\{\boldsymbol{x}_k\}_1^\infty$ in S, its limit point $\boldsymbol{x}$ belongs to S.*

PROOF. First assume that S is closed. Let $\{\boldsymbol{x}_k\}_1^\infty$ be a sequence in S with limit point $\boldsymbol{x}$. Since $\boldsymbol{x}$ obviously is not an outer point of S it belongs to $\mathrm{cl}(S)$, which by assumption is equal to S.

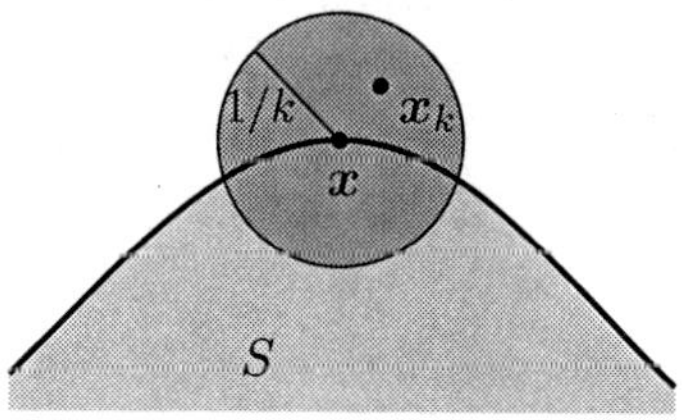

To prove the converse let $\boldsymbol{x}$ be a boundary point of S. Then every ball $B_{1/k}(\boldsymbol{x})$ with center $\boldsymbol{x}$ and radius $1/k$ contains a point, call it $\boldsymbol{x}_k$, in S. The sequence $\{\boldsymbol{x}_k\}_1^\infty$ converges to $\boldsymbol{x}$, hence by assumption $\boldsymbol{x}$ belongs to S. This proves that S is closed. □

Example 4. Let $\boldsymbol{A}$ be an $m \times n$ matrix and $\boldsymbol{b}$ a column vector with m elements. In the future we will often have occasion to consider the set S in $\mathbf{R}^n$ defined by the *vector inequality*

$$\boldsymbol{A}\boldsymbol{x} \le \boldsymbol{b}.$$

The inequality should be interpreted as inequalities between the individual m components, in other words, it is a *system of inequalities*[1]

$$\begin{cases} a_{11}x_1 + \ldots + a_{n1}x_n \le b_1 \\ \ldots \\ a_{m1}x_1 + \ldots + a_{mn}x_n \le b_m. \end{cases}$$

Geometrically S is an intersection of closed half-spaces. Later we will call it a *polyhedral* set. In linear programming (chapter 5) we study minimization of linear functions over polyhedral sets.

The set S defined by these inequalities is closed. This is easily seen using the lemma. If the elements of a convergent sequence $\{\boldsymbol{x}_k\}_1^\infty$ in $\mathbf{R}^n$ satisfy $\boldsymbol{A}\boldsymbol{x}_k \le \boldsymbol{b}$, then by continuity the limit point $\boldsymbol{x}$ satisfies $\boldsymbol{A}\boldsymbol{x} \le \boldsymbol{b}$. □

DEFINITION 4. *A set S in $\mathbf{R}^n$ is called* **bounded** *if it is contained in some ball $\|\boldsymbol{x}\| \le R$ with large enough radius R.*

A set S in $\mathbf{R}^n$ is called **compact** *if it is closed and bounded.*

4.1.2 Weierstrass' theorem

For a given sequence $\{\boldsymbol{x}_k\}_1^\infty$ (indexed as usual by the positive integers $\mathbf{Z}^+$) one often has reason to consider one or more *subsequences*. This means that we pick out a suitable infinite subset $\mathcal{K}$ of $\mathbf{Z}^+$, and allow only those $\boldsymbol{x}_k$ whose index k is taken from $\mathcal{K}$. For example, one may consider the subsequence of those $\boldsymbol{x}_k$ for which k is an even number (we could denote this subsequence by $\{\boldsymbol{x}_{2p}\}_1^\infty$), or the subsequence of all elements with an odd index ($\{\boldsymbol{x}_{2p-1}\}_1^\infty$). In the general case a subsequence is denoted by $\{\boldsymbol{x}_k\}_{k\in\mathcal{K}}$. The elements of $\mathcal{K}$ are assumed to have the ordering inherited from $\mathbf{Z}^+$.

Example 5. Consider the sequence $\{x_k\}_1^\infty$ of real numbers defined by

$$x_k = \frac{e^{-3k} + 1}{4k - (-1)^k}, \qquad k = 1, 2, \ldots .$$

[1]The reader should beware to handle vector inequalities too carelessly. For example, $\boldsymbol{a} \not\le \boldsymbol{b}$ is not equivalent to $\boldsymbol{a} > \boldsymbol{b}$. (However, one of these statements implies the other. Which one?)

From this sequence we may pick out the subsequence $\{x_{2p}\}_1^\infty$ of those elements which are indexed by an even number,

$$x_{2p} = \frac{e^{-6p} + 1}{8p - 1}, \qquad p = 1, 2, \ldots .$$

We may also pick the subsequence with all odd positive integers as index, $\{x_{2p-1}\}_1^\infty$, where

$$x_{2p-1} = \frac{e^{-6p+3} + 1}{8p - 3}, \qquad p = 1, 2, \ldots .$$

Other examples are the subsequence consisting of every third element of $\{x_k\}_1^\infty$, or the subsequence of those elements which have a prime number as index. □

THEOREM 1. (BOLZANO-WEIERSTRASS). *Let S be a compact subset of $\mathbf{R}^n$, and assume that $\{\boldsymbol{x}_k\}_1^\infty$ is a sequence of elements in S. Then it has a subsequence $\{\boldsymbol{x}_k\}_{k \in \mathcal{K}}$ which converges (to an element of S).*

Example 6. For example, in $\mathbf{R}$ the sequence $\{x_k\}_1^\infty$ defined by

$$x_k = \frac{1 + (-1)^k}{1 + \dfrac{1}{k}}$$

is not convergent. However, all elements x_k belong to the compact interval $[0, 2]$. Hence by the theorem it has a convergent subsequence. In this case it is easy to exhibit one; the subsequence of all elements with an odd index is $0, 0, 0, \ldots$, which is certainly convergent. The reader might verify that the subsequence of all elements with an even index converges as well. □

Example 7. Let $x_k = \sin 2^k$. Then $\{x_k\}_1^\infty$ is a sequence in the compact interval $[-1, 1]$ in $\mathbf{R}$. By the theorem it has a convergent subsequence.

Note that this is a purely theoretical result. The theorem gives no indication of how such a sequence might be constructed. □

Theorem 1 will be used occasionally in the course. One application is in the proof of the following theorem, which of course plays a very important role in optimization.

THEOREM 2. (WEIERSTRASS) *Assume that S is a compact subset of $\mathbf{R}^n$ and that f is a continuous function defined on S. Then there is an $\bar{\boldsymbol{x}}$ in S such that*

$$f(\bar{\boldsymbol{x}}) = \min_{\boldsymbol{x}\in S} f(\boldsymbol{x}).$$

Example 8. Let

$$f(x_1, x_2) = e^{x_1 x_2}\, \frac{x_1 + \sin x_2}{x_1^4 + x_2^4 + 1}, \qquad x_1^2 + x_2^2 \le 1.$$

This is a continuous function on the closed and bounded unit disc. Hence it has a minimum at some point there, by the theorem. How to find that point is another question. □

The proofs of theorems 1 and 2 are omitted.

Exercises

4.1 Identify the closure, interior, and boundary of each of the following sets:

a) $S = \{\,\boldsymbol{x} \in \mathbf{R}^2\,;\, x_1^2 \le x_2 < 1\,\}$,

b) $S = \{\,\boldsymbol{x} \in \mathbf{R}^2\,;\, 0 \le x_1 \le 1,\ 0 \le x_2 \le 1 - x_1\,\}$,

c) $S = \{\,\boldsymbol{x} \in \mathbf{R}^2\,;\, x_2 = 2x_1,\ 0 \le x_1 < 2\,\}$,

d) $S = \{\,\boldsymbol{x} \in \mathbf{R}^3\,;\, x_1 + x_3 = 3,\ x_1 + x_2 + 2x_3 \le 5\,\}$,

e) $S = \{\,\boldsymbol{x} \in \mathbf{R}^3\,;\, x_1^2 + x_2^2 + x_3^2 \le 5,\ x_1 + x_3 = 1\,\}$.

4.2 Prove that if S_1 and S_2 are closed sets then their intersection $S_1 \cap S_2$ is closed. What about the intersection of infinitely many closed sets?

S 4.3 Prove that the closure of a set S is a closed set.

4.4 Prove that $\mathrm{cl}(S) = \displaystyle\bigcap_{\substack{T \text{ closed} \\ T \supseteq S}} T.$

4.2 Convexity

4.2.1 Convex sets

DEFINITION 5. *A set S in* $\mathbf{R}^n$ *is said to be* **convex** *if*

(1) $\quad \boldsymbol{x}_1, \boldsymbol{x}_2 \in S \quad \Longrightarrow \quad \lambda \boldsymbol{x}_1 + (1-\lambda)\boldsymbol{x}_2 \in S$ *for all* $0 < \lambda < 1$.

In geometrical language: S is convex if for any pair of points $\boldsymbol{x}_1, \boldsymbol{x}_2$ in S the whole line segment between these points also belongs to S.

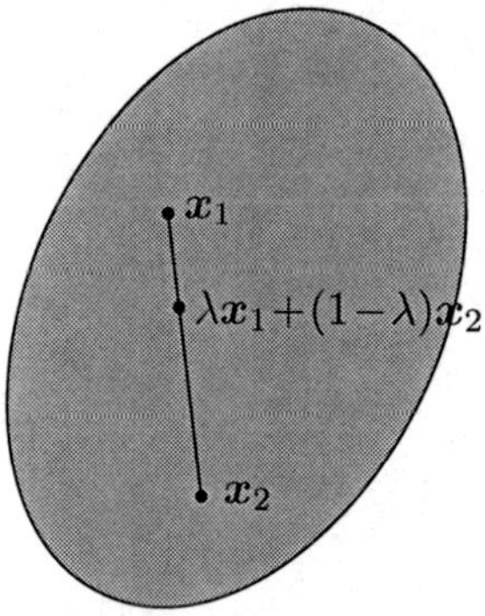

convex set

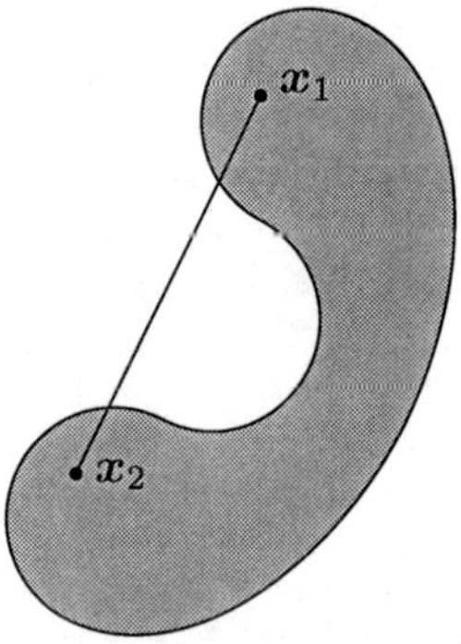

not convex set

Example 9. A hyperplane $\boldsymbol{p}^T\boldsymbol{x} = \alpha$ is a convex set, since $\boldsymbol{p}^T\boldsymbol{x}_1 = \alpha$ and $\boldsymbol{p}^T\boldsymbol{x}_2 = \alpha$ imply that

$$\boldsymbol{p}^T(\lambda \boldsymbol{x}_1 + (1-\lambda)\boldsymbol{x}_2) = \lambda \boldsymbol{p}^T\boldsymbol{x}_1 + (1-\lambda)\boldsymbol{p}^T\boldsymbol{x}_2 = \lambda\alpha + (1-\lambda)\alpha = \alpha.$$

That a half-space $\boldsymbol{p}^T\boldsymbol{x} \leq \alpha$ in $\mathbf{R}^n$ is a convex set is proved analogously. (Exercise!) □

Example 10. An open ball $B = \{\, \boldsymbol{x}\,;\, \|\boldsymbol{x} - \boldsymbol{a}\| < r \,\}$ is convex. To prove this, assume that $\boldsymbol{x}_1, \boldsymbol{x}_2 \in B$. Then, for $0 < \lambda < 1$, by the triangle inequality

$$\|(\lambda \boldsymbol{x}_1 + (1-\lambda)\boldsymbol{x}_2) - \boldsymbol{a}\| = \|\lambda(\boldsymbol{x}_1 - \boldsymbol{a}) + (1-\lambda)(\boldsymbol{x}_2 - \boldsymbol{a})\| \leq$$

$$\leq \lambda\|\boldsymbol{x}_1 - \boldsymbol{a}\| + (1-\lambda)\|\boldsymbol{x}_2 - \boldsymbol{a}\| <$$

$$< \lambda r + (1-\lambda) r = r,$$

proving that $\lambda \boldsymbol{x}_1 + (1-\lambda)\boldsymbol{x}_2 \in B$.

Also a closed ball is a convex set. A sphere $\|\boldsymbol{x} - \boldsymbol{a}\| = r$ is *not* a convex set. □

Example 11. Consider with notation from example 4 a polyhedral set

$$S = \{ \boldsymbol{x} \in \mathbf{R}^n \,;\, \boldsymbol{A}\boldsymbol{x} \le \boldsymbol{b} \}.$$

This set is convex. To prove this assume that $\boldsymbol{A}\boldsymbol{x}_1 \le \boldsymbol{b}$ and $\boldsymbol{A}\boldsymbol{x}_2 \le \boldsymbol{b}$. Then, if $0 < \lambda < 1$,

$$\boldsymbol{A}(\lambda\boldsymbol{x}_1 + (1-\lambda)\boldsymbol{x}_2) = \lambda\boldsymbol{A}\boldsymbol{x}_1 + (1-\lambda)\boldsymbol{A}\boldsymbol{x}_2 \le \lambda\boldsymbol{b} + (1-\lambda)\boldsymbol{b} = \boldsymbol{b}.$$

Another way of proving that S is convex is to note that it is the intersection of finitely many half-spaces, which we already know to be convex, and then use lemma 2 a) below. □

Example 12. The set in the plane

$$\{ (x_1, x_2) \in \mathbf{R}^2 \,;\, x_2 \ge |x_1| \}$$

is convex. It is left to the reader to verify this. □

Example 13. Let S be the set of all $\boldsymbol{x} \in \mathbf{R}^n$ which solve the problem of finding the minimum of

$$\boldsymbol{c}^T\boldsymbol{x}$$

subject to the constraints

$$\boldsymbol{A}\boldsymbol{x} = \boldsymbol{b}, \quad \boldsymbol{x} \ge \boldsymbol{0}.$$

Here $\boldsymbol{A}$ and $\boldsymbol{b}$ are given matrices of types $m \times n$ and $m \times 1$ respectively; $\boldsymbol{c}$ is of type $n \times 1$.

For the proof that S is convex assume that both $\boldsymbol{x}_1$ and $\boldsymbol{x}_2$ solve the minimization problem. For $0 < \lambda < 1$ we then have

$$\boldsymbol{A}(\lambda\boldsymbol{x}_1 + (1-\lambda)\boldsymbol{x}_2) = \lambda\boldsymbol{A}\boldsymbol{x}_1 + (1-\lambda)\boldsymbol{A}\boldsymbol{x}_2 = \lambda\boldsymbol{b} + (1-\lambda)\boldsymbol{b} = \boldsymbol{b},$$

and of course $\lambda\boldsymbol{x}_1 + (1-\lambda)\boldsymbol{x}_2 \ge \boldsymbol{0}$. Thus $\lambda\boldsymbol{x}_1 + (1-\lambda)\boldsymbol{x}_2$ satisfies the constraints. Furthermore, let v denote the minimum value. Then $\boldsymbol{c}^T\boldsymbol{x}_1 = v$, $\boldsymbol{c}^T\boldsymbol{x}_2 = v$ and consequently

$$\boldsymbol{c}^T(\lambda\boldsymbol{x}_1 + (1-\lambda)\boldsymbol{x}_2) = \lambda\boldsymbol{c}^T\boldsymbol{x}_1 + (1-\lambda)\boldsymbol{c}^T\boldsymbol{x}_2 = \lambda v + (1-\lambda)v = v.$$

Thus $\lambda\boldsymbol{x}_1 + (1-\lambda)\boldsymbol{x}_2$ solves the minimization problem.

Hence the convexity of S is proved. □

Example 14. On the real line, a set is convex if and only if it is an interval. It is left to the reader to check this. □

LEMMA 2. *Assume that the sets S_1 and S_2 are convex subsets of $\mathbf{R}^n$. Then the following sets are convex.*

a) $S_1 \cap S_2$,

b) $S_1 \oplus S_2 = \{\, \boldsymbol{x}_1 + \boldsymbol{x}_2 \,;\, \boldsymbol{x}_1 \in S_1, \boldsymbol{x}_2 \in S_2 \,\}$,

c) $S_1 \ominus S_2 = \{\, \boldsymbol{x}_1 - \boldsymbol{x}_2 \,;\, \boldsymbol{x}_1 \in S_1, \boldsymbol{x}_2 \in S_2 \,\}$.

PROOF. We prove b), and leave a) and c) as exercises for the reader.

Assume that $\boldsymbol{x}_1 + \boldsymbol{x}_2$ and $\boldsymbol{y}_1 + \boldsymbol{y}_2$ are elements of $S_1 \oplus S_2$, with $\boldsymbol{x}_1, \boldsymbol{y}_1 \in S_1$ and $\boldsymbol{x}_2, \boldsymbol{y}_2 \in S_2$. Rearranging we get

$$\lambda(\boldsymbol{x}_1 + \boldsymbol{x}_2) + (1-\lambda)(\boldsymbol{y}_1 + \boldsymbol{y}_2) = (\lambda \boldsymbol{x}_1 + (1-\lambda)\boldsymbol{y}_1) + (\lambda \boldsymbol{x}_2 + (1-\lambda)\boldsymbol{y}_2)\,.$$

By the convexity of S_1 and S_2 this is for $0 < \lambda < 1$ a sum of one element from each set, thus an element of $S_1 \oplus S_2$. □

It is easily seen, as in the proof of a) of the lemma, that the intersection of any number (finite or infinite) of convex sets is convex.

We now introduce some terminology: for real numbers $\lambda_1, \ldots, \lambda_k$ and vectors $\boldsymbol{x}_1, \ldots, \boldsymbol{x}_k$ in $\mathbf{R}^n$ we call

$\sum_{i=1}^k \lambda_i \boldsymbol{x}_i$ a **linear combination** of $\boldsymbol{x}_1, \ldots, \boldsymbol{x}_k$,

$\sum_{i=1}^k \lambda_i \boldsymbol{x}_i$ an **affine combination** of $\boldsymbol{x}_1, \ldots, \boldsymbol{x}_k$ if $\sum_{i=1}^k \lambda_i = 1$,

$\sum_{i=1}^k \lambda_i \boldsymbol{x}_i$ a **convex combination** of $\boldsymbol{x}_1, \ldots, \boldsymbol{x}_k$ if $\sum_{i=1}^k \lambda_i = 1$ and all $\lambda_i \geq 0$.

In definition 5 above of a convex set the condition (1) can be reformulated as

$$\left.\begin{array}{r} \boldsymbol{x}_1, \boldsymbol{x}_2 \in S \\ \lambda_1 + \lambda_2 = 1, \quad \lambda_1, \lambda_2 \geq 0 \end{array}\right\} \quad \Longrightarrow \quad \lambda_1 \boldsymbol{x}_1 + \lambda_2 \boldsymbol{x}_2 \in S.$$

It is left to the reader to prove by induction the following more general statement: *a set S is convex if and only if for any set of elements $\boldsymbol{x}_1, \ldots, \boldsymbol{x}_k$ in S every convex combination of them also belongs to S.*

4.2.2 Convex hull

DEFINITION 6. *Let S be a subset of $\mathbf{R}^n$. The* **convex hull** *of S is the set defined as*

$$H(S) = \{\, \textit{all convex combinations of elements in } S \,\}.$$

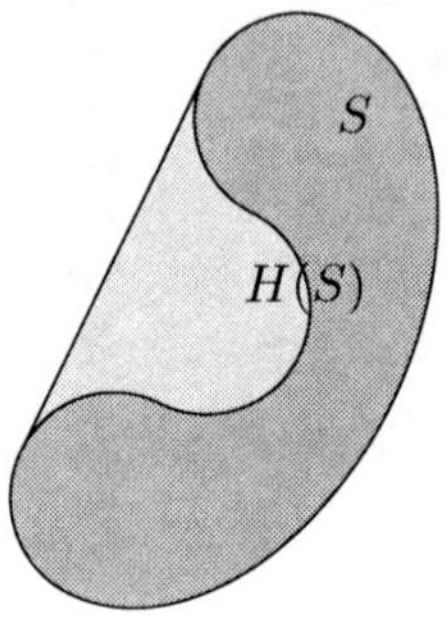

Example 15. If $S = \{\boldsymbol{a}, \boldsymbol{b}, \boldsymbol{c}\}$, a set containing three points in $\mathbf{R}^n$, then $H(S)$ is the closed triangle with corners $\boldsymbol{a}$, $\boldsymbol{b}$, $\boldsymbol{c}$. See the left figure below.

The situation is of course analogous for any finite set of points. The convex hull of finitely many points is called a **polytope**. The right figure gives an example of the convex hull of six points in a plane.

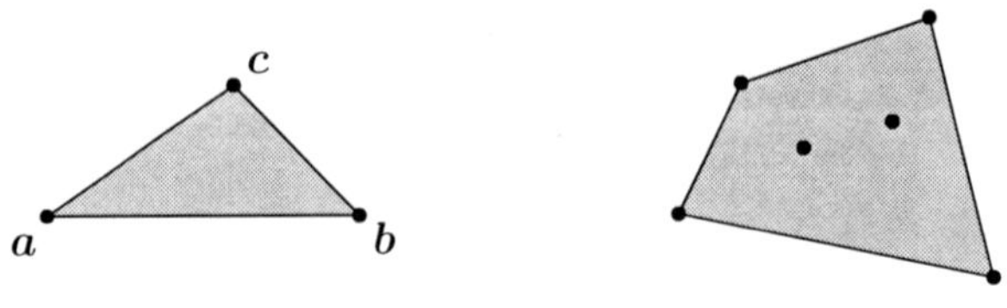

□

It is clear from the definition that $H(S) \supseteq S$ always.

LEMMA 3. *The set $H(S)$ is convex.*

PROOF. Assume that $\boldsymbol{x}_1$ and $\boldsymbol{x}_2$ belong to $H(S)$, i.e., that they are convex combinations of elements in S. Since new elements with zero coefficient can be added to a convex combination without destroying this property, it is no restriction to assume that $\boldsymbol{x}_1$ and $\boldsymbol{x}_2$ are convex combinations of the *same* elements $\boldsymbol{z}_1, \dots, \boldsymbol{z}_k$ in S:

$$\boldsymbol{x}_1 = \sum_{j=1}^{k} \mu_j \boldsymbol{z}_j, \qquad \boldsymbol{x}_2 = \sum_{j=1}^{k} \nu_j \boldsymbol{z}_j,$$

where $\sum_1^k \mu_j = \sum_1^k \nu_j = 1$ and $\mu_j \geq 0$, $\nu_j \geq 0$. Now consider the convex combination $\lambda \boldsymbol{x}_1 + (1-\lambda)\boldsymbol{x}_2$ of $\boldsymbol{x}_1$ and $\boldsymbol{x}_2$, with $0 < \lambda < 1$. We have

$$\lambda \boldsymbol{x}_1 + (1-\lambda)\boldsymbol{x}_2 = \lambda \sum_{j=1}^{k} \mu_j \boldsymbol{z}_j + (1-\lambda) \sum_{j=1}^{k} \nu_j \boldsymbol{z}_j = \sum_{j=1}^{k} (\lambda\mu_j + (1-\lambda)\nu_j)\boldsymbol{z}_j.$$

The coefficients here are non-negative and their sum is

$$\sum_{j=1}^{k} (\lambda\mu_j + (1-\lambda)\nu_j) = \lambda \sum_{j=1}^{k} \mu_j + (1-\lambda) \sum_{j=1}^{k} \nu_j = \lambda \cdot 1 + (1-\lambda) \cdot 1 = 1.$$

Hence $\lambda \boldsymbol{x}_1 + (1-\lambda)\boldsymbol{x}_2$ is a convex combination of elements in S, thus belonging to $H(S)$. □

LEMMA 4. *The convex hull of a set S is the least convex set containing S, i.e.*

$$H(S) = \bigcap_{\substack{T \text{ convex} \\ T \supseteq S}} T.$$

PROOF. Denote the right hand side set by R. Obviously R is the least (when ordering is by set inclusion) convex set containing S.

Since $H(S)$ is one of the sets T occuring in the definition av R we have $R \subseteq H(S)$. On the other hand, any $\boldsymbol{x}$ in $H(S)$ is a convex combination of elements in S. Those elements belong to any set T appearing in the definition of R, and since these T are convex, they contain $\boldsymbol{x}$ as well. Consequently $\boldsymbol{x}$ belongs to their intersection, in other words $\boldsymbol{x} \in R$. Thus $H(S) \subseteq R$. □

Analogously to the convex hull one defines the *affine hull* and the *linear hull* of a set S in $\mathbf{R}^n$. There are counterparts to lemma 4 for these cases. The details are left to the reader.

THEOREM 3. (CARATHÉODORY) *Let S be a subset of $\mathbf{R}^n$. Any point $\boldsymbol{x}$ of $H(S)$ can be written*

$$\boldsymbol{x} = \sum_{j=1}^{n+1} \lambda_j \boldsymbol{x}_j \qquad \textit{with} \quad \sum_{j=1}^{n+1} \lambda_j = 1 \textit{ and all } \lambda_j \geq 0$$

for suitable $\boldsymbol{x}_1, \ldots, \boldsymbol{x}_{n+1}$ in S (depending on $\boldsymbol{x}$). In other words, any $\boldsymbol{x} \in H(S)$ is a convex combination of at most *$n+1$ points in S.*

The reader may skip the proof without loss of continuity.

PROOF. By definition $\boldsymbol{x}$ is a convex combination of some elements $\boldsymbol{x}_1, \ldots, \boldsymbol{x}_k$ of S:

$$\boldsymbol{x} = \sum_{j=1}^{k} \lambda_j \boldsymbol{x}_j, \qquad \text{where} \quad \sum_{j=1}^{k} \lambda_j = 1, \quad \lambda_j \geq 0.$$

If $k \leq n+1$ we are finished.

Thus assume that $k > n+1$. Then the vectors

$$\boldsymbol{x}_2 - \boldsymbol{x}_1,\ \boldsymbol{x}_3 - \boldsymbol{x}_1,\ \ldots,\ \boldsymbol{x}_k - \boldsymbol{x}_1$$

are linearly dependent, since their number is larger than n. This means that there exist real numbers μ_j, not all of them equal to zero, such that

$$\sum_{j=2}^{k} \mu_j (\boldsymbol{x}_j - \boldsymbol{x}_1) = \boldsymbol{0} \quad \Longleftrightarrow \quad \sum_{j=2}^{k} \mu_j \boldsymbol{x}_j - \Big(\sum_{j=2}^{k} \mu_j\Big) \boldsymbol{x}_1 = \boldsymbol{0}.$$

Putting $\mu_1 = -\sum_{j=2}^{k} \mu_j$ we have

$$\sum_{j=1}^{k} \mu_j \boldsymbol{x}_j = \boldsymbol{0} \qquad \text{with} \quad \sum_{j=1}^{k} \mu_j = 0 \quad \text{and some } \mu_j > 0.$$

Now we can eliminate one of the $\boldsymbol{x}_j$:s as follows. Denote by α a real number to be chosen in a minute. We have

$$\boldsymbol{x} = \sum_{j=1}^{k} \lambda_j \boldsymbol{x}_j + \boldsymbol{0} = \sum_{j=1}^{k} \lambda_j \boldsymbol{x}_j - \alpha \sum_{j=1}^{k} \mu_j \boldsymbol{x}_j = \sum_{j=1}^{k} (\lambda_j - \alpha \mu_j) \boldsymbol{x}_j.$$

Choose α as the least of the numbers $\dfrac{\lambda_j}{\mu_j}$ with μ_j positive. (Such numbers μ_j exist.) Assume that $\alpha = \dfrac{\lambda_i}{\mu_i}$ and put $\lambda'_j = \lambda_j - \alpha \mu_j$ for all j. Then $\lambda'_i = 0$ and the choice of α implies that $\lambda'_j \geq 0$ when $j \neq i$. Consequently

$$\boldsymbol{x} = \sum_{\substack{j=1 \\ j\neq i}}^{k} \lambda'_j \boldsymbol{x}_j, \qquad \text{where} \qquad \sum_{\substack{j=1 \\ j\neq i}}^{k} \lambda'_j = 1, \quad \lambda'_j \geq 0.$$

This exhibits $\boldsymbol{x}$ as a convex combination of at most $k-1$ points. Continue this process until $n+1$ points remain, and the proof is finished. □

4.2.3 Connections between convexity and topological concepts like interior, closure, ...

LEMMA 5. *Let S be a convex set in R^n. Assume that $\boldsymbol{x}_1 \in \mathrm{cl}(S)$ and $\boldsymbol{x}_2 \in \mathrm{int}(S)$. Then*

$$\lambda \boldsymbol{x}_1 + (1-\lambda)\boldsymbol{x}_2 \in \mathrm{int}(S) \qquad \textit{when} \ \ 0 \leq \lambda < 1.$$

PROOF. Let

$$\boldsymbol{x} = \lambda \boldsymbol{x}_1 + (1-\lambda)\boldsymbol{x}_2.$$

Since $\boldsymbol{x}_2$ is an interior point of S we can find an open ball with center $\boldsymbol{x}_2$ which is contained in S; denote its radius by ε.

We intend to prove that the open ball with center $\boldsymbol{x}$ and radius $(1-\lambda)\varepsilon$ is contained in S; this will show that $\boldsymbol{x}$ is an interior point of S and the proof will be finished. What we need to prove then is that an arbitrary point $\boldsymbol{z}$ in this ball belongs to S. See the figure below.

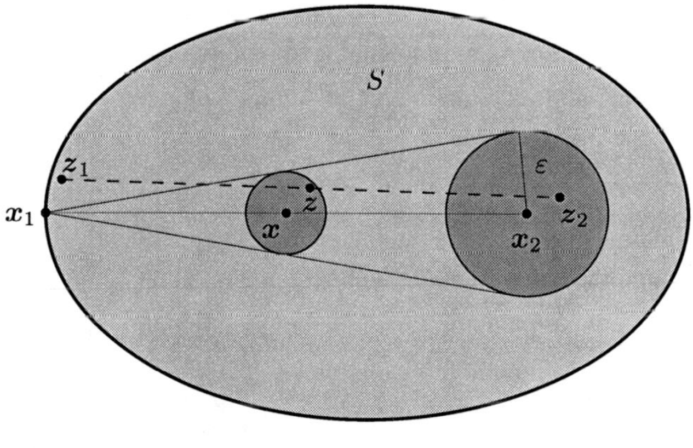

Since $\boldsymbol{x}_1$ belongs to the closure of S there is for every positive r a point $\boldsymbol{z}_1 \in S$ such that $\|\boldsymbol{z}_1 - \boldsymbol{x}_1\| < r$. We will choose a value for r shortly. Let $\boldsymbol{z}_2$ be the point in $\mathbf{R}^n$ defined by

$$\boldsymbol{z} = \lambda \boldsymbol{z}_1 + (1-\lambda)\boldsymbol{z}_2,$$

i.e., let

$$\boldsymbol{z}_2 = \frac{1}{1-\lambda}(\boldsymbol{z} - \lambda \boldsymbol{z}_1).$$

Then

$$\|\boldsymbol{z}_2 - \boldsymbol{x}_2\| = \left\|\frac{1}{1-\lambda}(\boldsymbol{z} - \lambda \boldsymbol{z}_1) - \frac{1}{1-\lambda}(\boldsymbol{x} - \lambda \boldsymbol{x}_1)\right\| =$$

$$= \frac{1}{1-\lambda}\left\|(\boldsymbol{z} - \boldsymbol{x}) - \lambda(\boldsymbol{z}_1 - \boldsymbol{x}_1)\right\| \leq$$

$$\leq \frac{1}{1-\lambda}\Big(\|\boldsymbol{z} - \boldsymbol{x}\| + \lambda\|\boldsymbol{z}_1 - \boldsymbol{x}_1\|\Big) < \frac{1}{1-\lambda}\Big(\|\boldsymbol{z} - \boldsymbol{x}\| + \lambda r\Big),$$

where the first inequality is the triangle inequality. Choosing

$$r = \frac{(1-\lambda)\varepsilon - \|\boldsymbol{z} - \boldsymbol{x}\|}{\lambda},$$

which is a positive number (check this!), we obtain the inequality

$$\|\boldsymbol{z}_2 - \boldsymbol{x}_2\| < \varepsilon.$$

In particular, this shows that $\boldsymbol{z}_2 \in S$. Since also $\boldsymbol{z}_1 \in S$, by the convexity of S we conclude that $\boldsymbol{z}$ belongs to S.

The proof is finished. □

The lemma provides the essential part of the proofs of most of the following statements.

THEOREM 4.

1. *If S is convex then* $\operatorname{int}(S)$ *is convex.*

2. *If S is convex then* $\operatorname{cl}(S)$ *is convex.*

3. *If S is convex with nonempty interior then* $\operatorname{cl}(\operatorname{int}(S)) = \operatorname{cl}(S)$.

4. *If S is convex with nonempty interior then* $\operatorname{int}(\operatorname{cl}(S)) = \operatorname{int}(S)$.

PROOF.

1. If $\boldsymbol{x}_1, \boldsymbol{x}_2 \in \operatorname{int}(S)$, then by the lemma any convex combination of these points is in $\operatorname{int}(S)$.

2. Let $\boldsymbol{x}_1, \boldsymbol{x}_2 \in \operatorname{cl}(S)$, and let $\boldsymbol{x} = \lambda\boldsymbol{x}_1 + (1-\lambda)\boldsymbol{x}_2$ be a convex combination of these points. By the definition of closure there is a sequence $\{\boldsymbol{y}_k\}_1^\infty$ of points in S with limit point $\boldsymbol{x}_1$, and another sequence $\{\boldsymbol{z}_k\}_1^\infty$ in S with limit point $\boldsymbol{x}_2$. Let

$$\boldsymbol{\xi}_k = \lambda\boldsymbol{y}_k + (1-\lambda)\boldsymbol{z}_k.$$

Since S is convex $\boldsymbol{\xi}_k$ belongs to S. Furthermore, the limit of $\boldsymbol{\xi}_k$ as $k \to \infty$ is $\lambda\boldsymbol{x}_1 + (1-\lambda)\boldsymbol{x}_2 = \boldsymbol{x}$. Consequently $\boldsymbol{x}$ belongs to $\operatorname{cl}(S)$. This proves that $\operatorname{cl}(S)$ is a convex set.

3. We only have to prove that $\operatorname{cl}(\operatorname{int}(S)) \supseteq \operatorname{cl}(S)$, since the reverse inequality is obvious. Let $\boldsymbol{x}$ be an element of $\operatorname{cl}(S)$, and choose a point $\boldsymbol{z}$ in $\operatorname{int}(S)$. Then by the lemma

$$\lambda\boldsymbol{x} + (1-\lambda)\boldsymbol{z} \in \operatorname{int}(S).$$

Taking the limit when $\lambda \to 1$ we conclude that $\boldsymbol{x} \in \operatorname{cl}(\operatorname{int}(S))$.

4. We need only prove that $\operatorname{int}(\operatorname{cl}(S)) \subseteq \operatorname{int}(S)$, the reverse inequality being obvious. Let $\boldsymbol{x} \in \operatorname{int}(\operatorname{cl}(S))$ and choose a point $\boldsymbol{z}$ in $\operatorname{int}(S)$. Define $\boldsymbol{y}$ so that

$$\boldsymbol{x} = \lambda\boldsymbol{z} + (1-\lambda)\boldsymbol{y},$$

i.e., let

$$\boldsymbol{y} = \frac{1}{1-\lambda}(\boldsymbol{x} - \lambda\boldsymbol{z}).$$

If λ is sufficiently close to zero the point $\boldsymbol{y}$ is in such a small neighbourhood of $\boldsymbol{x}$ that it belongs to $\operatorname{cl}(S)$, since $\boldsymbol{x}$ is in the interior of this set. Thus by the lemma $\boldsymbol{x} \in \operatorname{int}(S)$. □

The reader should provide examples showing that without the assumption of convexity of S the conclusions in 3. and 4. may not be true.

Exercises

4.5 Decide whether the following sets are convex.

a) $\{\,(x_1,\, x_2)\,;\, x_1^2 + x_2^2 = 1\,\}$,

b) $\{ (x_1, x_2) ; x_1 - x_2 \leq 1 \}$,

c) $\{ (x_1, x_2) ; x_1 < x_2^2 \}$,

d) $\{ (x_1, x_2, x_3) ; x_1 \geq x_2^2 \}$.

S 4.6 Let S be a convex set in $\mathbf{R}^n$ and $\boldsymbol{A}$ an $m \times n$-matrix. Show that the set

$$\boldsymbol{A}S = \{ \boldsymbol{y} ; \boldsymbol{y} = \boldsymbol{A}\boldsymbol{x}, \ \boldsymbol{x} \in S \}$$

is convex.

4.7 a) Let $\boldsymbol{p} \in \mathbf{R}^n$, $\alpha \in \mathbf{R}$. Verify that the half-space in $\mathbf{R}^n$

$$H = \{ \boldsymbol{x} \in \mathbf{R}^n ; \boldsymbol{p}^T \boldsymbol{x} \leq \alpha \}$$

is a convex set.

b) Verify that the closed unit ball in $\mathbf{R}^n$,

$$K = \{ \boldsymbol{x} \in \mathbf{R}^n ; \|\boldsymbol{x}\| \leq 1 \},$$

is a convex set.

4.8 Prove that if S_1 and S_2 are convex sets then $S_1 \cap S_2$ is convex. What about $S_1 \cup S_2$?

4.9 Let S be a nonempty set in $\mathbf{R}$. Show that S is convex if and only if for each integer $k \geq 2$, the following holds true: $\boldsymbol{x}_1, \boldsymbol{x}_2, \ldots, \boldsymbol{x}_k \in S$ implies that $\sum_{i=1}^k \lambda_i \boldsymbol{x}_i \in S$, where $\sum_{i=1}^k \lambda_i = 1$ and $\lambda_i \geq 0$ for $i = 1, 2, \ldots, k$.

4.10 Describe $S_1 \oplus S_2$ and $S_1 \ominus S_2$ when

$$S_1 = \{ \boldsymbol{x} ; x_1 = 0, \ 0 \leq x_2 \leq 1 \} \quad \text{and} \quad S_2 = \{ \boldsymbol{x} ; 0 \leq x_1 \leq 1, \ x_2 = 2 \}.$$

4.11 Draw a picture in the plane of the set S and its convex hull $H(S)$ in the following cases.

a) $S = \{ (0,0) \} \cup \{ (1,1) \} \cup \{ (0,1) \} \cup \{ (\frac{1}{2}, \frac{2}{3}) \}$,

b) $S = \{ (0,y) ; |y| \leq 1 \} \cup \{ (1,2) \}$,

c) $S = \{ (x,y) ; x^2 + y^2 = 1, \ x \leq 0 \}$,

d) $S = \{ (x,y) ; 0 \leq xy \leq 1, \ 0 \leq x \leq 2, \ 0 \leq y \leq 2 \}$.

(1997–04–04, 2d)

4.12 Find those values of a for which $(1, a, 1)$ belongs to the convex hull of

$$(0,0,0), \qquad (1,1,2), \qquad (2,4,-6), \qquad (1,3,8).$$

(1997–12–17, 1)

4.13 Let S_1 and S_2 be nonempty sets in $\mathbf{R}^n$. Show that

$$H(S_1 \cap S_2) \subseteq H(S_1) \cap H(S_2).$$

Is there in fact equality between these sets? If yes, give a proof. If not, give a counterexample.

4.3 Separation. Support planes

THEOREM 5. *Let S be a non-empty closed and convex set in $\mathbf{R}^n$. Assume that $\boldsymbol{y} \notin S$. Then there exists a unique point $\bar{\boldsymbol{x}} \in S$ such that*

(2) $$\|\boldsymbol{y} - \bar{\boldsymbol{x}}\| = \min_{\boldsymbol{x} \in S} \|\boldsymbol{y} - \boldsymbol{x}\|.$$

Furthermore, $\bar{\boldsymbol{x}}$ solves this minimum problem if and only if

(3) $$(\boldsymbol{y} - \bar{\boldsymbol{x}})^T(\boldsymbol{x} - \bar{\boldsymbol{x}}) \le 0 \quad \textit{for all } \boldsymbol{x} \in S.$$

Remark. The theorem is concerned with the problem of finding a point $\bar{\boldsymbol{x}}$ in S with least distance to $\boldsymbol{y}$. The number $\|\boldsymbol{y} - \bar{\boldsymbol{x}}\|$ is often called the **distance** from $\boldsymbol{y}$ to S. — Condition (3) has the geometrical interpretation that the angle between the vector $\boldsymbol{y} - \bar{\boldsymbol{x}}$ and any $\boldsymbol{x} - \bar{\boldsymbol{x}}$ with $\boldsymbol{x} \in S$ is *obtuse*.

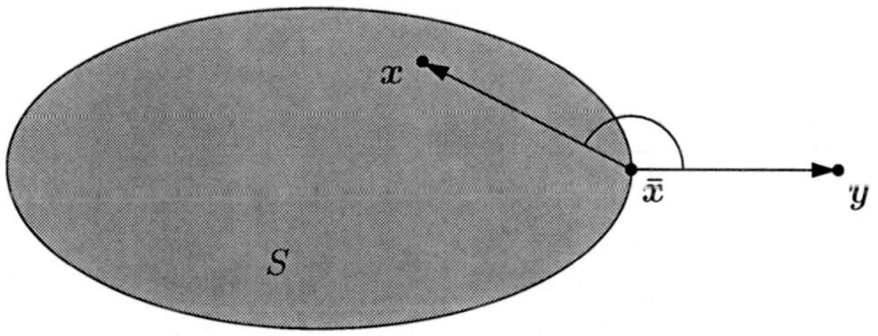

The reader should convince himself, by drawing a suitable figure, that without the assumption of convexity the point $\bar{\boldsymbol{x}}$ might not be unique.

PROOF. First we prove the *existence* of a minimum. If S is bounded this follows directly from theorem 2, page 118, since the function

$$S \ni \boldsymbol{x} \longmapsto \|\boldsymbol{y} - \boldsymbol{x}\| \in \mathbf{R}$$

is continuous. (Exercise: prove that.) If S is non-bounded we cut off with a big ball, that is, we consider $S' = S \cap \{\boldsymbol{x}\,;\, \|\boldsymbol{x}\| \leq R\}$ for R large enough. See the figure below. We then have

$$\min_{\boldsymbol{x}\in S} \|\boldsymbol{y} - \boldsymbol{x}\| = \min_{\boldsymbol{x}\in S'} \|\boldsymbol{y} - \boldsymbol{x}\|,$$

where the last minimum certainly exists since S' is compact.

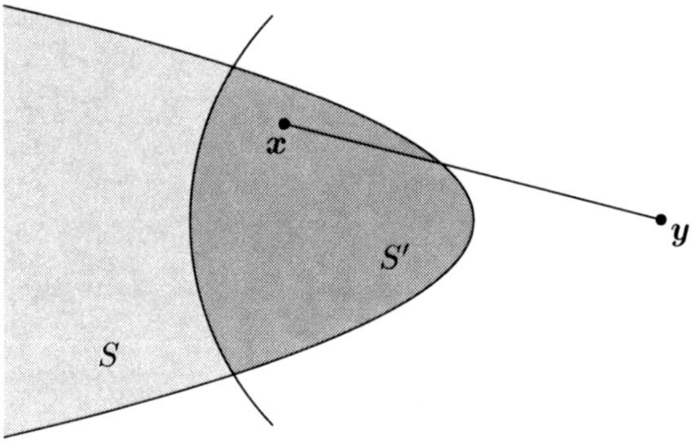

We now prove condition (3) for $\bar{\boldsymbol{x}}$ to have minimum distance to $\boldsymbol{y}$. Assume first that $\bar{\boldsymbol{x}}$ satisfies (3) and let $\boldsymbol{x} \in S$. Then

$$\begin{aligned}\|\boldsymbol{y} - \boldsymbol{x}\|^2 &= \|\boldsymbol{y} - \bar{\boldsymbol{x}} + \bar{\boldsymbol{x}} - \boldsymbol{x}\|^2 = \\ &= \Big((\boldsymbol{y}-\bar{\boldsymbol{x}}) + (\bar{\boldsymbol{x}}-\boldsymbol{x})\Big)^T \Big((\boldsymbol{y}-\bar{\boldsymbol{x}}) + (\bar{\boldsymbol{x}}-\boldsymbol{x})\Big) = \\ &= \|\boldsymbol{y} - \bar{\boldsymbol{x}}\|^2 + \|\bar{\boldsymbol{x}} - \boldsymbol{x}\|^2 + 2(\boldsymbol{y} - \bar{\boldsymbol{x}})^T(\bar{\boldsymbol{x}} - \boldsymbol{x}).\end{aligned}$$

Using (3) and that $\|\bar{\boldsymbol{x}} - \boldsymbol{x}\|^2 \geq 0$ we obtain

$$\|\boldsymbol{y} - \boldsymbol{x}\|^2 \geq \|\boldsymbol{y} - \bar{\boldsymbol{x}}\|^2.$$

Thus $\|\boldsymbol{y} - \boldsymbol{x}\|$ has a minimum for $\boldsymbol{x} = \bar{\boldsymbol{x}}$.

For the proof in the opposite direction we assume that $\bar{\boldsymbol{x}} \in S$ is a closest point to $\boldsymbol{y}$. Let $\boldsymbol{x}$ belong to S. By the convexity of S

$$\lambda \boldsymbol{x} + (1-\lambda)\bar{\boldsymbol{x}} \in S, \qquad 0 < \lambda < 1.$$

Hence by the assumption on $\bar{\boldsymbol{x}}$,

$$\|\boldsymbol{y} - \bar{\boldsymbol{x}}\|^2 \le \|\boldsymbol{y} - (\lambda \boldsymbol{x} + (1-\lambda)\bar{\boldsymbol{x}})\|^2 = \|\boldsymbol{y} - \bar{\boldsymbol{x}} - \lambda(\boldsymbol{x} - \bar{\boldsymbol{x}})\|^2 =$$

$$= \|\boldsymbol{y} - \bar{\boldsymbol{x}}\|^2 + \lambda^2\|\boldsymbol{x} - \bar{\boldsymbol{x}}\|^2 - 2\lambda(\boldsymbol{y} - \bar{\boldsymbol{x}})^T(\boldsymbol{x} - \bar{\boldsymbol{x}}).$$

Cancelling $\|\boldsymbol{y} - \bar{\boldsymbol{x}}\|^2$ and dividing by λ results in

$$0 \le \lambda\|\boldsymbol{x} - \bar{\boldsymbol{x}}\|^2 - 2(\boldsymbol{y} - \bar{\boldsymbol{x}})^T(\boldsymbol{x} - \bar{\boldsymbol{x}}).$$

Letting $\lambda \to 0$ we arrive at (3).

It remains to prove the *uniqueness* of the closest point $\bar{\boldsymbol{x}}$. Assume that both $\bar{\boldsymbol{x}}$ and $\bar{\boldsymbol{x}}'$ have minimum distance to $\boldsymbol{y}$. Then both points fulfil condition (3):

$$(\boldsymbol{y} - \bar{\boldsymbol{x}})^T(\boldsymbol{x} - \bar{\boldsymbol{x}}) \le 0 \quad \text{for all } \boldsymbol{x} \in S,$$

$$(\boldsymbol{y} - \bar{\boldsymbol{x}}')^T(\boldsymbol{x} - \bar{\boldsymbol{x}}') \le 0 \quad \text{for all } \boldsymbol{x} \in S.$$

Choosing $\boldsymbol{x} = \bar{\boldsymbol{x}}'$ in the first condition and $\boldsymbol{x} = \bar{\boldsymbol{x}}$ in the second we obtain after some rearrangement

$$(\boldsymbol{y} - \bar{\boldsymbol{x}})^T(\bar{\boldsymbol{x}}' - \bar{\boldsymbol{x}}) \le 0$$

$$(\bar{\boldsymbol{x}}' - \boldsymbol{y})^T(\bar{\boldsymbol{x}}' - \bar{\boldsymbol{x}}) \le 0.$$

Addition of these inequalities gives

$$(\bar{\boldsymbol{x}}' - \bar{\boldsymbol{x}})^T(\bar{\boldsymbol{x}}' - \bar{\boldsymbol{x}}) \le 0 \quad \Longleftrightarrow \quad \|\bar{\boldsymbol{x}}' - \bar{\boldsymbol{x}}\|^2 \le 0 \quad \Longleftrightarrow \quad \bar{\boldsymbol{x}}' = \bar{\boldsymbol{x}}.$$

This proves the uniqueness. □

DEFINITION 7. *Let A and B be subsets of $\mathbf{R}^n$. The hyperplane $\mathbf{p}^T\boldsymbol{x} = \alpha$ is said to* **separate** *these sets if*

$$\mathbf{p}^T\boldsymbol{x} > \alpha \quad \text{when } \boldsymbol{x} \in \text{cl}(A),$$

$$\mathbf{p}^T\boldsymbol{x} < \alpha \quad \text{when } \boldsymbol{x} \in \text{cl}(B)$$

(or vice versa). (See the figure on the next page.)

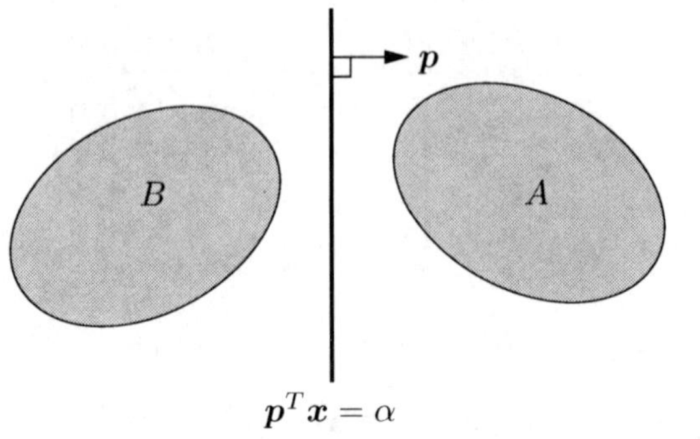

THEOREM 6. *Let S be a non-empty closed, convex set in $\mathbf{R}^n$. Assume that $\boldsymbol{y} \notin S$. Then $\boldsymbol{y}$ and S (more precisely: the sets $\{\boldsymbol{y}\}$ and S) have a separating hyperplane.*

PROOF. By theorem 5 there is a point $\bar{\boldsymbol{x}}$ in S with least distance to $\boldsymbol{y}$. Put

$$\boldsymbol{p} = \boldsymbol{y} - \bar{\boldsymbol{x}}.$$

Note that $\boldsymbol{p} \neq \boldsymbol{0}$. Consider the hyperplane with equation

$$\boldsymbol{p}^T\left(\boldsymbol{x} - \frac{1}{2}(\boldsymbol{y} + \bar{\boldsymbol{x}})\right) = 0 \quad \Longleftrightarrow \quad \boldsymbol{p}^T\boldsymbol{x} = \frac{1}{2}\boldsymbol{p}^T(\boldsymbol{y} + \bar{\boldsymbol{x}}).$$

This plane separates $\boldsymbol{y}$ and S since

$$1^\circ. \quad \boldsymbol{p}^T\boldsymbol{y} = \frac{1}{2}\boldsymbol{p}^T(\boldsymbol{y} - \bar{\boldsymbol{x}}) + \frac{1}{2}\boldsymbol{p}^T(\boldsymbol{y} + \bar{\boldsymbol{x}}) = \frac{1}{2}\|\boldsymbol{p}\|^2 + \frac{1}{2}\boldsymbol{p}^T(\boldsymbol{y} + \bar{\boldsymbol{x}}) >$$
$$> \frac{1}{2}\boldsymbol{p}^T(\boldsymbol{y} + \bar{\boldsymbol{x}}),$$

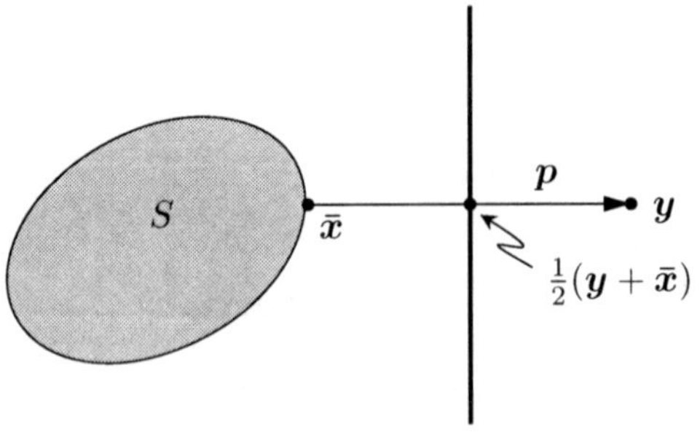

2°. if $\boldsymbol{x} \in S$ then using (3) we have

$$\boldsymbol{p}^T\boldsymbol{x} = \boldsymbol{p}^T(\boldsymbol{x} - \bar{\boldsymbol{x}}) + \boldsymbol{p}^T\bar{\boldsymbol{x}} \le 0 + \boldsymbol{p}^T\bar{\boldsymbol{x}} = \frac{1}{2}\boldsymbol{p}^T(\bar{\boldsymbol{x}} - \boldsymbol{y}) + \frac{1}{2}\boldsymbol{p}^T(\bar{\boldsymbol{x}} + \boldsymbol{y}) =$$

$$= -\frac{1}{2}\|\boldsymbol{p}\|^2 + \frac{1}{2}\boldsymbol{p}^T(\bar{\boldsymbol{x}} + \boldsymbol{y}) < \frac{1}{2}\boldsymbol{p}^T(\bar{\boldsymbol{x}} + \boldsymbol{y}).$$

□

COROLLARY 1. *Assume that S is a closed and convex set. Then*

$$S = \bigcap_{\substack{H \text{ half-space} \\ H \supseteq S}} H.$$

PROOF. Denote the right hand side set by R. That $S \subseteq R$ is obvious. To prove the reverse inequality assume that $\boldsymbol{y} \notin S$. Then by the theorem there is a hyperplane separating $\boldsymbol{y}$ and S. This hyperplane is the boundary of a half-space H containing S but not $\boldsymbol{y}$. It follows that $\boldsymbol{y} \notin R$. □

The following theorem is included for the sake of completeness, but will not be needed in the future.

THEOREM 7. *Let S_1 and S_2 be non-empty closed, convex sets, and assume that $S_1 \cap S_2 = \emptyset$. Then S_1 and S_2 have a separating hyperplane.*

PROOF. Let

$$S = S_1 \ominus S_2 = \{\, \boldsymbol{x}_1 - \boldsymbol{x}_2 \,;\, \boldsymbol{x}_1 \in S_1, \boldsymbol{x}_2 \in S_2 \,\}.$$

This set is closed and is convex by lemma 2, page 121. Since S_1 and S_2 are disjoint S does not contain $\boldsymbol{0}$. By theorem 6 there is a hyperplane separating $\boldsymbol{0}$ and S, i.e., there is a nonzero vector $\boldsymbol{p}$ and a number α such that

$$\boldsymbol{p}^T\boldsymbol{x} > \alpha > \boldsymbol{p}^T\boldsymbol{0} = \boldsymbol{0} \qquad \text{for all } \boldsymbol{x} \in S.$$

This implies that

$$\boldsymbol{p}^T\boldsymbol{x}_1 > \alpha + \boldsymbol{p}^T\boldsymbol{x}_2 \qquad \text{for all } \boldsymbol{x}_1 \in S_1,\ \boldsymbol{x}_2 \in S_2.$$

Put $\beta = \inf_{\boldsymbol{x} \in S_1} \boldsymbol{p}^T\boldsymbol{x}$ (exercise: this is a finite number) and let $\gamma = \beta - \frac{1}{2}\alpha$. The hyperplane $\boldsymbol{p}^T\boldsymbol{x} = \gamma$ separates S_1 and S_2; indeed we have

$$\boldsymbol{p}^T\boldsymbol{x}_1 > \gamma \quad \text{for all } \boldsymbol{x}_1 \in S_1,$$
$$\boldsymbol{p}^T\boldsymbol{x}_2 < \gamma \quad \text{for all } \boldsymbol{x}_2 \in S_2.$$

□

Definition 8. *Let A be a subset of $\mathbf{R}^n$. The hyperplane $\boldsymbol{p}^T\boldsymbol{x} = \alpha$ is called a* **support plane** *of A if*

$$\boldsymbol{p}^T\boldsymbol{x} \le \alpha \qquad \text{for all } \boldsymbol{x} \in A$$

(or $\boldsymbol{p}^T\boldsymbol{x} \ge \alpha$ for all $\boldsymbol{x} \in A$), with equality occuring for some $\bar{\boldsymbol{x}} \in \mathrm{cl}(A)$.

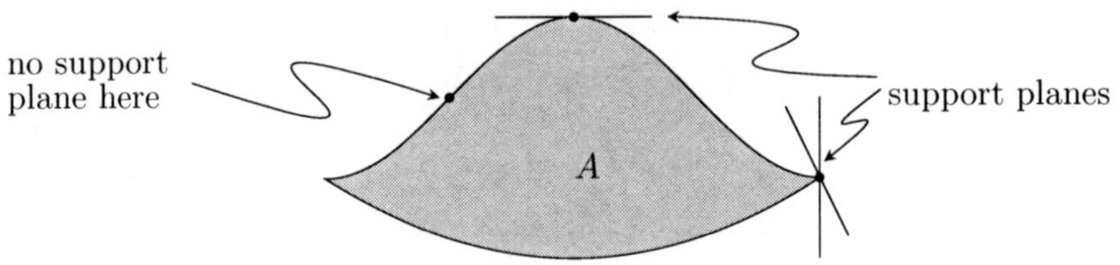

Theorem 8.*Let S be a non-empty convex set in $\mathbf{R}^n$, and assume that $\bar{\boldsymbol{x}} \in \partial S$. Then there is a support plane of S at $\bar{\boldsymbol{x}}$.*

Proof. As a consequence of the definition of a boundary point, there is a sequence $\{\boldsymbol{y}_k\}_1^\infty$ with $\boldsymbol{y}_k \notin \mathrm{cl}\,(S)$ and with limit point $\bar{\boldsymbol{x}}$. By theorem 6, for every $\boldsymbol{y}_k$ there is a hyperplane separating it from $\mathrm{cl}\,(S)$. In particular, there is a sequence $\{\boldsymbol{p}_k\}_1^\infty$ such that for all k

$$\boldsymbol{p}_k^T\boldsymbol{y}_k > \boldsymbol{p}_k^T\boldsymbol{x} \qquad \text{for all} \quad \boldsymbol{x} \in S. \tag{4}$$

It is no restriction to assume that the $\boldsymbol{p}_k$ are normalized, $\|\boldsymbol{p}_k\| = 1$. Then the sequence $\{\boldsymbol{p}_k\}_1^\infty$ belongs to the compact set $\|\boldsymbol{x}\| = 1$ in $\mathbf{R}^n$, and thus has by Bolzano-Weierstrass' theorem (theorem 1, page 117) a convergent subsequence $\{\boldsymbol{p}_k\}_{k\in\mathcal{K}}$. Denote by $\boldsymbol{p}$ the limit of this subsequence. Taking the limit in (4) for $k \in \mathcal{K}$ we get

$$\boldsymbol{p}^T\bar{\boldsymbol{x}} \ge \boldsymbol{p}^T\boldsymbol{x} \qquad \text{for all} \quad \boldsymbol{x} \in S.$$

Let $\alpha = \boldsymbol{p}^T\bar{\boldsymbol{x}}$. Then the plane $\boldsymbol{p}^T\boldsymbol{x} = \alpha$ is a support plane of S at $\bar{\boldsymbol{x}}$. □

4.4 Systems of inequalities. Farkas' theorem

Let $\boldsymbol{A} = (a_{ij})$ be an $m \times n$-matrix and $\boldsymbol{c}$ an n-vector (column matrix). We are going to consider two systems of inequalities,

$$(*) \qquad \boldsymbol{Ax} \leq \boldsymbol{0}, \quad \boldsymbol{c}^T\boldsymbol{x} > 0, \qquad \text{i.e.} \quad \begin{cases} a_{11}x_1 + \ldots + a_{1n}x_n \leq 0 \\ \ldots \\ a_{m1}x_1 + \ldots + a_{mn}x_n \leq 0 \\ c_1x_1 + \ldots + c_nx_n > 0 \end{cases}$$

and

$$(**) \qquad \boldsymbol{A}^T\boldsymbol{y} = \boldsymbol{c}, \quad \boldsymbol{y} \geq \boldsymbol{0}, \qquad \text{i.e.} \quad \begin{cases} a_{11}y_1 + \ldots + a_{m1}y_m = c_1 \\ \ldots \\ a_{1n}y_1 + \ldots + a_{mn}y_m = c_n \\ y_1 \geq 0, \ \ldots, \ y_m \geq 0. \end{cases}$$

Both systems can not have a solution, since in that case

$$\boldsymbol{y}^T\boldsymbol{Ax} = \sum_{i=1}^{m} y_i(\boldsymbol{Ax})_i \leq 0$$

and

$$\boldsymbol{y}^T\boldsymbol{Ax} = (\boldsymbol{A}^T\boldsymbol{y})^T\boldsymbol{x} - \boldsymbol{c}^T\boldsymbol{x} > 0,$$

resulting in a contradiction.

The geometrical interpretation of $(*)$ is the following. The inequalities $\boldsymbol{Ax} \leq \boldsymbol{0}$ define an intersection C of closed half-spaces, all with $\boldsymbol{0}$ as a boundary point. It is a closed, convex set[2]. The plane

$$\boldsymbol{c}^T\boldsymbol{x} = 0$$

contains the origin; if $(*)$ has no solution it is a support plane of C.

[2] In fact, it is a *cone*, as we shall see later.

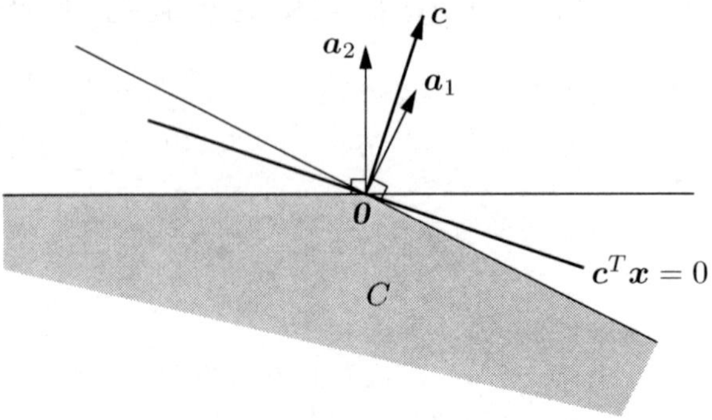

Denote the normals of the bounding planes of these half-spaces by $a_1, \ldots, a_m$; they are the rows of $\boldsymbol{A}$ and columns of $\boldsymbol{A}^T$. The figure indicates the situation when $(*)$ has no solution in the case $n=2$.

System $(**)$ is equivalent to

$$\boldsymbol{c} = y_1 \boldsymbol{a}_1 + \ldots + y_m \boldsymbol{a}_m \quad \text{with all} \quad y_j \geq 0.$$

We call $\boldsymbol{c}$ a *positive linear combination* of $\boldsymbol{a}_1, \ldots, \boldsymbol{a}_m$.

THEOREM 9. (FARKAS) *Exactly one of the systems* $(*)$ *and* $(**)$ *has a solution.*

PROOF. Above we showed that both systems can not have a solution. This can be phrased as: if $(**)$ has a solution then $(*)$ has not. It remains to show that if $(**)$ has no solution then $(*)$ has one.

Thus assume that $(**)$ has no solution. Let

$$S = \{ \boldsymbol{x} \in \mathbf{R}^n \, ; \, \boldsymbol{x} = \boldsymbol{A}^T \boldsymbol{y}, \quad \boldsymbol{y} \geq \boldsymbol{0} \}.$$

This is the set of positive linear combinations of the rows $\boldsymbol{a}_1, \ldots, \boldsymbol{a}_m$ of $\boldsymbol{A}$. It is easily seen to be a closed convex set. By assumption $\boldsymbol{c}$ does not belong to S. Consequently, by theorem 6 there is a hyperplane $\boldsymbol{p}^T \boldsymbol{x} = \alpha$ separating $\boldsymbol{c}$ and S. Assume that

$$\begin{aligned} &\boldsymbol{p}^T \boldsymbol{c} > \alpha, \\ &\boldsymbol{p}^T \boldsymbol{x} < \alpha \quad \text{when } \boldsymbol{x} \in S. \end{aligned}$$

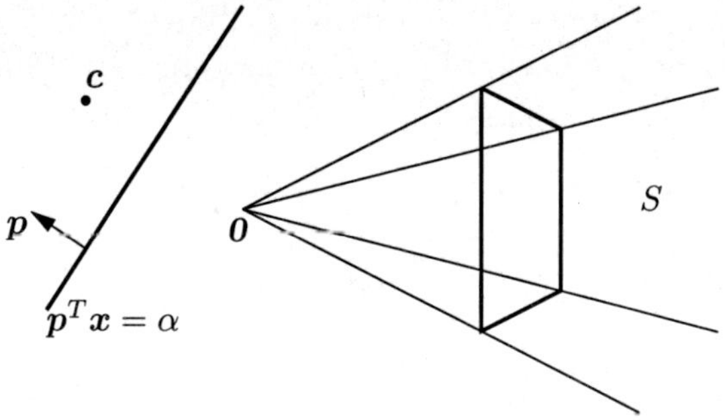

In particular, since $\mathbf{0} \in S$ the second condition implies that $\alpha > 0$. Furthermore, by the second condition and the definition of S

$$\alpha > \mathbf{p}^T(\mathbf{A}^T\mathbf{y}) = \mathbf{y}^T\mathbf{A}\mathbf{p} = \sum_{i=1}^{m} y_i(\mathbf{A}\mathbf{p})_i \qquad \text{for } all \ \ \mathbf{y} \geq \mathbf{0}.$$

If some $(\mathbf{Ap})_i$ were positive we would get a contradiction by choosing the corresponding y_i large enough and the others equal to zero. Hence $(\mathbf{Ap})_i \leq 0$ for all i.

We have now shown that $\mathbf{p}$ satisfies

$$\mathbf{Ap} \leq \mathbf{0}, \qquad \mathbf{c}^T\mathbf{p} > \alpha > 0.$$

i.e., that $\mathbf{p}$ solves $(*)$. This finishes the proof. □

Example 16. For what numbers a is the following implication valid?

$$\begin{cases} x_1 + 3x_2 - \ \ x_3 \leq 0 \\ 2x_1 + 7x_2 + \ \ x_3 \leq 0 \\ -x_1 \qquad\quad + 9x_3 \leq 0 \end{cases} \quad \Longrightarrow \quad x_1 + 3x_2 + ax_3 \leq 0.$$

Solution: With obvious notation we write the above implication as

$$\mathbf{Ax} \leq \mathbf{0} \quad \Longrightarrow \quad \mathbf{c}^T\mathbf{x} \leq 0.$$

This implication is equivalent to the statement that the system

$$\mathbf{Ax} \leq \mathbf{0}, \qquad \mathbf{c}^T\mathbf{x} > 0$$

has no solution. Here we have system (∗) above. Thus by Farkas' theorem we can transform our problem to a question concerning system (∗∗): for what values of a is the system

$$\boldsymbol{A}^T \boldsymbol{y} = \boldsymbol{c}, \qquad \boldsymbol{y} \geq \boldsymbol{0}$$

solvable? We now solve the system of equations $\boldsymbol{A}^T \boldsymbol{y} = \boldsymbol{c}$ by Gauss elimination:

$$\begin{cases} y_1 + 2y_2 - y_3 = 1 \\ 3y_1 + 7y_2 = 3 \\ -y_1 + y_2 + 9y_3 = a \end{cases} \iff \quad \ldots \quad \iff \begin{cases} y_1 = -6 - 7a \\ y_2 = 3 + 3a \\ y_3 = -1 - a. \end{cases}$$

Applying the conditions $y_i \geq 0$ results in $a \leq -\frac{6}{7}$, $a \geq -1$ and $a \leq -1$ respectively. The only possibility to fulfil all three conditions is letting $a = -1$, which thus is the answer. □

Example 17. In the applications Farkas' theorem may be used in more complicated settings than above. System (∗) may involve equalities as well as inequalities, there may be strict inequalities somewhere, the inequalities may be reversed, ... The proofs of these more involved cases are usually done by some transformation to the situation in theorem 9. We give one example, which we will have occasion to use later on.

We consider a mixed system of equalites and inequalities of the form

$$\boldsymbol{A}\boldsymbol{x} \leq \boldsymbol{0}, \quad \boldsymbol{B}\boldsymbol{x} = \boldsymbol{0}, \quad \boldsymbol{x}' \geq \boldsymbol{0}, \quad \boldsymbol{c}^T\boldsymbol{x} > 0. \tag{5}$$

The matrices $\boldsymbol{A}$ and $\boldsymbol{B}$ are of type $m \times n$ and $k \times n$ respectively. The $\boldsymbol{x}$-variables are partitioned into two kinds, $\boldsymbol{x} = (\boldsymbol{x}', \boldsymbol{x}'')$, where $\boldsymbol{x}'$ consists of the first r variables, which are the ones having a sign restriction in (5).

The equalities $\boldsymbol{B}\boldsymbol{x} = \boldsymbol{0}$ in (5) are equivalent to the inequalities $\boldsymbol{B}\boldsymbol{x} \leq \boldsymbol{0}$ and $-\boldsymbol{B}\boldsymbol{x} \leq \boldsymbol{0}$. Thus (5) is equivalent to a system of inequalities

$$\mathcal{A}\boldsymbol{x} \leq \boldsymbol{0}, \qquad \boldsymbol{c}^T\boldsymbol{x} > 0, \tag{6}$$

where $\mathcal{A}$ is a block matrix

$$\mathcal{A} = \begin{pmatrix} \boldsymbol{A} \\ \boldsymbol{B} \\ -\boldsymbol{B} \\ -\widetilde{\boldsymbol{I}} \end{pmatrix}.$$

Here $\widetilde{\boldsymbol{I}}$ denotes the $r \times n$ block matrix $(\boldsymbol{I} \quad \boldsymbol{O})$, in which $\boldsymbol{I}$ is an $r \times r$ identity matrix and $\boldsymbol{O}$ an $r \times (n-r)$ zero matrix.

System (6) is of the form (∗) above. The corresponding system (∗∗) is

$$\mathcal{A}^T \begin{pmatrix} \boldsymbol{y} \\ \boldsymbol{u} \\ \boldsymbol{v} \\ \boldsymbol{w} \end{pmatrix} = \boldsymbol{c}, \qquad \begin{pmatrix} \boldsymbol{y} \\ \boldsymbol{u} \\ \boldsymbol{v} \\ \boldsymbol{w} \end{pmatrix} \geq \boldsymbol{0}, \tag{7}$$

where $\boldsymbol{y}$, $\boldsymbol{u}$, $\boldsymbol{v}$, $\boldsymbol{w}$ are column matrices of size m, k, k and r respectively, and

$$\mathcal{A}^T = \begin{pmatrix} \boldsymbol{A}^T & \boldsymbol{B}^T & -\boldsymbol{B}^T & -\widetilde{\boldsymbol{I}}^T \end{pmatrix}.$$

Of course (7) can be written

$$\boldsymbol{A}^T\boldsymbol{y} + \boldsymbol{B}^T\boldsymbol{u} - \boldsymbol{B}^T\boldsymbol{v} - \widetilde{\boldsymbol{I}}^T\boldsymbol{w} = \boldsymbol{c}, \qquad \boldsymbol{y}, \boldsymbol{u}, \boldsymbol{v}, \boldsymbol{w} \geq \boldsymbol{0}.$$

Putting $\boldsymbol{z} = \boldsymbol{u} - \boldsymbol{v}$, there is no sign condition on $\boldsymbol{z}$ and (7) is equivalent to

$$\boldsymbol{A}^T\boldsymbol{y} + \boldsymbol{B}^T\boldsymbol{z} - \widetilde{\boldsymbol{I}}^T\boldsymbol{w} = \boldsymbol{c}, \qquad \boldsymbol{y}, \boldsymbol{w} \geq \boldsymbol{0}.$$

This can be further simplified if we introduce some more notation. We partition the rows of $\boldsymbol{A}^T$ and $\boldsymbol{B}^T$ (i.e., the columns of $\boldsymbol{A}$ and $\boldsymbol{B}$) and divide $\boldsymbol{c}$ as well in correspondence with the partition of the $\boldsymbol{x}$-variables, using the notation

$$\boldsymbol{A}^T = \begin{pmatrix} \boldsymbol{A}'^T \\ \boldsymbol{A}''^T \end{pmatrix}, \quad \boldsymbol{B}^T = \begin{pmatrix} \boldsymbol{B}'^T \\ \boldsymbol{B}''^T \end{pmatrix}, \quad \widetilde{\boldsymbol{I}}^T = \begin{pmatrix} \boldsymbol{I} \\ \boldsymbol{O} \end{pmatrix}, \quad \boldsymbol{c} = \begin{pmatrix} \boldsymbol{c}' \\ \boldsymbol{c}'' \end{pmatrix}$$

(where $\boldsymbol{I}$ is still an $r \times r$ identity matrix). With this notation (7) is equivalent to

$$\begin{cases} \boldsymbol{A}'^T\boldsymbol{y} + \boldsymbol{B}'^T\boldsymbol{z} \geq \boldsymbol{c}' \\ \boldsymbol{A}''^T\boldsymbol{y} + \boldsymbol{B}''^T\boldsymbol{z} = \boldsymbol{c}'', \end{cases} \qquad \boldsymbol{y} \geq \boldsymbol{0}. \tag{8}$$

By Farkas' theorem exactly one of the systems (5) and (8) have a solution.

□

The next example is known in economic literature as *the arbitrage theorem.*

Example 18. Let a_{ij}, $1 \le i \le n$, $1 \le j \le m$, be real numbers. Prove that exactly one of the following statements is true.

- There exist numbers $p_1, \ldots, p_m$ such that

$$\sum_{j=1}^{m} a_{ij} p_j = 0, \quad i = 1, \ldots, n,$$
$$\sum_{j=1}^{m} p_j = 1,$$
$$p_j \ge 0 \quad \text{for all } j.$$

- There exist numbers $x_1, \ldots, x_n$ such that

$$\sum_{i=1}^{n} a_{ij} x_i > 0, \quad j = 1, \ldots, m.$$

Solution: Let $\boldsymbol{A} = (a_{ij})$, an $n \times m$ matrix, and let $\boldsymbol{e}$ denote the column vector of length m whose all entries are equal to 1. Then the equations $\sum_{j=1}^{m} a_{ij} p_j = 0$ can be written $\boldsymbol{Ap} = \boldsymbol{0}$ and $\sum_{j=1}^{m} p_j = 1$ is equivalent to $\boldsymbol{e}^T \boldsymbol{p} = 1$. Now introduce the $(n+1) \times m$ and $(n+1) \times 1$ matrices

$$\mathcal{A} = \begin{pmatrix} \boldsymbol{A} \\ \boldsymbol{e}^T \end{pmatrix}, \qquad \boldsymbol{c} = \begin{pmatrix} 0 \\ \vdots \\ 0 \\ 1 \end{pmatrix}.$$

The conditions in the first statement in the theorem can be formulated as

$$\mathcal{A}\boldsymbol{p} = \boldsymbol{c}, \qquad \boldsymbol{p} \ge \boldsymbol{0}.$$

Thus we have in other notation exactly statement (**) in Farkas' theorem.

The corresponding statement (*) is

$$\mathcal{A}^T \begin{pmatrix} \boldsymbol{x} \\ z \end{pmatrix} \le \boldsymbol{0}, \qquad \boldsymbol{c}^T \begin{pmatrix} \boldsymbol{x} \\ z \end{pmatrix} > \boldsymbol{0}.$$

Here $\boldsymbol{x}$ is an n-vector and z a real number. Considering the form of $\boldsymbol{c}$ this can be restated as

$$\boldsymbol{A}^T \boldsymbol{x} + z\boldsymbol{e} \le 0, \qquad z > 0. \tag{9}$$

The system (9) is obviously solvable if and only if the system of strict inequalities

$$\mathbf{A}^T x < \mathbf{0}$$

is solvable. Rewriting the system as $\mathbf{A}^T(-x) > \mathbf{0}$ we see that this is equivalent to the second statement above.

Hence the result follows from Farkas' theorem. □

We close this section with an application of Farkas' theorem to probability theory. We prove a theorem of Markov stating that any stochastic matrix has eigenvalue 1. For this it is enough to prove that its transpose has eigenvalue 1.

Example 19. Let $\mathbf{P} = (p_{ij})$ be an $n \times n$ matrix with the properties that $p_{ij} \geq 0$ for all i, j and

$$\sum_{j=1}^{n} p_{ij} = 1, \qquad i = 1, \ldots, n.$$

We want to prove that there is a column vector y satisfying

$$\mathbf{P}^T y = y, \qquad y \geq \mathbf{0},$$

and with $\sum_1^n y_i = 1$.[3] Introducing the column vector $e = (1, \ldots, 1)^T$, our conditions can be stated

$$(10) \qquad (\mathbf{P}^T - \mathbf{I})y = \mathbf{0}, \qquad e^T y = 1, \qquad y \geq \mathbf{0}.$$

This is a system of the form (**) in Farkas' theorem, with

$$\mathbf{A}^T = \begin{pmatrix} \mathbf{P}^T - \mathbf{I} \\ e^T \end{pmatrix}, \qquad c = \begin{pmatrix} \mathbf{0} \\ 1 \end{pmatrix}$$

(matrices of type $(n+1) \times n$ and $(n+1) \times 1$ respectively).

The corresponding system (*) is

$$(11) \qquad \begin{pmatrix} \mathbf{P} - \mathbf{I} & e \end{pmatrix} x \leq \mathbf{0}, \qquad x_{n+1} > 0,$$

[3]This normalization condition ensures that the y_i can be interpreted as probabilities.

where $\boldsymbol{x}$ is an $(n+1)$-vector. Our goal is to prove that system (10) has a solution. By Farkas' theorem this is equivalent to proving that (11) has no solution.

Assume then that (11) has a solution $\boldsymbol{x}$, and let $x_k = \min_{1\le j\le n} x_j$. From the kth row in (11) we conclude that[4]

$$0 \ge \sum_{j=1}^{n}(p_{kj} - \delta_{kj})x_j + x_{n+1} = \sum_{j=1}^{n} p_{kj}x_j - x_k + x_{n+1} \ge$$
$$\ge x_k \sum_{j=1}^{n} p_{kj} - x_k + x_{n+1} = x_k \cdot 1 - x_k + x_{n+1} = x_{n+1}.$$

This is a contradiction to the last condition in (11). Thus (11) has no solution, and the proof is finished. □

Exercises

S 4.14 For what values of the real constant a is it true that

$$\begin{cases} x_1 + x_2 + 2x_3 \le 0 \\ 3x_1 + 4x_2 - x_3 \le 0 \\ -3x_1 - 4x_2 + ax_3 \le 0 \end{cases} \implies -x_1 - 2x_2 + 4x_3 \le 0\,?$$

4.15 Is it true that

a)

$$\begin{cases} 3x_1 + 2x_2 + 5x_3 \le 0 \\ -2x_1 + x_2 - x_3 \le 0 \\ 2x_1 - 3x_2 + 2x_3 \le 0 \end{cases} \implies 7x_1 - 8x_2 + 11x_3 \le 0\,?$$

b)

$$\begin{cases} 3x_1 + 2x_2 - x_3 \le 0 \\ 5x_1 - 3x_2 + x_3 \le 0 \\ 4x_1 - x_2 + 5x_3 \le 0 \end{cases} \implies 12x_1 + 7x_2 + 6x_3 \le 0\,?$$

[4]By δ_{jk} we denote the *Kronecker delta*, defined as 1 if $j = k$ and 0 otherwise.

c)

$$\begin{cases} x_1 + x_2 - x_3 \le 0 \\ 2x_1 + x_2 - 6x_3 \le 0 \end{cases} \implies 3x_1 - 8x_2 + 3x_3 \le 0\,?$$

4.16 For what values of a is it true that

$$\begin{cases} x_1 + 2x_2 + 3x_3 \le 0 \\ -x_1 + 2x_2 - x_3 \le 0 \\ 2x_1 + x_2 - 3x_3 \le 0 \end{cases} \implies ax_1 + 6x_2 + x_3 \le 0\,?$$

4.17 For what values of a does the following implication hold?

$$\begin{cases} x_1 + x_2 + 2x_3 \le 0 \\ x_1 - 3x_2 \qquad x_3 \le 0 \end{cases} \implies ax_1 + 5x_3 \le 0.$$

(1999–04–09, 4c)

S 4.18 Prove, using Farkas' theorem, that exactly one of the following two systems has a solution:

- $\boldsymbol{A}\boldsymbol{x} \le \boldsymbol{0}, \quad \boldsymbol{x} \le \boldsymbol{0}, \quad \boldsymbol{c}^T\boldsymbol{x} > 0,$
- $\boldsymbol{A}^T\boldsymbol{y} \le \boldsymbol{c}, \quad \boldsymbol{y} \ge \boldsymbol{0}.$

4.19 Let $\boldsymbol{A}$ be an $m \times n$-matrix and $\boldsymbol{c}$ be an m-vector. Show that exactly one of the following two systems has a solution:

- $\boldsymbol{A}\boldsymbol{x} = \boldsymbol{c},$
- $\boldsymbol{A}^T\boldsymbol{y} = \boldsymbol{0}, \quad \boldsymbol{c}^T\boldsymbol{y} = 1.$

4.5 Cones

DEFINITION 9. *A subset S of $\mathbf{R}^n$ is called a* **cone** *if*

$$\boldsymbol{x} \in S, \quad \lambda \ge 0 \quad \implies \quad \lambda\boldsymbol{x} \in S.$$

In geometrical language, a set S is a cone if $\boldsymbol{x} \in S$ implies that S contains the whole half-ray from the origin through $\boldsymbol{x}$. Taking $\lambda = 0$ in the definition we see that the origin belongs to any cone.

Example 20. A half-space with $\boldsymbol{0}$ as a boundary point is a cone. In the plane the set defined by $x_2 \geq |x_1|$ is a cone, as is the set $x_2 \geq -|x_1|$. See the figures below.

The sets defined by strict inequalities $x_2 > |x_1|$ and $x_2 > -|x_1|$ are not cones. However, if we adjoin $\boldsymbol{0}$ to these sets we obtain cones! These statements should be checked by the reader.

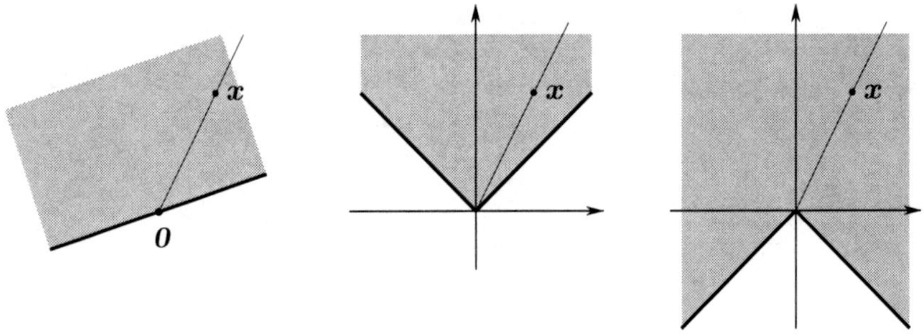

□

Example 21. Let $\boldsymbol{A}$ be an $m \times n$-matrix. The set C defined by the vector inequality $\boldsymbol{Ax} \leq \boldsymbol{0}$ is a (convex) cone. This is easily seen by the definition above: if $\boldsymbol{Ax} \leq \boldsymbol{0}$ and λ is a non-negative number then

$$\boldsymbol{A}(\lambda \boldsymbol{x}) = \lambda \boldsymbol{Ax} \leq \boldsymbol{0}.$$

In fact this set is an intersection of closed half-planes, as we already know. □

Example 22. Let $\boldsymbol{a}_1, \ldots, \boldsymbol{a}_m$ be vectors in $\mathbf{R}^n$, and let C^* denote the set of all positive linear combinations of these vectors, i.e., the set of all vectors $\sum_1^m \lambda_i \boldsymbol{a}_i$ with all λ_i non-negative. This set is a cone, since obviously multiplying an element with a non-negative number λ results in another positive linear combination of $\boldsymbol{a}_1, \ldots, \boldsymbol{a}_m$. □

DEFINITION 10. *Let C be a cone in $\mathbf{R}^n$. The* ***dual cone*** *of C is the set*[5]

$$C^* = \{\boldsymbol{p} \in \mathbf{R}^n \,;\, \boldsymbol{p}^T \boldsymbol{x} \leq 0 \;\; \textit{for all} \;\; \boldsymbol{x} \in C\}.$$

[5]For any set S a set S^* can be defined in the corresponding way. Even in this more general case S^* becomes a closed, convex cone; it is called the *polar cone* of S.

LEMMA 6. *The set C^* is a closed, convex cone.*

LEMMA 7. *If $C_1 \subseteq C_2$ then the opposite inequality is valid for the duals: $C_1^* \supseteq C_2^*$.*

The proofs of these lemmas are left as exercises.

Example 23. The dual of the cone

$$x_2 \geq 2|x_1|$$

in $\mathbf{R}^2$ is the cone

$$x_2 \leq -\frac{1}{2}|x_1|.$$

□

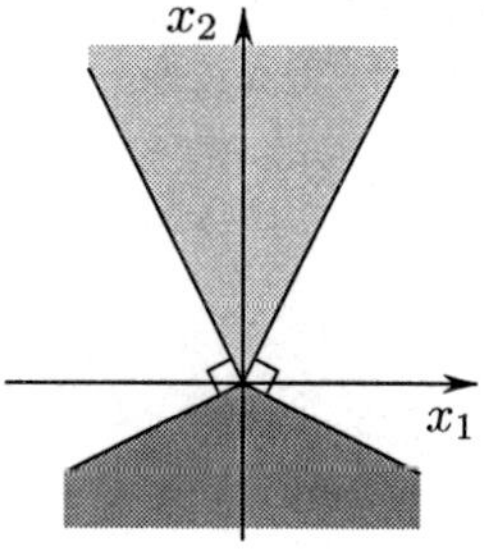

Example 24. Consider again the cone C of example 21. Using the definition of dual cone and Farkas' theorem we have

$$\boldsymbol{c} \in C^* \iff (*) \text{ admits no solution} \iff (**) \text{ has a solution.}$$

Consequently, the dual of C is exactly the cone C^* in example 22, if $\boldsymbol{a}_1, \ldots, \boldsymbol{a}_m$ are the rows of $\boldsymbol{A}$.

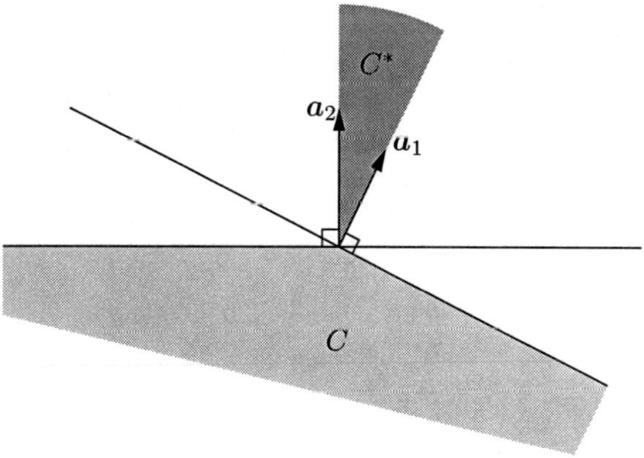

□

The dual of the dual of C is of course denoted by C^{**}.

THEOREM 10. *If C is a non-empty closed, convex cone then $C = C^{**}$.*

PROOF. That $C \subseteq C^{**}$ is a direct consequence of the definition of dual cone. We prove the reverse inequality by proving that if $\boldsymbol{y} \notin C$ then $\boldsymbol{y} \notin C^{**}$.

If $\boldsymbol{y} \notin C$ then, since C is closed and convex, $\boldsymbol{y}$ and C have a separating hyperplane by theorem 6, page 132. Thus there exist a vector $\boldsymbol{p}$ and a scalar α such that

$$\boldsymbol{p}^T\boldsymbol{y} > \alpha$$
$$\boldsymbol{p}^T\boldsymbol{x} < \alpha \quad \text{when } \boldsymbol{x} \in C.$$

Since $\boldsymbol{0} \in C$ we have $\alpha > 0$, implying that

$$\boldsymbol{p}^T\boldsymbol{y} > 0. \tag{12}$$

Furthermore $\boldsymbol{p} \in C^*$, since otherwise there would be an $\bar{\boldsymbol{x}} \in C$ such that $\boldsymbol{p}^T\bar{\boldsymbol{x}} > 0$. However, since C is a cone $\lambda\bar{\boldsymbol{x}} \in C$ for all $\lambda \geq 0$. Thus

$$\boldsymbol{p}^T(\lambda\bar{\boldsymbol{x}}) = \lambda\boldsymbol{p}^T\bar{\boldsymbol{x}} < \alpha,$$

and we would get a contradiction when $\lambda \to \infty$.

Knowing that $\boldsymbol{p} \in C^*$ we conclude from (12) that $\boldsymbol{y} \notin C^{**}$. The proof is finished. □

The reader might try to give an independent proof of Farkas' theorem based on theorem 10.

Exercises

S 4.20 Let C be a nonempty set in $\mathbf{R}^n$. Show that C is a convex cone if and only if $\boldsymbol{x}_1, \boldsymbol{x}_2 \in C$ implies that $\lambda_1\boldsymbol{x}_1 + \lambda_2\boldsymbol{x}_2 \in C$ for all $\lambda_1, \lambda_2 \geq 0$.

4.21 Derive an explicit form of the dual cone C^* of the following cones:

a) $C = \{\,(x_1, x_2)\,;\, 0 \leq x_2 \leq x_1\,\}$,
b) $C = \{\,(x_1, x_2)\,;\, x_2 \geq -|x_1|\,\}$,
c) $C = \{\,\boldsymbol{x} \in \mathbf{R}^n\,;\, \boldsymbol{x} = \boldsymbol{A}\boldsymbol{p},\ \boldsymbol{p} \geq \boldsymbol{0}\,\}$.

4.22 Prove that the dual C^* of a cone C is a closed, convex cone.

Miscellanous exercises

4.23 Find those values of a for which $(-a, 3a, 5a)$ belongs to the convex hull of

$$(-4,0,8), \qquad (2,6,2), \qquad (-1,27,29).$$

(1999–10–21, 1)

4.24 For what values of a is the following implication valid?

$$\begin{cases} x_1 + x_2 + x_3 \geq 0 \\ x_1 + 3x_2 + 10x_3 \geq 0 \\ -x_1 + x_2 + 5x_3 \geq 0 \end{cases} \implies 3x_1 + 5x_2 + ax_3 \geq 0.$$

(1998–04–16, 3)

4.25 Let $\boldsymbol{A}$ be a $p \times n$-matrix and $\boldsymbol{B}$ be a $q \times n$-matrix. Show that exactly one of the following systems has a solution:

- $\boldsymbol{Ax} < \boldsymbol{0}, \quad \boldsymbol{Bx} = \boldsymbol{0}$ where $\boldsymbol{x} \in \mathbf{R}^n$,
- $\boldsymbol{A}^T\boldsymbol{u} + \boldsymbol{B}^T\boldsymbol{v} = \boldsymbol{0}$ where $\boldsymbol{u} \in \mathbf{R}^p$, $\boldsymbol{v} \in \mathbf{R}^q$, $\boldsymbol{u} \neq \boldsymbol{0}$, $\boldsymbol{u} \geq \boldsymbol{0}$.

4.26 a) Give an example showing that the convex hull $H(S)$ of a closed set S is not necessarily closed.

b) Prove that if the set S is compact then $H(S)$ *is* closed.

Hint: Carathéodory's theorem might be useful.

4.27 For what values of a is $-x + 10y + az = 0$ a support plane at $\boldsymbol{0}$ to the convex set defined by

$$\begin{cases} x + 3y - 5z \leq 0 \\ x - y + 3z \leq 0 \\ -x + 2y \leq 0 \text{ ?} \end{cases}$$

(1998–12–14, 4)

4.28 Does the vector $\boldsymbol{b} = (-1,0,0)^T$ belong to the convex hull of the columns of the matrix

$$\begin{pmatrix} 2 & -5 & -3 & -6 \\ 0 & 2 & 1 & 1 \\ 2 & 2 & -2 & -4 \end{pmatrix} \text{ ?}$$

(2007–04–10, 3b)

Chapter 5

LINEAR PROGRAMMING

One of the most important classes of optimization problems is called *linear programming*. Here the objective function is linear and its domain is defined by linear constraints, that is, geometrically the domain is a set bounded by hyperplanes. These problems are often solved by the *simplex algorithm*, which is discussed in section 5.2. In section 5.3 we present the theory of duality; to every linear programming problem there is another one, the *dual problem*, and information about one of the two can be turned into information on the other one.

However, first we need to study the properties of the sets which appear as the domain of the objective function in these problems, the *polyhedral sets*.

5.1 Polyhedral sets

5.1.1 Extreme points

DEFINITION 1. *The intersection of finitely many closed half-spaces in* $\mathbf{R}^n$ *is called a* ***polyhedral set****.*

A polyhedral set is closed and convex, since it is an intersection of closed and convex sets. Obviously an intersection of polyhedral sets is itself a polyhedral set.

Example 1. A hyperplane $\boldsymbol{p}^T\boldsymbol{x} = \alpha$ is the intersection of two closed half-spaces

$$\boldsymbol{p}^T\boldsymbol{x} \leq \alpha \quad \text{and} \quad \boldsymbol{p}^T\boldsymbol{x} \geq \alpha\,;$$

hence a hyperplane is a polyhedral set. □

Example 2. The figures show two examples of polyhedral sets in the plane.

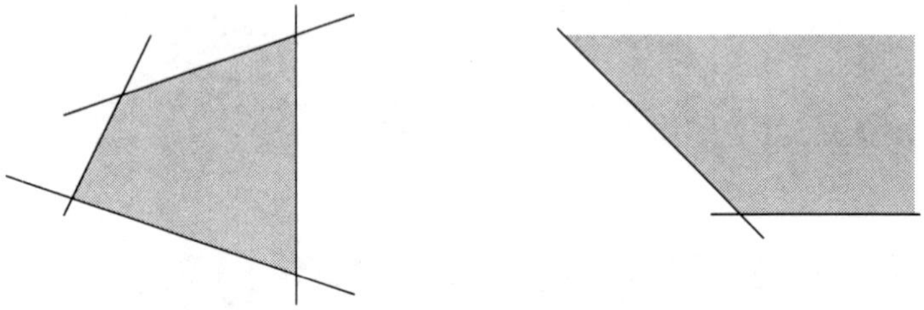

□

Example 3. Let $\boldsymbol{A}$ be an $m \times n$-matrix and $\boldsymbol{b}$ an $m \times 1$-matrix. The set

$$S = \{ \boldsymbol{x} \in \mathbf{R}^n ;\ \boldsymbol{A}\boldsymbol{x} \leq \boldsymbol{b} \}$$

is a polyhedral set, and so is

$$T = \{ \boldsymbol{x} \in \mathbf{R}^n ;\ \boldsymbol{A}\boldsymbol{x} \leq \boldsymbol{b},\ \boldsymbol{x} \geq \boldsymbol{0} \}.$$

□

We state the next definition for general convex sets, although we will need it for polyhedral sets only.

DEFINITION 2. *Let S be a convex set. A point $\boldsymbol{x}$ in S is called an* ***extreme point*** *of S if $\boldsymbol{x}$ is not strictly between two points of S.*

In other words, $\boldsymbol{x}$ is an extreme point of S if no two points $\boldsymbol{y}$ and $\boldsymbol{z}$, both in S, can be found such that

$$\boldsymbol{x} = \lambda \boldsymbol{y} + (1-\lambda)\boldsymbol{z}, \qquad 0 < \lambda < 1.$$

Obviously an extreme point of S is a boundary point of S.

Example 4. The extreme points of the polyhedral sets from example 2 are marked with dots.

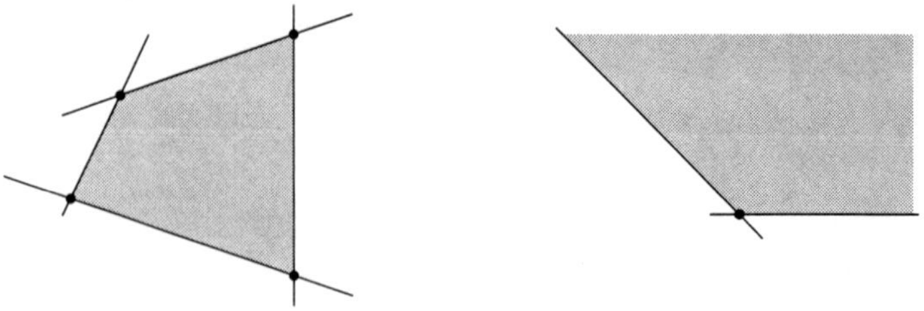

□

Example 5. For the half circular disc in the figure, the whole half-circle consists of extreme points. □

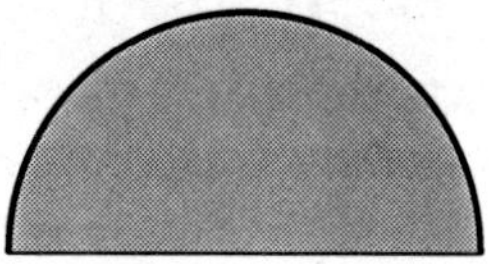

Example 6. A half-space has *no* extreme point. □

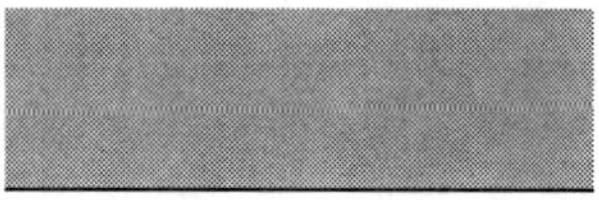

THEOREM 1.

1. *Assume that $P \neq \emptyset$ is a bounded polyhedral set in $\mathbf{R}^n$. Then P has only finitely many extreme points $\boldsymbol{x}_1, \ldots, \boldsymbol{x}_N$, and in fact P is the convex hull of its extreme points.*

2. *Assume that $P \neq \emptyset$ is a polyhedral set in $\mathbf{R}^n$ that contains no line. Then P has at least one extreme point.*

PROOF. We use induction over the dimension[1] k of P.

In the case $k = 1$ the polyhedral set is an interval on a line in $\mathbf{R}^n$. Then the theorem is obvious, since the extreme points of an interval are its endpoints.

For the induction step, assume the theorem to be true for all polyhedral sets of dimension $k-1$ or less. Let P be a polyhedral set of dimension k. It is then no restriction to assume P to be contained in $\mathbf{R}^k$. Since P is an intersection of half-spaces it is defined by a number of inequalities

$$\boldsymbol{p}_1^T \boldsymbol{x} \leq \alpha_1$$

$$\ldots$$

$$\boldsymbol{p}_l^T \boldsymbol{x} \leq \alpha_l.$$

Denote by H_i the corresponding hyperplanes $\boldsymbol{p}_i^T \boldsymbol{x} = \alpha_i$, $i = 1, \ldots, l$, in $\mathbf{R}^k$.

[1]This is defined as the dimension of the least plane variety that contains P.

We now make three observations.

1°. If $\boldsymbol{x}_0$ is an extreme point of P then $\boldsymbol{x}_0$ is an extreme point of $P \cap H_i$ for some bounding hyperplane H_i. This follows since $\boldsymbol{x}_0$ must be a boundary point of P, which implies that for some i

$$\boldsymbol{p}_i^T \boldsymbol{x}_0 = \alpha_i.$$

Then for this i $\boldsymbol{x}_0$ is an extreme point of $P \cap H_i$.

2°. If $\boldsymbol{x}_0$ is an extreme point of $P \cap H_i$ for some H_i then $\boldsymbol{x}_0$ is an extreme point of P. Because suppose otherwise, i.e., that $\boldsymbol{x}_0 = \lambda \boldsymbol{y} + (1-\lambda)\boldsymbol{z}$, $0 < \lambda < 1$, where $\boldsymbol{y}, \boldsymbol{z} \in P$. Then

$$\alpha_i = \boldsymbol{p}_i^T \boldsymbol{x}_0 = \lambda \boldsymbol{p}_i^T \boldsymbol{y} + (1-\lambda)\boldsymbol{p}_i^T \boldsymbol{z} \le \lambda \alpha_i + (1-\lambda)\alpha_i = \alpha_i.$$

Thus we have equality everywhere, implying that $\boldsymbol{y}, \boldsymbol{z} \in H_i$. We have arrived at at contradiction.

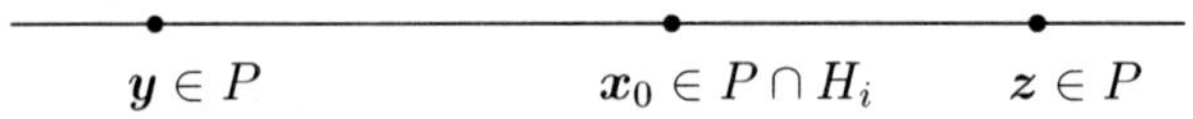

3°. All the sets $P \cap H_i$ are polyhedral sets of dimension at most $k-1$. They are of course bounded if P is.

Using these observations we conclude the following. The H_i are finitely many, and in the case of bounded P the sets $P \cap H_i$ all have finitely many extreme points by the induction hypothesis. These points together constitute all extreme points of P. Thus P has finitely many extreme points, $\boldsymbol{x}_1, \ldots, \boldsymbol{x}_N$.

To show that P is the convex hull of these points we consider a point $\boldsymbol{x} \in P$. We want to prove that $\boldsymbol{x}$ is a convex combination of $\boldsymbol{x}_1, \ldots, \boldsymbol{x}_N$.

Let ℓ be a line through $\boldsymbol{x}$. The set $P \cap \ell$ is convex, bounded and of dimension 1, i.e., an interval. Denote its endpoints by $\boldsymbol{y}_1$ and $\boldsymbol{y}_2$ respectively (see the figure on the opposite page) and assume that

$$\boldsymbol{y}_1 \in P \cap H_i, \qquad \boldsymbol{y}_2 \in P \cap H_j.$$

By the induction hypothesis the points $\boldsymbol{y}_1$ and $\boldsymbol{y}_2$ are convex combinations of $\boldsymbol{x}_1, \ldots, \boldsymbol{x}_N$. Then the same goes for $\boldsymbol{x}$, which is a convex combination of $\boldsymbol{y}_1$ and $\boldsymbol{y}_2$.

The first statement of the theorem is now proved.

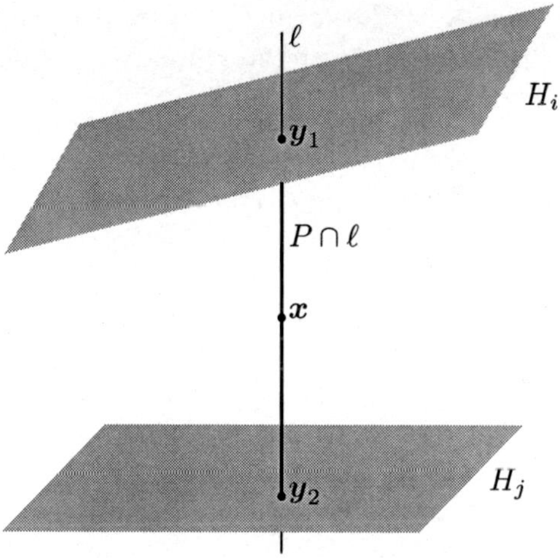

For the second statement let as above $\boldsymbol{x} \in P$ and let ℓ be a line through $\boldsymbol{x}$. Since by assumption P contains no line, the set $P \cap \ell$ has at least one boundary point $\boldsymbol{y}_1$. Assume that $\boldsymbol{y}_1 \in P \cap H_i$.

By the induction hypothesis $P \cap H_i$ has at least one extreme point $\boldsymbol{z}$. But then $\boldsymbol{z}$ is an extreme point of P as well. Thus the proof is finished. □

We now specialize to the case when the polyhedral set is defined by

(1) $$S = \{\, \boldsymbol{x} \in \mathbf{R}^n \,;\; \boldsymbol{A}\boldsymbol{x} = \boldsymbol{b},\ \boldsymbol{x} \geq \boldsymbol{0} \,\}.$$

Here $\boldsymbol{A}$ is an $m \times n$-matrix and $\boldsymbol{b}$ an m-vector. Assuming that $S \neq \emptyset$ we can discard excess equations in the system $\boldsymbol{A}\boldsymbol{x} = \boldsymbol{b}$ and assume that $\operatorname{rank} \boldsymbol{A} = m$. Then $m \leq n$. Reordering the columns of $\boldsymbol{A}$ if necessary, there is a partition of the columns such that

$$\boldsymbol{A} = (\boldsymbol{B} \quad \boldsymbol{N}),$$

where $\boldsymbol{B}$ is an invertible $m \times m$-matrix and $\boldsymbol{N}$ an $m \times (n-m)$-matrix. Letting

$$\boldsymbol{x} = \begin{pmatrix} \boldsymbol{x}_B \\ \boldsymbol{x}_N \end{pmatrix},$$

be the corresponding partition of the rows of $\boldsymbol{x}$, we have

$$\boldsymbol{A}\boldsymbol{x} = \boldsymbol{b} \quad \Longleftrightarrow \quad \boldsymbol{B}\boldsymbol{x}_B + \boldsymbol{N}\boldsymbol{x}_N = \boldsymbol{b} \quad \Longleftrightarrow \quad \boldsymbol{x}_B = \boldsymbol{B}^{-1}\boldsymbol{b} - \boldsymbol{B}^{-1}\boldsymbol{N}\boldsymbol{x}_N.$$
(2)

For a non-empty set S of the form (1) we will in the future always use this notation. The reordering of the columns of $\boldsymbol{A}$ is implicitly understood.

THEOREM 2. *A point $\boldsymbol{x} \in \mathbf{R}^n$ is an extreme point of the set S in (1) if and only if there is a partition of the variables such that*

$$\text{(3)} \qquad \boldsymbol{x} = \begin{pmatrix} \boldsymbol{x}_B \\ \boldsymbol{x}_N \end{pmatrix} = \begin{pmatrix} \boldsymbol{B}^{-1}\boldsymbol{b} \\ \boldsymbol{0} \end{pmatrix} \quad \text{and} \quad \boldsymbol{B}^{-1}\boldsymbol{b} \geq \boldsymbol{0}.$$

PROOF. Assume first that $\boldsymbol{x}$ has the form (3). That $\boldsymbol{x}$ belongs to S is easily verified. We prove that it is an extreme point. Suppose to the contrary that

$$\boldsymbol{x} = \lambda \boldsymbol{y} + (1-\lambda)\boldsymbol{z}, \qquad 0 < \lambda < 1,$$

with points $\boldsymbol{y}, \boldsymbol{z}$ in S. This means that

$$\begin{pmatrix} \boldsymbol{B}^{-1}\boldsymbol{b} \\ \boldsymbol{0} \end{pmatrix} = \lambda \begin{pmatrix} \boldsymbol{y}_B \\ \boldsymbol{y}_N \end{pmatrix} + (1-\lambda) \begin{pmatrix} \boldsymbol{z}_B \\ \boldsymbol{z}_N \end{pmatrix}.$$

In particular, $\boldsymbol{0} = \lambda \boldsymbol{y}_N + (1-\lambda)\boldsymbol{z}_N$, which implies that $\boldsymbol{y}_N = \boldsymbol{z}_N = \boldsymbol{0}$, since we know a priori that $\boldsymbol{y}_N, \boldsymbol{z}_N \geq \boldsymbol{0}$. Using (2) we now find that $\boldsymbol{y}_B = \boldsymbol{z}_B = \boldsymbol{B}^{-1}\boldsymbol{b}$. Thus

$$\boldsymbol{y} = \boldsymbol{z} = \boldsymbol{x}.$$

From this contradiction it follows that $\boldsymbol{x}$ is an extreme point of S.

To prove the converse statement, assume that $\boldsymbol{x}$ is an extreme point of S. After a reordering of the variables we can assume that

$$\boldsymbol{x} = (x_1, \ldots, x_k,\ 0, \ldots, 0),$$

where $x_1, \ldots, x_k$ are strictly positive. We are going to show that the corresponding columns $\boldsymbol{a}_1, \ldots, \boldsymbol{a}_k$ in $\boldsymbol{A}$ are linearly independent. Accepting that for a moment, we can complete them with more columns to a basis $\boldsymbol{a}_1, \ldots, \boldsymbol{a}_m$ in $\mathbf{R}^m$, since rank $\boldsymbol{A} = m$. Choosing $\boldsymbol{B}$ as the submatrix with columns $\boldsymbol{a}_1, \ldots, \boldsymbol{a}_m$ defines the partition and makes $\boldsymbol{B}$ invertible. Furthermore, for this partition $\boldsymbol{x}_N = \boldsymbol{0}$ and (3) follows from (2).

We now give the proof that $\boldsymbol{a}_1, \ldots, \boldsymbol{a}_k$ are linearly independent. Assume to the contrary that there exist numbers $\lambda_1, \ldots, \lambda_k$, not all of them zero, such that

$$\sum_{j=1}^{k} \lambda_j \boldsymbol{a}_j = \boldsymbol{0}.$$

Denote by $\boldsymbol{\lambda}$ the column vector in $\mathbf{R}^n$ having the first k components equal to the λ_j:s and the other ones equal to 0. Then the condition above can be written

$$\boldsymbol{A\lambda} = \boldsymbol{0}.$$

This implies that for any t we have

$$\boldsymbol{A}(\boldsymbol{x} \pm t\boldsymbol{\lambda}) = \boldsymbol{b}.$$

Furthermore $\boldsymbol{x} \pm t\boldsymbol{\lambda} \geq \boldsymbol{0}$ for t small, since $x_1, \ldots, x_k$ are strictly positive. Hence the points $\boldsymbol{x} \pm t\boldsymbol{\lambda}$ belong to S for these t. This is a contradiction to the assumption that $\boldsymbol{x}$ is an extreme point. □

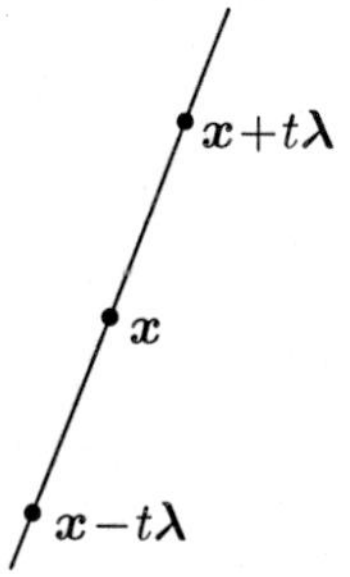

Since there is only a finite number of ways to choose the $n-m$ components of $\boldsymbol{x}$ that should take the value 0 in (3), one consequence of theorem 2 is that there are only finitely many extreme points of S. This of course we already know from theorem 1. However we can now say more. The number of ways to make the partition in (3) is $\binom{n}{n-m} = \binom{n}{m}$, hence this number provides an upper bound on the number of extreme points of S. Note that in practice $\binom{n}{m}$ is a very large number.

5.1.2 Directions

For non-bounded convex sets we now introduce the concept of direction.

DEFINITION 3. *Let S be a closed convex set. A vector $\mathbf{d} \neq \boldsymbol{0}$ is called a* ***direction*** *of S if*

$$\boldsymbol{x} \in S \quad \Longrightarrow \quad \boldsymbol{x} + \lambda \boldsymbol{d} \in S \quad \text{for all } \lambda \geq 0.$$

We consider two directions to be different if they are not parallel.

The concept of **extreme direction** is defined analogously to that of extreme point.

Example 7. Let the polyhedral set S in the plane be defined by $x_2 \geq |x_1|$. Among others, the vectors $\boldsymbol{d}_1$, $\boldsymbol{d}_2$, $\boldsymbol{d}_3$ in the figure are directions of S; $\boldsymbol{d}_1$ and $\boldsymbol{d}_3$ are extreme directions.

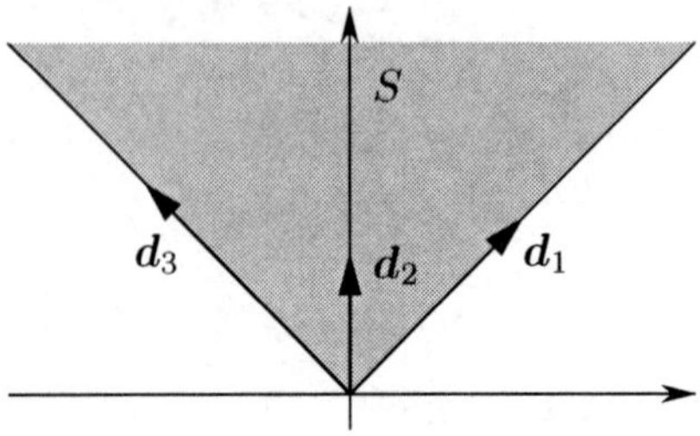

□

Example 8. Let $S = \{\boldsymbol{x} \in \mathbf{R}^n\,;\, \boldsymbol{A}\boldsymbol{x} = \boldsymbol{b},\ \boldsymbol{x} \geq \boldsymbol{0}\}$. Then $\boldsymbol{d} \neq \boldsymbol{0}$ is a direction of S if and only if

$$\boldsymbol{A}\boldsymbol{d} = \boldsymbol{0}, \qquad \boldsymbol{d} \geq \boldsymbol{0}.$$

This follows because by definition $\boldsymbol{d}$ is a direction if and only if for any $\boldsymbol{x} \in S$ and for all $\lambda \geq 0$

$$\boldsymbol{A}(\boldsymbol{x} + \lambda\boldsymbol{d}) = \boldsymbol{b} \quad \text{and} \quad \boldsymbol{x} + \lambda\boldsymbol{d} \geq 0.$$

The first condition is equivalent to $\boldsymbol{A}\boldsymbol{d} = \boldsymbol{0}$, the second one to $\boldsymbol{d} \geq \boldsymbol{0}$.

It can be shown that S has only finitely many extreme directions, but we abstain from this. □

We have the following representation theorem.

THEOREM 3. *Let $S = \{\boldsymbol{x} \in \mathbf{R}^n\,;\, \boldsymbol{A}\boldsymbol{x} = \boldsymbol{b},\ \boldsymbol{x} \geq \boldsymbol{0}\}$. Denote by $\boldsymbol{x}_1, \ldots, \boldsymbol{x}_k$ the extreme points of S, and by $\boldsymbol{d}_1, \ldots, \boldsymbol{d}_l$ the extreme directions of S. Then a point $\boldsymbol{x}$ in $\mathbf{R}^n$ belongs to S if and only if it can be written*

$$\boldsymbol{x} = \sum_{i=1}^{k} \lambda_i \boldsymbol{x}_i + \sum_{j=1}^{l} \mu_j \boldsymbol{d}_j \qquad \textit{with} \quad \sum_{i=1}^{k} \lambda_i = 1, \quad \textit{all} \ \lambda_i \geq 0, \quad \textit{all} \ \mu_j \geq 0.$$

In the case when S is bounded there are no directions, and we are back at theorem 1. The proof in the general case is omitted. The theorem will not be used in the sequel.

Exercises

5.1 Draw a figure in the plane of the set

$$S = \{\, \boldsymbol{x} \in \mathbf{R}^2\,;\; -x_1 + 2x_2 \le 3,\; x_1 + x_2 \le 2,\; x_2 \le 1,\; x_1 \ge 0,\; x_2 \ge 0 \,\}.$$

Verify that it is bounded, and identify all extreme points. Represent the point $(1, \frac{1}{2})$ as a convex combination of the extreme points.

5.2 Draw a figure and find the set of directions for each of the following convex sets:

a) $S = \{\, (x_1, x_2)\,;\; x_1 \le x_2 \le \sqrt{3}\, x_1 \,\}$,
b) $S = \{\, (x_1, x_2)\,;\; x_2 \ge x_1^2 \,\}$,
c) $S = \{\, (x_1, x_2)\,;\; x_1 x_2 \ge 1,\; x_1 > 0 \,\}$,
d) $S = \{\, (x_1, x_2)\,;\; x_1^2 + 3x_2^2 \le 1 \,\}$.

Which are the extreme directions?

5.3 Let S be a closed convex set in $\mathbf{R}^n$ and let $\bar{\boldsymbol{x}}$ be a fixed point of S. Suppose that $\boldsymbol{d}$ is a nonzero vector in $\mathbf{R}^n$ and that $\bar{\boldsymbol{x}} + \lambda \boldsymbol{d} \in S$ for all $\lambda \ge 0$. Show that $\boldsymbol{d}$ is a direction of S.

5.2 Linear programming and the simplex algorithm

5.2.1 Introduction

A linear programming (LP for short) minimization problem has the following form:

Find the minimum of

$$\boldsymbol{c}^T \boldsymbol{x} = c_1 x_1 + \ldots + c_n x_n$$

when $\boldsymbol{x}$ belongs to a polyhedral set S in $\mathbf{R}^n$.

Example 9. In the case $n = 2$ there is an important geometrical interpretation of an LP problem. Suppose that we wish to find the minimum of

$$f(\boldsymbol{x}) = a x_1 + b x_2 \qquad (a, b \text{ positive constants})$$

over the polyhedral set S defined by

$$\begin{aligned} 2x_1 + \ \ x_2 &\geq 3 \\ x_1 + 2x_2 &\geq 3 \\ x_1 \geq 0,\ x_2 &\geq 0. \end{aligned}$$

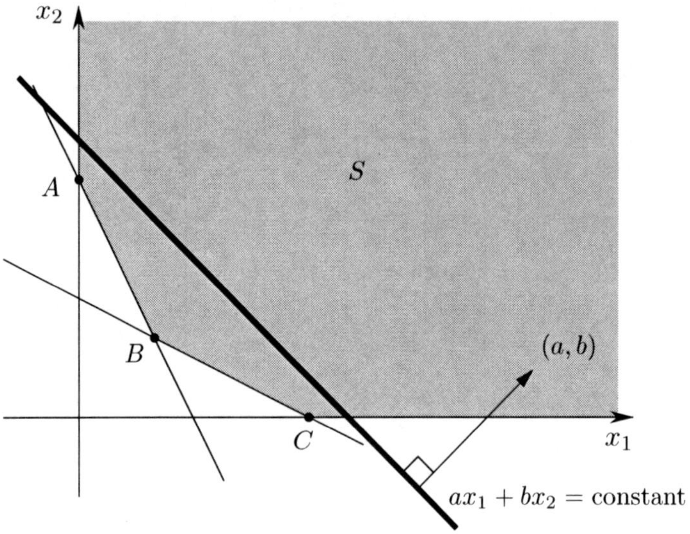

The picture shows the set S in the plane. Also shown is a level curve of f, which of course is a straight line. The values of f grow in the northeast direction. Thus it is obvious that the minimum of f occurs at one of the points A: $(0, 3)$, B: $(1, 1)$ or C: $(3, 0)$. Which of the points that actually is the minimum one depends on the slope of the level lines, that is, on the values of a and b.

The values at A, B and C are $f(A) = 3b$, $f(B) = a+b$, $f(C) = 3a$. It is left to the reader to prove that if $2b \leq a$ the minimum is at A, if $2a \leq b$ then C is the solution, and otherwise it is point B. □

We now give an example of the kind of context in which LP problems appear.

Example 10. Consider a company running two factories (F_1 and F_2), both producing goods of two kinds (G_1 and G_2). The production capacities of the factories are given in the following table (unit e.g. tons per day).

	F_1	F_2
G_1	2	1
G_2	1	2

The cost per day to operate factory F_1 is a kiloeuros, and for factory F_2 the cost is b kiloeuros.

The company's objective is to produce (at least) 3 tons of G_1 and 3 tons of G_2. Assume the production to take place x_1 days at factory F_1, x_2 days at F_2. Then the total production of G_1 is (using the table of capacities above) $2x_1 + x_2$ and the production of G_2 is $x_1 + 2x_2$. Thus we have the requirements

$$2x_1 + x_2 \geq 3, \qquad x_1 + 2x_2 \geq 3.$$

We wish to choose x_1 and x_2 so that the total production cost $ax_1 + bx_2$ is minimized.

This problem is exactly the one we stated and solved geometrically in example 9 above. □

An LP problem is said to be in *canonical form* if the polyhedral set S has the form (1), i.e.

$$S = \{\boldsymbol{x} \in \mathbf{R}^n\,;\, \boldsymbol{A}\boldsymbol{x} = \boldsymbol{b},\ \boldsymbol{x} \geq \boldsymbol{0}\}.$$

Any LP problem can be put in canonical form:

- An inequality $a_{11}x_1 + \ldots + a_{1n}x_n \leq b_1$ can be transformed to an equality by the introduction of a *slack variable* $s_1 \geq 0$, making

$$a_{11}x_1 + \ldots + a_{1n}x_n + s_1 = b_1.$$

- A *free* variable x_j, that is, a variable with no sign condition, can be replaced by two non-negative variables by putting

$$x_j = x_j' - x_j'', \qquad x_j', x_j'' \geq 0.$$

Using these operations we can turn any LP problem into canonical form, of course at the expense of having to increase the number of variables, and as a consequence to consider a polyhedral set in a space of higher dimension.

In the sequel we assume whenever necessary that S is defined in the above manner with $\boldsymbol{A}$ of type $m \times n$ and of rank m. We assume that S is non-empty.[2]

First we prove that if $\mathbf{c}^T\boldsymbol{x}$ is bounded from below on S then the minimum actually exists.

LEMMA 1. *Assume that the set S defined above is non-empty, and that*

$$\beta = \inf_{\boldsymbol{x} \in S} \mathbf{c}^T\boldsymbol{x}$$

is a finite number. Then there is an $\bar{\boldsymbol{x}} \in S$ such that $\mathbf{c}^T\bar{\boldsymbol{x}} = \beta$.

PROOF. Assume to the contrary that there is no such $\bar{\boldsymbol{x}}$. This implies that the system

$$\begin{cases} \boldsymbol{A}\boldsymbol{x} = \boldsymbol{b} \\ \mathbf{c}^T\boldsymbol{x} = \beta \\ \boldsymbol{x} \geq \boldsymbol{0} \end{cases}$$

has no solution. Here we have the system (∗∗) in Farkas' theorem (page 136) in the form

$$\begin{pmatrix} \boldsymbol{A} \\ \mathbf{c}^T \end{pmatrix} \boldsymbol{x} = \begin{pmatrix} \boldsymbol{b} \\ \beta \end{pmatrix}, \qquad \boldsymbol{x} \geq \boldsymbol{0}.$$

The corresponding system (∗) is obtained by transposing the matrices, and is

$$\begin{pmatrix} \boldsymbol{A}^T & \mathbf{c} \end{pmatrix} \begin{pmatrix} y_1 \\ y_2 \end{pmatrix} \leq \boldsymbol{0}, \qquad \begin{pmatrix} \boldsymbol{b}^T & \beta \end{pmatrix} \begin{pmatrix} y_1 \\ y_2 \end{pmatrix} > 0,$$

[2]Later we will address the issue of how to find out if this is really the case. In practice, when the number of variables and constraints is large, it is quite possible that so many constraints have been imposed that it is impossible to satisfy them all.

or equivalently

$$\begin{cases} \boldsymbol{A}^T \boldsymbol{y}_1 + \boldsymbol{c}\, y_2 \leq \boldsymbol{0} \\ \boldsymbol{b}^T \boldsymbol{y}_1 + \beta\, y_2 > 0. \end{cases}$$

By Farkas' theorem this system *has* a solution $(\boldsymbol{y}_1, y_2)$. Note that y_2 here is a real number. In fact $y_2 \neq 0$, since $y_2 = 0$ would imply (use Farkas' theorem in the direction from (∗) to (∗∗)) that S is empty. Furthermore, we may scale the solution $(\boldsymbol{y}_1, y_2)$ to make

$$\boldsymbol{b}^T \boldsymbol{y}_1 + \beta\, y_2 = 1.$$

Assume this scaling to be done and put $\boldsymbol{z} = \dfrac{1}{y_2} \boldsymbol{y}_1$.

First assume that $y_2 > 0$. Then

$$\boldsymbol{A}^T \boldsymbol{z} + \boldsymbol{c} \leq \boldsymbol{0} \qquad \text{and} \quad \boldsymbol{b}^T \boldsymbol{z} + \beta = \frac{1}{y_2}.$$

The first inequality can be rewritten $\boldsymbol{c}^T \leq -(\boldsymbol{A}^T \boldsymbol{z})^T = -\boldsymbol{z}^T \boldsymbol{A}$. This implies that for $\boldsymbol{x} \in S$

$$\boldsymbol{c}^T \boldsymbol{x} \leq -\boldsymbol{z}^T \boldsymbol{A} \boldsymbol{x} = -\boldsymbol{z}^T \boldsymbol{b} = \beta - \frac{1}{y_2},$$

which is a contradiction to the definition of β as $\inf_{\boldsymbol{x} \in S} \boldsymbol{c}^T \boldsymbol{x}$.

Now assume that $y_2 < 0$. Then

$$\boldsymbol{A}^T \boldsymbol{z} + \boldsymbol{c} \geq \boldsymbol{0} \qquad \text{and} \quad \boldsymbol{b}^T \boldsymbol{z} + \beta = \frac{1}{y_2},$$

implying that for any $\boldsymbol{x} \in S$

$$\boldsymbol{c}^T \boldsymbol{x} \geq -\boldsymbol{z}^T \boldsymbol{A} \boldsymbol{x} = -\boldsymbol{z}^T \boldsymbol{b} = \beta - \frac{1}{y_2}.$$

Since $y_2 < 0$ we have once again obtained a contradiction to the definition of β.

From this contradiction we conclude that the system (∗∗) is in fact solvable. The proof is finished. □

The following theorem is fundamental.

THEOREM 4. *If* $\min_S \boldsymbol{c}^T\boldsymbol{x}$ *exists, there is an extreme point of* S *at which the minimum occurs.*

PROOF. Assume that $\min_S \boldsymbol{c}^T\boldsymbol{x} = \beta$. Then the hyperplane

$$H: \quad \boldsymbol{c}^T\boldsymbol{x} = \beta$$

is a support plane of the convex set S. Since $\boldsymbol{x} \geq \boldsymbol{0}$ in S the set $H \cap S$ contains no line. Thus it has at least one extreme point $\bar{\boldsymbol{x}}$ by theorem 1. This is an extreme point of S as well, and since $\boldsymbol{c}^T\bar{\boldsymbol{x}} = \beta$ the minimum occurs at $\bar{\boldsymbol{x}}$. □

In the case of an unbounded domain S we also have

THEOREM 5. *The minimum* $\min_S \boldsymbol{c}^T\boldsymbol{x}$ *exists if and only if* $\boldsymbol{c}^T\boldsymbol{d} \geq 0$ *for all directions* $\boldsymbol{d}$ *of* S.

PROOF. Assume first that there is a direction $\boldsymbol{d}$ such that $\boldsymbol{c}^T\boldsymbol{d} < 0$. Let $\boldsymbol{x} \in S$. By the definition of a direction $\boldsymbol{x} + \lambda\boldsymbol{d} \in S$ for all $\lambda \geq 0$. Furthermore

$$\boldsymbol{c}^T(\boldsymbol{x} + \lambda\boldsymbol{d}) = \boldsymbol{c}^T\boldsymbol{x} + \lambda\boldsymbol{c}^T\boldsymbol{d} \to -\infty \quad \text{as} \quad \lambda \to \infty.$$

Thus no minimum exists.

To prove the converse assume that $\boldsymbol{c}^T\boldsymbol{d} \geq 0$ for all directions $\boldsymbol{d}$ of S. By example 8 the directions of S are characterized by $\boldsymbol{A}\boldsymbol{d} = \boldsymbol{0}$, $\boldsymbol{d} \geq \boldsymbol{0}$. Thus the assumption is that

$$\boldsymbol{A}\boldsymbol{d} = \boldsymbol{0}, \quad \boldsymbol{d} \geq \boldsymbol{0} \quad \Longrightarrow \quad \boldsymbol{c}^T\boldsymbol{d} \geq 0.$$

This statement is equivalent to the system

$$\begin{cases} \boldsymbol{A}\boldsymbol{d} \leq \boldsymbol{0} \\ (-\boldsymbol{A})\boldsymbol{d} \leq \boldsymbol{0} \\ -\boldsymbol{d} \leq \boldsymbol{0} \\ (-\boldsymbol{c})^T\boldsymbol{d} > 0 \end{cases}$$

having no solution. Here we have system (∗) of Farkas' theorem, page 136. The corresponding system (∗∗) is

$$\begin{pmatrix} \boldsymbol{A}^T & -\boldsymbol{A}^T & -\boldsymbol{I} \end{pmatrix} \begin{pmatrix} \boldsymbol{y}_1 \\ \boldsymbol{y}_2 \\ \boldsymbol{y}_3 \end{pmatrix} = -\boldsymbol{c}, \qquad \begin{pmatrix} \boldsymbol{y}_1 \\ \boldsymbol{y}_2 \\ \boldsymbol{y}_3 \end{pmatrix} \geq \boldsymbol{0}.$$

Thus the assumption implies that this system is solvable. This in turn means that there exist $\boldsymbol{y}_1, \boldsymbol{y}_2, \boldsymbol{y}_3 \geq \boldsymbol{0}$ such that

$$\boldsymbol{A}^T\boldsymbol{y}_1 - \boldsymbol{A}^T\boldsymbol{y}_2 - \boldsymbol{y}_3 = -\boldsymbol{c}.$$

Putting $\boldsymbol{z} = \boldsymbol{y}_2 - \boldsymbol{y}_1$ we have $\boldsymbol{A}^T\boldsymbol{z} + \boldsymbol{y}_3 = \boldsymbol{c}$, with no sign restriction on $\boldsymbol{z}$. Consequently there is a $\boldsymbol{z}$ such that

$$\boldsymbol{A}^T\boldsymbol{z} \leq \boldsymbol{c}.$$

For the objective function this implies that when $\boldsymbol{x} \in S$

$$\boldsymbol{c}^T\boldsymbol{x} \geq (\boldsymbol{A}^T\boldsymbol{z})^T\boldsymbol{x} = \boldsymbol{z}^T\boldsymbol{A}\boldsymbol{x} = \boldsymbol{z}^T\boldsymbol{b}.$$

Hence the function $\boldsymbol{c}^T\boldsymbol{x}$ is bounded from below on S. By lemma 1 this proves that $\min_S \boldsymbol{c}^T\boldsymbol{x}$ exists. □

Theorem 4 tells us that if the minimum exists it occurs at an extreme point of the polyhedral set S. Hence a conceivable strategy might be to compile a list of all extreme points and then compare the values of the objective function at these. However, as we have seen, the number of extreme points is bounded above by the number $\binom{n}{m}$, which in realistic problems is a *very* large number. Thus this idea is no good. The simplex algorithm aims at improving it by, given any extreme point, only consider extreme points with lesser function values. However, before discussing this further we introduce some terminology and give an example.

5.2.2 Terminology

In connection with the problem

Find the minimum of

$$\boldsymbol{c}^T\boldsymbol{x} = c_1x_1 + \ldots + c_nx_n$$

when

$$\boldsymbol{A}\boldsymbol{x} = \boldsymbol{b}, \qquad \boldsymbol{x} \geq \boldsymbol{0}$$

we as usual call $f(\boldsymbol{x}) = \boldsymbol{c}^T\boldsymbol{x}$ the *objective function*. A point $\boldsymbol{x} \in \mathbf{R}^n$ is called a *solution* if $\boldsymbol{A}\boldsymbol{x} = \boldsymbol{b}$, and a *feasible solution* if $\boldsymbol{A}\boldsymbol{x} = \boldsymbol{b}$ and $\boldsymbol{x} \geq \boldsymbol{0}$.

Now assume that $\boldsymbol{A}$ is partitioned as $\boldsymbol{A} = (\boldsymbol{B}\ \boldsymbol{N})$ with $\boldsymbol{B}$ invertible and let

$$\boldsymbol{x} = \begin{pmatrix} \boldsymbol{x}_B \\ \boldsymbol{x}_N \end{pmatrix}$$

be the corresponding partition of the variables. Here $\boldsymbol{x}_B$ are called the *basic* variables, and $\boldsymbol{x}_N$ the *non-basic* ones (of the partition). Remember that by theorem 2 $\boldsymbol{x}$ is an extreme point (i.e., a candidate for a minimum) if and only if a partition can be found with $\boldsymbol{x}_N = \boldsymbol{0}$ and $\boldsymbol{x}_B = \boldsymbol{B}^{-1}\boldsymbol{b} \geq \boldsymbol{0}$. We call $\boldsymbol{x}$ a *basic solution* if $\boldsymbol{A}\boldsymbol{x} = \boldsymbol{b}$ and $\boldsymbol{x}_N = \boldsymbol{0}$, and a *feasible basic solution* if $\boldsymbol{x}_B \geq \boldsymbol{0}$ as well. If $\boldsymbol{x}_B > \boldsymbol{0}$ the solution is called *non-degenerate.*

Example 11. Let the polyhedral set S in $\mathbf{R}^4$ be defined by the conditions

$$\begin{cases} x_1 + x_2 + 3x_3 - 3x_4 = 3 \\ \qquad x_2 + \ x_3 - \ x_4 = 1 \\ x_i \geq 0, \quad i = 1, 2, 3, 4. \end{cases}$$

Then

$$\boldsymbol{A} = \begin{pmatrix} 1 & 1 & 3 & -3 \\ 0 & 1 & 1 & -1 \end{pmatrix}, \qquad \boldsymbol{b} = \begin{pmatrix} 3 \\ 1 \end{pmatrix}.$$

A basic solution $\boldsymbol{x} = (x_1, x_2, x_3, x_4)$ is recognized by at least two variables being zero. There are $\binom{4}{2} = 6$ ways to choose two variables from four, corresponding to the 6 different ways of partitioning $\boldsymbol{A}$ into $\boldsymbol{B}$ and $\boldsymbol{N}$. All six cases are displayed in the table on the next page.

By theorem 2 a point $\boldsymbol{x} \in \mathbf{R}^n$ is an extreme point of S if and only if it is a feasible basic solution. Furthermore, by theorem 4, if the minimum of $\mathbf{c}^T\boldsymbol{x}$ in S exists it occurs at some extreme point of S. Consequently, for our S the only possibilities for a minimum are $(2, 1, 0, 0)$ and $(0, 0, 1, 0)$, whatever the objective function $f(\boldsymbol{x}) = \mathbf{c}^T\boldsymbol{x}$. □

columns in $\boldsymbol{B}$	columns in $\boldsymbol{N}$	equations for $\boldsymbol{x}_B$	result
1, 2	3, 4	$\begin{cases} x_1 + x_2 = 3 \\ x_2 = 1 \end{cases}$	$\boldsymbol{x} = (2,1,0,0)$. A basic feasible, non-degenerate solution.
1, 3	2, 4	$\begin{cases} x_1 + 3x_3 = 3 \\ x_3 = 1 \end{cases}$	$\boldsymbol{x} = (0,0,1,0)$. A basic feasible, degenerate solution.
1, 4	2, 3	$\begin{cases} x_1 - 3x_4 = 3 \\ -x_4 = 1 \end{cases}$	$\boldsymbol{x} = (0,0,0,-1)$. A non-feasible solution.
2, 3	1, 4	$\begin{cases} x_2 + 3x_3 = 3 \\ x_2 + x_3 = 1 \end{cases}$	$\boldsymbol{x} = (0,0,1,0)$. Same as above.
2, 4	1, 3	$\begin{cases} x_2 - 3x_4 = 3 \\ x_2 - x_4 = 1 \end{cases}$	$\boldsymbol{x} = (0,0,0,-1)$. Same as above.
3, 4	1, 2	$\begin{cases} 3x_3 - 3x_4 = 3 \\ x_3 - x_4 = 1 \end{cases}$	No solution. The matrix $\begin{pmatrix} 3 & -3 \\ 1 & -1 \end{pmatrix}$ is not invertible.

5.2.3 The simplex algorithm

As we have said earlier, any interesting LP problem is too large (in respect of the number of variables and constraints) to make it possible, in a reasonable amount of time, to examine all extreme points. Instead, the *simplex algorithm*[3] makes use of the following strategy.

- Find one extreme point of S, i.e., a feasible basic solution $\bar{\boldsymbol{x}} = \begin{pmatrix} \bar{\boldsymbol{x}}_B \\ \bar{\boldsymbol{x}}_N \end{pmatrix} = \begin{pmatrix} \bar{\boldsymbol{x}}_B \\ \boldsymbol{0} \end{pmatrix}$. We postpone the discussion of how this is done.
- Try to find another extreme point of S with a lesser value of the objective function, by finding a new set of basic variables, including

[3]The name derives from an early geometrical interpretation which we do not discuss here.

The concept of linear programming and the simplex algorithm were conceived around 1947 by above all *George B. Dantzig*. In the years since, the impact on the applications of the simplex algorithm has been tremendous. Lately some alternatives have appeared, notably the *Karmarkar algorithm*.

one of the $\boldsymbol{x}_N$ and all but one of the $\boldsymbol{x}_B$. We call the new extreme point a *neighbour* of the old one.

- Repeat the process as long as possible. When (if) it terminates we have found the minimum.

To proceed with the discussion we need the notation introduced earlier. Corresponding to the initial extreme point $\bar{\boldsymbol{x}}$ there is a partition of the variables into basic and non-basic ones, and corresponding partitions of the matrices:

$$\mathsf{A} = \begin{pmatrix} \mathsf{B} & \mathsf{N} \end{pmatrix}, \qquad \boldsymbol{x} = \begin{pmatrix} \boldsymbol{x}_B \\ \boldsymbol{x}_N \end{pmatrix}, \qquad \boldsymbol{c} = \begin{pmatrix} \boldsymbol{c}_B \\ \boldsymbol{c}_N \end{pmatrix}.$$

The constraints are $\boldsymbol{x} \geq \boldsymbol{0}$ and

$$(4) \qquad \mathsf{A}\boldsymbol{x} = \boldsymbol{b} \quad \Longleftrightarrow \quad \mathsf{B}\boldsymbol{x}_B + \mathsf{N}\boldsymbol{x}_N = \boldsymbol{b},$$

and the value of the objective function, which we call f, can be written

$$(5) \quad f = \boldsymbol{c}^T\boldsymbol{x} = \boldsymbol{c}_B^T\boldsymbol{x}_B + \boldsymbol{c}_N^T\boldsymbol{x}_N \quad \Longleftrightarrow \quad f - \boldsymbol{c}_B^T\boldsymbol{x}_B - \boldsymbol{c}_N^T\boldsymbol{x}_N = \boldsymbol{0}.$$

It is convenient to put this information into a tableau. (Of course this is exactly how it is implemented on a computer.) The tableau contains the matrices appearing in (4) and (5), while the $\boldsymbol{x}$-variables are implicitly understood; for convenience we have put them in a head line above the tableau. RHS of course means the right hand side. The leftmost column is never used and usually omitted; we have put it there to help the reader remember what the first row stands for.

(6)

f	$\boldsymbol{x}_B$	$\boldsymbol{x}_N$	RHS
1	$-\boldsymbol{c}_B^T$	$-\boldsymbol{c}_N^T$	0
$\boldsymbol{0}$	B	N	$\boldsymbol{b}$

Remark. In other literature the simplex tableau may appear in modified form. As we have said already, the left column is usually omitted. The first row may have been put as the last row; it also sometimes appears with the opposite sign in every entry.

The reordering of the columns rendering the first m variables the basic ones, which is implicitly understood in the tableau, is never done in practice.

Equation (4) can be solved for $\boldsymbol{x}_B$, since $\boldsymbol{B}$ is invertible. Multiplying by $\boldsymbol{B}^{-1}$ from the left gives

$$\boldsymbol{x}_B + \boldsymbol{B}^{-1}\boldsymbol{N}\boldsymbol{x}_N = \boldsymbol{B}^{-1}\boldsymbol{b}, \tag{7}$$

or $\boldsymbol{x}_B = \boldsymbol{B}^{-1}\boldsymbol{b} - \boldsymbol{B}^{-1}\boldsymbol{N}\boldsymbol{x}_N$. Putting this expression into f, that is, into (5), the result is

$$f = \boldsymbol{c}_B^T\boldsymbol{B}^{-1}\boldsymbol{b} - \left(-\boldsymbol{c}_N^T + \boldsymbol{c}_B^T\boldsymbol{B}^{-1}\boldsymbol{N}\right)\boldsymbol{x}_N. \tag{8}$$

This shows how f can be considered as a function of the $\boldsymbol{x}_N$-variables alone, with no constraints except all variables being non-negative. In particular, in our extreme point $\bar{\boldsymbol{x}}$, where $\bar{\boldsymbol{x}}_N = \boldsymbol{0}$, the value is

$$f(\bar{\boldsymbol{x}}) = \boldsymbol{c}_B^T\boldsymbol{B}^{-1}\boldsymbol{b} = \boldsymbol{c}_B^T\bar{\boldsymbol{x}}_B = \boldsymbol{c}^T\bar{\boldsymbol{x}}$$

as it should. Equation (8) can be rewritten as

$$f + \left(-\boldsymbol{c}_N^T + \boldsymbol{c}_B^T\boldsymbol{B}^{-1}\boldsymbol{N}\right)\boldsymbol{x}_N = \boldsymbol{c}^T\bar{\boldsymbol{x}}. \tag{9}$$

All these calculations can of course be performed as row operations in the tableau. The result is the following new tableau. Note the identity matrix under the basic variables.

(10)

f	$\boldsymbol{x}_B$	$\boldsymbol{x}_N$	RHS
1	$\boldsymbol{0}^T$	$-\boldsymbol{c}_N^T + \boldsymbol{c}_B^T\boldsymbol{B}^{-1}\boldsymbol{N}$	$\boldsymbol{c}^T\bar{\boldsymbol{x}}$
$\boldsymbol{0}$	$\boldsymbol{I}$	$\boldsymbol{B}^{-1}\boldsymbol{N}$	$\boldsymbol{B}^{-1}\boldsymbol{b}$

After this rewriting of the objective function and the constraints we face the problem of finding a new (better) extreme point $\bar{\boldsymbol{x}}'$ to replace $\bar{\boldsymbol{x}}$, i.e., to find another partition $\left(\boldsymbol{B}' \ \boldsymbol{N}'\right)$ of $\boldsymbol{A}$. The choice of a new partition brings us back to the situation in (6).

We first consider some examples.

Example 12. Find the minimum of $f = -5x_1 - 4x_2$ when

$$\begin{cases} x_1 + 2x_2 \leq 8 \\ 3x_1 + 2x_2 \leq 9 \\ x_1, x_2 \geq 0. \end{cases}$$

Solution: Since there are only two variables in this problem it can be given a geometrical interpretation and solved graphically as in example 9. The figure shows the polyhedral set defined by the constraints, and a level line $-5x_1 - 4x_2 =$ constant of f. The minimum obviously occurs at the point P, whose coordinates are $(\frac{1}{2}, \frac{15}{4})$.

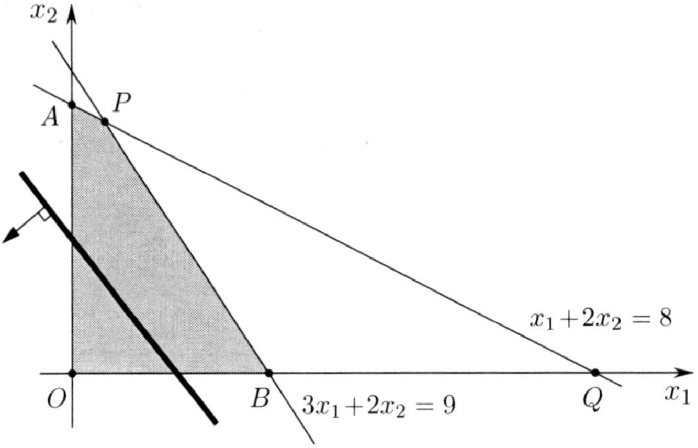

We now turn to an algebraic solution of the problem. Introduce *slack variables* $s_1, s_2 \geq 0$ transforming the constraints into equalities:

$$\begin{aligned} x_1 + 2x_2 + s_1 \quad\quad &= 8 \\ 3x_1 + 2x_2 \quad\quad + s_2 &= 9. \end{aligned}$$

Together with the conditions $x_1, x_2, s_1, s_2 \geq 0$ these equations define a polyhedral set in $\mathbf{R}^4$ (whose projection on the x_1x_2-plane is exactly the shadowed set in the figure above). We use the notation $\boldsymbol{x} = (x_1, x_2, s_1, s_2)$. The problem is now in canonical form, and we can identify matrices:

$$\mathbf{A} = \left(\begin{array}{cc|cc} 1 & 2 & 1 & 0 \\ 3 & 2 & 0 & 1 \end{array}\right), \qquad \mathbf{b} = \begin{pmatrix} 8 \\ 9 \end{pmatrix}, \qquad \mathbf{c} = \begin{pmatrix} -5 \\ -4 \\ 0 \\ 0 \end{pmatrix}.$$

A partition of $\boldsymbol{A}$ into $\boldsymbol{B} = \begin{pmatrix} 1 & 0 \\ 0 & 1 \end{pmatrix}$ and $\boldsymbol{N} = \begin{pmatrix} 1 & 2 \\ 3 & 2 \end{pmatrix}$ can be seen at once (luck!), however not in the standard order. The corresponding feasible basic solution is $\boldsymbol{x}_1 = (\underbrace{0,0}_{\boldsymbol{x}_N}, \underbrace{8,9}_{\boldsymbol{x}_B})$. The first simplex tableau is[4]

This row indicates $f - \boldsymbol{c}^T\boldsymbol{x} = 0$, i.e. in this case $f + 5x_1 + 4x_2 = 0$

A $*$ denotes a basic variable

	↓		*	*		
f	x_1	x_2	s_1	s_2	RHS	
1	5	4	0	0	0	
0	1	2	1	0	8	8/1=8
0	3	2	0	1	9	9/3=3 ←

These rows state that $\boldsymbol{Ax} = \boldsymbol{b}$

Arrows indicate row and column for the next pivot element

Here is an $\boldsymbol{I}$ already!

Since the tableau already contains an identity matrix, we are in the situation (10) above. Our aim now is to identify another basic feasible solution with a smaller value of the objective function.

The coefficients of the non-basic variables x_1 and x_2 in the expression for f are negative, since the first row contains positive elements. This implies that the values of f can be decreased by increasing the values of x_1 or x_2 from 0, of course at the expense of s_1 and s_2, since the constraints must be observed. However, only one of the non-basic variables should be made basic. Let us choose x_1. (See *Note 1* below for further comments on this choice.)

We have thus decided to introduce x_1 as a new basic variable. We must also choose one of s_1 and s_2 to leave the basis. Let s_2 be the choice. (See *Note 2* below).

Our choices imply that a new partition of the variables has been chosen. We are then in the position of a tableau like (6). Everything should now be expressed in the non-basic variables x_2 and s_2, i.e., the tableau should be transformed so that columns x_1 and s_1 make up an identity matrix. In particular we need the number 1 in the last entry of

[4]Some details of the tableau will be explained later in the example.

column x_1, where we now have 3 (the *pivot element*). The appropriate row and column are indicated with small arrows in the tableau above.

To achieve this end we make suitable row operations on the tableau. In principle this is multiplication with the inverse of $\begin{pmatrix} 1 & 1 \\ 3 & 0 \end{pmatrix}$. At the same time we express f in the new basic variables by making the topmost row entry of the x_1 column equal to zero. The net result is the following tableau. (The reader should check on the computations.)

	*	↓	*			
f	x_1	x_2	s_1	s_2	RHS	
1	0	$\frac{2}{3}$	0	$-\frac{5}{3}$	-15	$f = -\frac{2}{3}x_2 + \frac{5}{3}s_2 - 15$
0	0	$\frac{4}{3}$	1	$-\frac{1}{3}$	5	15/4 ←
0	1	$\frac{2}{3}$	0	$\frac{1}{3}$	3	9/2

We can now identify a new feasible basic solution $\boldsymbol{x}_2 = (3, 0, 5, 0)$, corresponding to the new partition of the variables. The non-zero coordinates of $\boldsymbol{x}_2$ appear in the RHS column. The value of f at $\boldsymbol{x}_2$ is -15; it appears in the first entry of the RHS column. As it should, it is less than the value 0 of f at the starting point $\boldsymbol{x}_1$.

In the projection onto the x_1x_2-plane in the figure above this solution corresponds to point B.

We now repeat the procedure. Since the first row in the new tableau contains a positive entry in the x_2 column, we can decrease f further by increasing x_2 from 0, that is, making x_2 a basic variable. The basic variables are now x_1, x_2, and s_1 becomes non-basic. Once again we perform the process of converting tableau (6) into tableau (10). The pivot element is $\frac{4}{3}$. With row operations we obtain a new tableau with an identity matrix in the x_2 and x_1 columns.

	*	*				
f	x_1	x_2	s_1	s_2	RHS	
1	0	0	$-\frac{1}{2}$	$-\frac{3}{2}$	$-\frac{35}{2}$	$f = \frac{1}{2}s_1 + \frac{3}{2}s_2 - \frac{35}{2}$
0	0	1	$\frac{3}{4}$	$-\frac{1}{4}$	$\frac{15}{4}$	
0	1	0	$-\frac{1}{2}$	$\frac{1}{2}$	$\frac{1}{2}$	

The tableaux indicates how f and x_1, x_2 are expressed in the non-basic varibles s_1,s_2. All coefficients in the first row are now non-positive. Hence any increase in the variables s_1 and s_2 will result in an increase in f. There is no way in which we can decrease the value of f further. Thus we have found the minimum, which equals $-\frac{35}{2}$, at the point $\bar{\boldsymbol{x}} = (\frac{1}{2}, \frac{15}{4}, 0, 0)$. — As expected, the projection of $\bar{\boldsymbol{x}}$ onto the x_1x_2-plane is the point P.

Note 1. In the first tableau above the coefficient of x_2 is also positive. You might as well decide to introduce x_2 and not x_1 as a basic variable. This will lead to different computations in the future (in the projection you move from O to point A), but the end result will be the same. In practice the following rule of thumb is often used: choose the variable with the largest coefficient.

Note 2. It is however not at all obvious that variable s_2 is the one that should be turned non-basic. But if we try s_1 instead, we end up att $\boldsymbol{x} = (8, 0, 0, -15)$, which is unfeasible. (In the figure we move to point Q outside the feasible region.)

When choosing which variable to leave the basis you must for every constraint find out how much the elected new basic variable may increase without making the variable corresponding to the constraint negative. The numbers in the right margin of the tableaux above state the limit case when this variable becomes zero. We must ensure that nonc of these limits are exceeded, otherwise some constraint $x_j \geq 0$ or $s_i \geq 0$ will be violated. That variable (constraint) must be chosen which provides the least limit. In the first step above s_2 should be chosen. — Sometimes you get a negative number in the margin, indicating that there are no limits for the variable in question (see the next example). If all margin numbers are negative the minimum is $-\infty$. □

Here is another example.

Example 13. Find the minimum of $f = x_1 + x_2 - x_3 - x_4 - x_5$ when

$$\begin{cases} x_1 + 2x_2 - x_3 + 3x_4 \qquad = 4 \\ \qquad x_2 - x_3 + 2x_4 - x_5 = 1 \\ \text{all } x_i \geq 0. \end{cases}$$

Solution: In this case the problem is already in canonical form, and there is no need for any slack variables. By trial and error (a general method will be discussed later, page 175) we find the feasible basic solution $x_1 = 2$, $x_2 = 1$, $x_3 = x_4 = x_5 = 0$. In the corresponding partition x_1 and x_2 are the basic variables, for which the constraints should be solved by Gauss elimination. The solution proceeds with the following tableaux.

*	*					
x_1	x_2	x_3	x_4	x_5		
-1	-1	1	1	1	0	
1	2	-1	3	0	4	
0	1	-1	2	-1	1	

This is tableau (6). The x_1 column already happens to contain $\begin{pmatrix}1\\0\end{pmatrix}$. We must arrange for $\begin{pmatrix}0\\1\end{pmatrix}$ in the x_2 column, and zeroes for x_1 och x_2 in the topmost row.

*	*		$\downarrow$			
0	0	1	2	2	3	
1	0	1	-1	2	2	< 0 no restriction
0	1	-1	$\textcircled{2}$	-1	1	$1/2 \leftarrow$

This is where the search for a new feasible basic solution starts. The present basic variables have been marked with a $*$. The pivot element 2 is encircled, and x_4 is to replace x_2 as a basic variable. Note that due to the minus sign the first constraint $x_1 + x_3 - x_4 + 2x_5 = 2$ will not be violated whatever the increase of x_4 (which can be compensated for by *increasing* x_1).

*			*	$\downarrow$		
0	-1	2	0	3	2	
1	$\frac{1}{2}$	$\frac{1}{2}$	0	$\textcircled{\frac{3}{2}}$	$\frac{5}{2}$	$5/3 \leftarrow$
0	$\frac{1}{2}$	$-\frac{1}{2}$	1	$-\frac{1}{2}$	$\frac{1}{2}$	< 0 no restriction

The new extreme point is at $(\frac{5}{2}, 0, 0, \frac{1}{2}, 0)$, where f has value 2. The search continues. New pivot element is $\frac{3}{2}$. New basic variables are x_4 and x_5.

		$\downarrow$	*	*		
-2	-2	1	0	0	-3	
$\frac{2}{3}$	$\frac{1}{3}$	$\textcircled{\frac{1}{3}}$	0	1	$\frac{5}{3}$	$5 \leftarrow$
$\frac{1}{3}$	$\frac{2}{3}$	$-\frac{1}{3}$	1	0	$\frac{4}{3}$	< 0 no restriction

The present value of f is -3 at the extreme point $(0, 0, 0, \frac{4}{3}, \frac{5}{3})$. New pivot element is $\frac{1}{3}$. New basic variables are x_3 and x_4.

		*	*		
−4	−3	0	0	−3	−8
2	1	1	0	3	5
1	1	0	1	1	3

End. All coefficients in the topmost row are non-positive.

The minimum equals -8 for $x_3 = 5$, $x_4 = 3$, $x_1 = x_2 = x_5 = 0$. □

We now return to the general case,in order to explain why things always work out as in the examples. Look again at tableau (10), where everything is expressed in the non-basic variables $\boldsymbol{x}_N$ of an extreme point $\bar{\boldsymbol{x}}$ of S.

(10)

f	$\boldsymbol{x}_B$	$\boldsymbol{x}_N$	RHS
1	$\boldsymbol{0}^T$	$-\boldsymbol{c}_N^T + \boldsymbol{c}_B^T\boldsymbol{B}^{-1}\boldsymbol{N}$	$\boldsymbol{c}^T\bar{\boldsymbol{x}}$
$\boldsymbol{0}$	$\boldsymbol{I}$	$\boldsymbol{B}^{-1}\boldsymbol{N}$	$\boldsymbol{B}^{-1}\boldsymbol{b}$

Consider first the case when all coefficients in the topmost row are non-positive, i.e. when

$$-\boldsymbol{c}_N^T + \boldsymbol{c}_B^T\boldsymbol{B}^{-1}\boldsymbol{N} \leq \boldsymbol{0}^T.$$

Then by (8)

$$f = \boldsymbol{c}^T\bar{\boldsymbol{x}} - \left(-\boldsymbol{c}_N^T + \boldsymbol{c}_B^T\boldsymbol{B}^{-1}\boldsymbol{N}\right)\boldsymbol{x}_N \geq \boldsymbol{c}^T\bar{\boldsymbol{x}} \qquad \text{for all } \boldsymbol{x} \geq \boldsymbol{0}.$$

Thus f attains its minimum at $\bar{\boldsymbol{x}}$.

Now assume that some coefficient in the first row is positive. More specifically, let the coefficient of x_j (a non-basic variable) be positive. Denote the corresponding column of $\boldsymbol{N}$ by $\boldsymbol{a}_j$. Then the coefficient in question is

$$-c_j + \boldsymbol{c}_B^T\boldsymbol{B}^{-1}\boldsymbol{a}_j > 0.$$

Put

$$\boldsymbol{d}_j = \begin{pmatrix} -\boldsymbol{B}^{-1}\boldsymbol{a}_j \\ \boldsymbol{e}_j \end{pmatrix},$$

where $\boldsymbol{e}_j$ denotes an $n-m$ column vector with the entry corresponding to x_j equal to 1 and the others equal to 0. Then

$$\boldsymbol{A}\boldsymbol{d}_j = \begin{pmatrix} \boldsymbol{B} & \boldsymbol{N} \end{pmatrix} \begin{pmatrix} -\boldsymbol{B}^{-1}\boldsymbol{a}_j \\ \boldsymbol{e}_j, \end{pmatrix} = -\boldsymbol{a}_j + \boldsymbol{N}\boldsymbol{e}_j = -\boldsymbol{a}_j + \boldsymbol{a}_j = \boldsymbol{0}.$$

It follows from example 8, page 156, that $\boldsymbol{d}_j$ is a direction of S if and only if $\boldsymbol{d}_j \geq \boldsymbol{0}$, that is, if and only if

$$\boldsymbol{B}^{-1}\boldsymbol{a}_j \leq 0.$$

Note that $\boldsymbol{B}^{-1}\boldsymbol{a}_j$ is the x_j column of $\boldsymbol{B}^{-1}\boldsymbol{N}$ in (10).

Put

$$\boldsymbol{x} = \bar{\boldsymbol{x}} + \lambda \boldsymbol{d}_j = \begin{pmatrix} \bar{\boldsymbol{x}}_B - \lambda \boldsymbol{B}^{-1}\boldsymbol{a}_j \\ \lambda \boldsymbol{e}_j \end{pmatrix}.$$

When λ increases from 0 the point $\boldsymbol{x}$ moves from $\bar{\boldsymbol{x}}$ in direction $\boldsymbol{d}_j$. Since

$$f = \boldsymbol{c}^T(\bar{\boldsymbol{x}} + \lambda \boldsymbol{d}_j) = \boldsymbol{c}^T\bar{\boldsymbol{x}} + \lambda \boldsymbol{c}^T\boldsymbol{d}_j = \boldsymbol{c}^T\bar{\boldsymbol{x}} + \lambda(-\boldsymbol{c}_B^T\boldsymbol{B}^{-1}\boldsymbol{a}_j + c_j)$$

with the coefficient of λ negative, the value of f decreases with increasing λ. We now have two possibilities to consider.

First assume that $\boldsymbol{B}^{-1}\boldsymbol{a}_j \leq \boldsymbol{0}$, i.e., all entries (except the topmost one) in the x_j column of (10) are non-positive. Then, as we concluded above, $\boldsymbol{d}_j$ is a direction of S, we can let $\lambda \to \infty$ and f may attain as large negative values as we please. In this case there is no minimum. See the left figure on the next page.

Now assume that $\boldsymbol{B}^{-1}\boldsymbol{a}_j \not\leq \boldsymbol{0}$. Then some entry in the x_j column is positive. Also $\boldsymbol{d}_j$ is not a direction of S. It is not possible to let λ increase indefinitely, only as long as the $\boldsymbol{x}_B$ variables remain non-negative. Thus we find a new extreme point $\bar{\boldsymbol{x}} + \lambda_{\max}\boldsymbol{d}_j$ when one of the basic variables becomes zero. Simultaneously, x_j grows from zero to some positive value (in fact $\lambda_{\max}$). See the right figure.

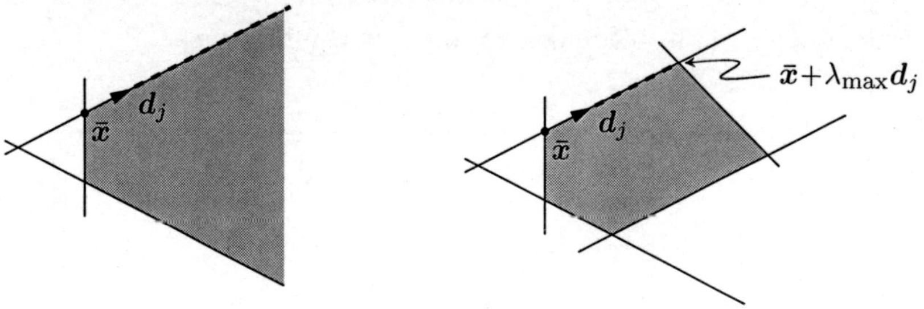

The move to the new (neighbouring) extreme point is of course accompanied by a new partition of the variables into basic and non-basic ones and a new tableau of the form (10); then the process starts all over again.

Remark. What is said above is not quite true. Unless all basic variables at an extreme point $\bar{x}$ are strictly positive we have *degeneracy*. Then you might have to change the partition in basic and non-basic variables while remaining at $\bar{x}$. (See example 11, page 164.) In itself this is no problem, if you can continue the normal routine afterwards. But we now run the risk of the algorithm ending up in an infinite loop, oscillating between the same two partitions of the variables. The phenomenon is called *cycling*. Of course any serious implementation of the simplex algorithm must be prepared for it. However, we will neglect this possibility in the present course.

5.2.4 How to find an initial point

We discuss one method, the *two-phase method*, for finding an initial extreme point from which to start the simplex algorithm.

The LP problem is assumed to be written in canonical form, with notation as before. By multiplying some constraints with -1, if necessary, we can ensure that $\boldsymbol{b} \geq \boldsymbol{0}$. In *phase I* we introduce *artificial variables* $u_1, \ldots, u_m$ and consider the following LP problem in $n+m$ variables:

Find the minimum of

$$g(\boldsymbol{x}, \boldsymbol{u}) = u_1 + \ldots + u_m$$

when

$$\boldsymbol{Ax} + \boldsymbol{Iu} = \boldsymbol{b}, \quad \boldsymbol{x} \geq \boldsymbol{0}, \quad \boldsymbol{u} \geq \boldsymbol{0}.$$

The simplex tableau for this problem is the following:

(11)

g	$\boldsymbol{x}$	$\boldsymbol{u}$	RHS
1	$0 \;\ldots\; 0$	$-1 \;\ldots\; -1$	0
$\boldsymbol{0}$	$\boldsymbol{A}$	$\boldsymbol{I}$	$\boldsymbol{b}$

For this problem there is no difficulty in finding a feasible basic solution: $\boldsymbol{x} = \boldsymbol{0}$, $\boldsymbol{u} = \boldsymbol{b}$ will do. Furthermore, the minimum certainly exists, since $g = \sum_1^m u_i$ is bounded from below by zero. Thus, if we apply the simplex algorithm to this problem it will yield a minimum. There are two possibilities.

1) Suppose the minimum turns out to be strictly positive. Then we can conclude that there is no $\boldsymbol{x} \geq \boldsymbol{0}$ such that $\boldsymbol{Ax} = \boldsymbol{b}$. In other words, the polyhedral set S defined by the original constraints is *empty*, and our original LP problem is insolvable.

2) The minimum is zero. Denote the minimum point by $(\boldsymbol{x}_0, \boldsymbol{u}_0)$; obviously $\boldsymbol{u}_0 = \boldsymbol{0}$. Since this is a feasible basic solution to the auxiliary problem, dropping the $\boldsymbol{u}$-variables shows that $\boldsymbol{x}_0$ is a feasible basic solution to the original problem.

Phase II of the method consists in applying the simplex algorithm to the original problem with $\boldsymbol{x}_0$ as initial point. In practice one puts both problems in the same tableau, by adding a row for the original objective function f in (11). Carrying out the row operations simultaneously in both problems gives a tableau with an identity matrix as in (10) at the end of phase I, when the u_i variables are dropped.

Exercises

5.4 Solve graphically the following LP problem:

$$\begin{aligned} \text{Maximize} \quad & x_1 + 5x_2 \\ \text{subject to} \quad & x_1 + x_2 \leq 5 \\ & 2x_1 + 3x_2 \leq 11 \\ & -x_1 + 2x_2 \leq 5 \\ & x_1 \geq 0,\ x_2 \geq 0. \end{aligned}$$

(1996–12–16, 3a)

5.5 Consider the linear program

$$\begin{aligned} \text{Minimize} \quad & -4x_1 - 2x_2 - 2x_3 \\ \text{subject to} \quad & 3x_1 + x_2 + x_3 = 12 \\ & x_1 - x_2 + x_3 = -8 \\ & x_1 \geq 0, x_2 \geq 0, x_3 \geq 0. \end{aligned}$$

Verify that the choice $B = \{1,2\}$, $N = \{3\}$ gives a basic feasible solution, and hence solve the problem. (Here B and N denote the index sets of basic and non-basic variables respectively.)

S 5.6 Use the simplex method to solve the following problem. Start with the solution $x_1 = 3$, $x_5 = 5$, $x_2 = x_3 = x_4 = 0$.

$$\begin{aligned} \text{Minimize} \quad & x_1 + 6x_2 - 7x_3 + x_4 + 5x_5 \\ \text{subject to} \quad & 5x_1 - 4x_2 + 13x_3 - 2x_4 + x_5 = 20 \\ & x_1 - x_2 + 5x_3 - x_4 + x_5 = 8 \\ & x_1, x_2, x_3, x_4, x_5 \geq 0. \end{aligned}$$

5.7 Consider the following LP problem in the plane.

$$\begin{aligned} \text{minimize} \quad & -2x_1 - 3x_2 \\ \text{subject to} \quad & x_1 - 3x_2 \leq 3 \\ & -x_1 + x_2 \leq 1 \\ & x_1 \geq 0, \; x_2 \geq 0. \end{aligned}$$

a) Solve it graphically.

b) Solve it using the simplex method starting at $(0,0)$. Remember to explain carefully what is happening.

(2010–08–27, 3)

5.8 Here is a tableau in the solution of a minimum LP problem with the simplex algorithm:

x_1	x_2	x_3	x_4	x_5	x_6	x_7	
2	0	3	−1	−4	0	0	−17/2
1	1	1	1	1	0	0	4
5	0	1	−1	−1	1	0	7/2
−2	0	2	2	−1	0	1	5

a) Identify the current basic variables. Find the values of all variables and of the objective function.

b) Perform one iteration step and then answer the same questions again.

c) Have you reached the optimum?

(1999–04–09, 2)

5.9 Solve the following LP problem using the simplex method:

$$\begin{array}{ll} \text{Minimize} & 100 - 5x_1 - 7x_2 - 11x_3 \\ \text{subject to} & 3x_1 + 4x_2 + x_3 \leq 25 \\ & 2x_1 + 3x_2 + 4x_3 \leq 20 \\ & x_1 \geq 0, x_2 \geq 0, x_3 \geq 0. \end{array}$$

5.10 Solve the following problem.

$$\begin{array}{ll} \text{Maximize} & 2x_1 + 4x_2 + x_3 + x_4 \\ \text{subject to} & x_1 + 3x_2 + \quad + x_4 \leq 4 \\ & 2x_1 + x_2 + \quad \leq 3 \\ & x_2 + 4x_3 + x_4 \leq 3 \\ & x_i \geq 0 \quad \text{for all } i. \end{array}$$

S 5.11 Consider the LP problem of exercise 5.5 and replace the objective function by the function $-4x_1 - 2x_3$. Show that the basic feasible solutions derived from $B = \{1, 2\}$, $N = \{3\}$ and $B = \{2, 3\}$, $N = \{1\}$ are both optimal. Show also that any convex combination of these solutions is a solution to the problem.

S 5.12 Apply the *Phase I* method to the problem of finding a feasible point to the set

$$\begin{aligned} 5x_1 + 5x_2 - 3x_3 &\leq 9 \\ -9x_1 - 8x_2 + 10x_3 &\geq 7 \\ -3x_1 + 3x_2 - 7x_3 &\geq 2 \\ \text{all } x_k &\geq 0 \end{aligned}$$

and provide an interpretation of the result obtained.

(2007–01–11, 2a)

5.3 Duality

To every minimization LP problem there is a *dual* maximization LP problem. The connections between these problems improves the understanding of linear optimization, and can be used to devise numerical methods for their solution. Also, as the reader will find for himself, the theory of duality is of great intrinsic beauty.

5.3.1 Notation. The primal and dual problems

We wish to study the minimization of $\boldsymbol{c}^T\boldsymbol{x}$ under general linear constraints, some of which are equalities and some inequalities. To keep track of the different types we partition the row index set $\{1,\ldots,m\}$ into two disjoint sets J' and J'', where indices in J' correspond to inequalities and indices in J'' to equalities. The corresponding partitions of the matrices are written

$$\mathsf{A} = \begin{pmatrix} \mathsf{A}_{J'} \\ \mathsf{A}_{J''} \end{pmatrix}, \qquad \boldsymbol{b} = \begin{pmatrix} \boldsymbol{b}_{J'} \\ \boldsymbol{b}_{J''} \end{pmatrix}.$$

Furthermore, we wish to allow for the possibility of both non-negative variables and *free* ones (no sign condition). For this purpose we divide the column index set $\{1,\ldots,n\}$ into disjoints sets K' and K'', and assume a variable x_k be free when $k \in K''$ and to be non-negative when $k \in K'$. For the matrices we have corresponding partitions

$$\mathsf{A} = \begin{pmatrix} \mathsf{A}_{K'} & \mathsf{A}_{K''} \end{pmatrix}, \qquad \boldsymbol{x} = \begin{pmatrix} \boldsymbol{x}_{K'} \\ \boldsymbol{x}_{K''} \end{pmatrix}, \qquad \boldsymbol{c} = \begin{pmatrix} \boldsymbol{c}_{K'} \\ \boldsymbol{c}_{K''} \end{pmatrix}.$$

In total we have divided A row-wise as well as column-wise, resulting in four blocks[5] which we denote

$$\begin{array}{rcc} & & \text{indices} \\ \mathsf{A} = & \begin{pmatrix} \mathsf{A}_{11} & \mathsf{A}_{12} \\ \mathsf{A}_{21} & \mathsf{A}_{22} \end{pmatrix} & \begin{array}{c} J' \\ J'' \end{array} \\ \text{indices} & \begin{array}{cc} K' & K'' \end{array} & \end{array}$$

[5]The notation assumes an appropriate reordering of rows and columns, which of course is never executed in practice.

To summarize: the problem we have introduced (the *primal* problem) is

Problem P: to minimize $\mathbf{c}^T \boldsymbol{x}$ *subject to the constraints*

$$\boldsymbol{A}_{J'}\boldsymbol{x} \geq \boldsymbol{b}_{J'}, \qquad \boldsymbol{A}_{J''}\boldsymbol{x} = \boldsymbol{b}_{J''}, \qquad \boldsymbol{x}_{K'} \geq \boldsymbol{0}, \qquad \boldsymbol{x}_{K''} \text{ free.}$$

Note the form of the inequality constraints ($\geq$). Of course any minimization LP problem can be transformed into this form. We denote by S the polyhedral subset of $\mathbf{R}^n$ defined by these constraints.

The **dual** problem of problem P is a maximum problem in m variables $y_1, \ldots, y_m$:

Problem D: to maximize $\boldsymbol{b}^T \boldsymbol{y}$ *subject to the constraints*

$$\boldsymbol{A}^T_{K'}\boldsymbol{y} \leq \boldsymbol{c}_{K'}, \qquad \boldsymbol{A}^T_{K''}\boldsymbol{y} = \boldsymbol{c}_{K''}, \qquad \boldsymbol{y}_{J'} \geq \boldsymbol{0}, \qquad \boldsymbol{y}_{J''} \text{ free.}$$

Let T denote the polyhedral set in $\mathbf{R}^m$ defined by the constraints.

The connections between problems P and D are as follows:

1. The matrix describing the constraints in problem D is the transpose of the corresponding one in problem P.

2. Conditions in the primal problem given by equalities correspond to free variables in the dual problem.

3. Free variables in the primal problem correspond to equality constraints in the dual problem.

4. The coefficients in the objective function of problem P appear in the right hand side of the constraints in problem D, and vice versa.

It is easy to see that the dual of problem D is problem P. (The reader should check this. Note that D trivially can be written in the form of P.)

Example 14. Let P be the problem to minimize $2x_1 - x_2 + x_3$ subject to

$$\begin{cases} x_1 + 2x_2 - x_3 = 3 \\ 4x_1 - 7x_2 - x_3 \geq 9 \\ x_1 \geq 0,\ x_2 \geq 0,\ x_3 \text{ free.} \end{cases}$$

Then the partition of the row indices is $J' = \{2\}$, $J'' = \{1\}$, and the partition of the column indices is $K' = \{1,2\}$, $K'' = \{3\}$.

For this problem the dual D is to maximize $3y_1 + 9y_2$ subject to

$$\begin{cases} y_1 + 4y_2 \leq 2 \\ 2y_1 - 7y_2 \leq -1 \\ -y_1 - \ \ y_2 = 1 \\ y_1 \text{ free},\ y_2 \geq 0. \end{cases}$$

□

5.3.2 The complementary slackness principle

DEFINITION 4. *The* **Lagrange function** *L connected to problems P and D above is a function of $n+m$ variables, defined by any of the following equal expressions:*

$$(12) \qquad \begin{aligned} L(\boldsymbol{x};\boldsymbol{y}) &= \boldsymbol{c}^T\boldsymbol{x} + \boldsymbol{y}^T(\boldsymbol{b} - \boldsymbol{A}\boldsymbol{x}) = \boldsymbol{c}^T\boldsymbol{x} + \boldsymbol{y}^T\boldsymbol{b} - \boldsymbol{y}^T\boldsymbol{A}\boldsymbol{x} = \\ &= \boldsymbol{b}^T\boldsymbol{y} + \boldsymbol{x}^T(\boldsymbol{c} - \boldsymbol{A}^T\boldsymbol{y}), \qquad \boldsymbol{x} \in \mathbf{R}^n,\ \boldsymbol{y} \in \mathbf{R}^m. \end{aligned}$$

(Note that all terms appearing in $L(\boldsymbol{x};\boldsymbol{y})$ are 1×1 matrices; hence they are equal to their transpose.)

From the different versions of $L(\boldsymbol{x};\boldsymbol{y})$ in (12) we deduce an important inequality. Using the conditions on sets S and T (here the form of the inequalities is important) it is easily seen that

$$(13) \qquad \boldsymbol{c}^T\boldsymbol{x} \geq L(\boldsymbol{x};\boldsymbol{y}) \geq \boldsymbol{b}^T\boldsymbol{y} \qquad \text{when } \boldsymbol{x} \in S,\ \boldsymbol{y} \in T.$$

This inequality will be used several times in the future. One immediate consequence is that, by lemma 1, page 160,

$$S \neq \emptyset,\ T \neq \emptyset \quad \Longrightarrow \quad \begin{cases} \min_S \boldsymbol{c}^T\boldsymbol{x} \quad \text{exists} \\ \max_T \boldsymbol{b}^T\boldsymbol{y} \quad \text{exists.} \end{cases}$$

Because putting any $\boldsymbol{y} \in T$ into (13) you see that $\boldsymbol{c}^T\boldsymbol{x}$ is bounded from below by $\boldsymbol{b}^T\boldsymbol{y}$, and correspondingly for problem D.

Example 15. Consider the primal and dual problems in example 14. It is easily seen that $\boldsymbol{y} = (-2, 1)$ is feasible for problem D, and gives the value 3 in its objective function. As a consequence, we obtain from (13) that the primal problem is bounded from below; more precisely

$$2x_1 - x_2 + x_3 \geq 3$$

for all $\boldsymbol{x}$ that are feasible for problem P. In particular the minimum value is greater than or equal to 3. (It should be verified that there *are* feasible solutions to problem P.) □

THEOREM 6. (THE COMPLEMENTARY SLACKNESS PRINCIPLE[6]) *For points $\hat{\boldsymbol{x}} \in S$ and $\hat{\boldsymbol{y}} \in T$ the following conditions are equivalent.*

(i) *$\hat{\boldsymbol{x}}$ and $\hat{\boldsymbol{y}}$ are optimal for P and D respectively.*

(ii) $\boldsymbol{c}^T\hat{\boldsymbol{x}} = \boldsymbol{b}^T\hat{\boldsymbol{y}}$.

(iii) $\hat{y}_j(b_j - (\boldsymbol{A}\hat{\boldsymbol{x}})_j) = 0, \quad j = 1, \ldots, m,$

$\hat{x}_k(c_k - (\boldsymbol{A}^T\hat{\boldsymbol{y}})_k) = 0, \quad k = 1, \ldots, n.$

Remark. Conditions (iii) could be put more compactly as

$$\hat{\boldsymbol{y}}^T(\boldsymbol{b} - \boldsymbol{A}\hat{\boldsymbol{x}}) = 0 \qquad \text{and} \qquad \hat{\boldsymbol{x}}^T(\boldsymbol{c} - \boldsymbol{A}^T\hat{\boldsymbol{y}}) = 0$$

respectively. They are of course interesting only for $j \in J'$ and $k \in K'$. In the first case they state that unless $\hat{y}_j$ is zero $\hat{\boldsymbol{x}}$ must give equality in the jth constraint of problem P, and in the second case correspondingly that if $\hat{x}_k > 0$ then $(\boldsymbol{A}^T\hat{\boldsymbol{y}})_k = c_k$.

PROOF. For a start, (ii) is equivalent to $\hat{\boldsymbol{x}}$ and $\hat{\boldsymbol{y}}$ giving equality in (13), which in turn is equivalent to (iii) by the definition of L.

Furthermore, (ii) implies (i), since by (13),

$$\boldsymbol{c}^T\boldsymbol{x} \geq \boldsymbol{b}^T\hat{\boldsymbol{y}} = \boldsymbol{c}^T\hat{\boldsymbol{x}} \qquad \text{for all } \boldsymbol{x} \in S$$

[6] CSP for short.

(with equality when $\boldsymbol{x} = \hat{\boldsymbol{x}}$). This shows that $\hat{\boldsymbol{x}}$ is optimal for problem P. That $\hat{\boldsymbol{y}}$ is optimal for D is proved in the same way.

Finally, that (i) implies (ii) is the so called *duality theorem*, which will be proved in the next subsection. □

Before turning to the duality theorem we give some examples.

Example 16. Show that $\hat{\boldsymbol{x}} = (0, 2, 1)$ is an optimal solution to the problem

$$\begin{array}{lll} \text{minimize} & 4x_1 + 5x_2 + x_3 & \\ \text{subject to} & x_1 + x_2 + x_3 \geq 3 & (1) \\ & 2x_1 + 2x_2 - x_3 \geq 2 & (2) \\ & x_1 + 2x_2 \geq 4 & (3) \\ & x_1, x_2, x_3 \geq 0 & \end{array}$$

Solution: The dual problem is

$$\begin{array}{lll} \text{maximize} & 3y_1 + 2y_2 + 4y_3 & \\ \text{subject to} & y_1 + 2y_2 + y_3 \leq 4 & (4) \\ & y_1 + 2y_2 + 2y_3 \leq 5 & (5) \\ & y_1 - y_2 \leq 1 & (6) \\ & y_1, y_2, y_3 \geq 0 & \end{array}$$

First we check that $\hat{\boldsymbol{x}}$ does satisfy constraints (1)–(3). (This is left to the reader.) Since also $\hat{\boldsymbol{x}} \geq \boldsymbol{0}$ it is a feasible solution. In particular, since we now know that S is nonempty, the dual problem must have an optimal solution (unless $T = \emptyset$).

By the CSP, $\hat{\boldsymbol{x}}$ is optimal for problem P if and only if there is a $\boldsymbol{y} \geq \boldsymbol{0}$ such that

$$\begin{cases} y_1 + 2y_2 + y_3 \leq 4 & \\ y_1 + 2y_2 + 2y_3 = 5 & \text{since } \hat{x}_2 = 2 > 0 \\ y_1 - y_2 = 1 & \text{since } \hat{x}_3 = 1 > 0 \\ y_2 = 0 & \text{since } \hat{\boldsymbol{x}} \text{ gives strict inequality in (2).} \end{cases}$$

The last three of these conditions constitute a system of equations, which is easily solved and has the unique solution $\hat{\boldsymbol{y}} = (1, 0, 2)$. Since $\hat{\boldsymbol{y}}$ satisfies the first condition as well, it is a feasible solution to the dual problem. Using (iii) $\Rightarrow$ (i) in theorem 6, we have shown that $\hat{\boldsymbol{x}}$ is an optimal solution to the given problem (and that $\hat{\boldsymbol{y}}$ is optimal for the dual problem).

As a further check we note that $\boldsymbol{c}^T\hat{\boldsymbol{x}} = 11 = \boldsymbol{b}^T\hat{\boldsymbol{y}}$, confirming (ii). □

Example 17. Is $\hat{\boldsymbol{x}}$ in example 16 the *unique* optimal solution to the problem?

Solution: We use the CSP again, this time starting from $\hat{\boldsymbol{y}}$, which we now know to be an optimal solution to the dual problem. By theorem 6, $\boldsymbol{x}$ is an optimal solution to the primal problem if and only if $\boldsymbol{x}$ is feasible and

$$\begin{cases} x_1 + x_2 + x_3 = 3 & \text{since } \hat{y}_1 = 1 > 0 \\ x_1 + 2x_2 \phantom{{}+x_3} = 4 & \text{since } \hat{y}_3 = 2 > 0 \\ x_1 \phantom{{}+2x_2+x_3} = 0 & \text{since } \hat{\boldsymbol{y}} \text{ gives strict inequality in (4).} \end{cases}$$

This system has the unique solution $x_1 = 0$, $x_2 = 2$, $x_3 = 1$, that is, $\hat{\boldsymbol{x}}$. Thus this solution to the minimization problem is unique. □

5.3.3 The duality theorem

We use the same notation as before.

THEOREM 7. (THE DUALITY THEOREM) *Assume that the sets S and T both are nonempty. Then*

$$\min_S \boldsymbol{c}^T\boldsymbol{x} = \max_T \boldsymbol{b}^T\boldsymbol{y}.$$

We have already made the observation that both optima exist as a consequence of inequality (13). What we have to prove is that they are equal. It is enough to do this in the case when S is given by

$$\boldsymbol{A}\boldsymbol{x} = \boldsymbol{b}, \quad \boldsymbol{x} \geq \boldsymbol{0}$$

(i.e., the index sets J' and K'' are empty), since any LP problem can be transformed to canonical form. (The dual problem is transformed accordingly.) For this S the dual feasible set T is defined by

$$\mathbf{A}^T \boldsymbol{y} \leq c, \quad \boldsymbol{y} \text{ free.}$$

We give two proofs.

FIRST PROOF OF THE DUALITY THEOREM. We know that by the assumption both optima exist. Then in particular there is an optimal solution $\bar{\boldsymbol{x}}$ to problem P which is an extreme point of S. By theorem 2, page 154, there corresponds to $\bar{\boldsymbol{x}}$ a partition of the variables into basic and non-basic ones such that

$$\bar{\boldsymbol{x}} = \begin{pmatrix} \bar{\boldsymbol{x}}_B \\ \bar{\boldsymbol{x}}_N \end{pmatrix} = \begin{pmatrix} \mathbf{B}^{-1}\boldsymbol{b} \\ \boldsymbol{0} \end{pmatrix}$$

where $\mathbf{A} = (\mathbf{B} \quad \mathbf{N})$. By (9), page 167,

$$f = \boldsymbol{c}^T \bar{\boldsymbol{x}} + \left(\boldsymbol{c}_N^T - \boldsymbol{c}_B^T \mathbf{B}^{-1} \mathbf{N}\right) \boldsymbol{x}_N,$$

with

(14) $$\boldsymbol{c}_N^T - \boldsymbol{c}_B^T \mathbf{B}^{-1} \mathbf{N} \geq \boldsymbol{0}^T$$

since $\bar{\boldsymbol{x}}$ is optimal. Put

$$\bar{\boldsymbol{y}} = (\boldsymbol{c}_B^T \mathbf{B}^{-1})^T.$$

We are going to prove that $\bar{\boldsymbol{y}}$ solves the dual problem.

For a start, $\bar{\boldsymbol{y}}$ is feasible for the dual problem since

$$\mathbf{A}^T \bar{\boldsymbol{y}} = \mathbf{A}^T (\boldsymbol{c}_B^T \mathbf{B}^{-1})^T = \left(\boldsymbol{c}_B^T \mathbf{B}^{-1} \mathbf{A}\right)^T =$$

$$= \left(\boldsymbol{c}_B^T \quad \boldsymbol{c}_B^T \mathbf{B}^{-1} \mathbf{N}\right)^T \leq \left(\boldsymbol{c}_B^T \quad \boldsymbol{c}_N^T\right)^T = \boldsymbol{c}.$$

The inequality is a consequence of (14).

Furthermore, $\bar{\boldsymbol{y}}$ is optimal for the dual problem, since

(15) $$\boldsymbol{b}^T \bar{\boldsymbol{y}} = \bar{\boldsymbol{y}}^T \boldsymbol{b} = \boldsymbol{c}_B^T \mathbf{B}^{-1} \boldsymbol{b} = \boldsymbol{c}_B^T \bar{\boldsymbol{x}}_B = \boldsymbol{c}^T \bar{\boldsymbol{x}} \geq \boldsymbol{b}^T \boldsymbol{y} \qquad \text{for all } \boldsymbol{y} \in T.$$

The inequality follows from (13).

In particular, (15) also shows that $\boldsymbol{c}^T \bar{\boldsymbol{x}} = \boldsymbol{b}^T \bar{\boldsymbol{y}}$. Thus the theorem is proved. □

Remark. Consider again the simplex tableau for the two-phase method, and suppose that we keep the artificial variables even after an initial point has been found. The initial tableau is a combination of (6) and (11). The final tableau looks like (only essential information is displayed)

g	f	$\boldsymbol{x}_B$	$\boldsymbol{x}_N$	$\boldsymbol{u}$	RHS
1	0			$\boldsymbol{0}^T$	0
0	1	$\boldsymbol{0}^T$	$-\boldsymbol{c}_N^T + \boldsymbol{c}_B^T \boldsymbol{B}^{-1}\boldsymbol{N}$	$\boldsymbol{c}_B^T \boldsymbol{B}^{-1}$	$\boldsymbol{c}^T \bar{\boldsymbol{x}}$
$\boldsymbol{0}$	$\boldsymbol{0}$	$\boldsymbol{I}$	$\boldsymbol{B}^{-1}\boldsymbol{N}$	$\boldsymbol{B}^{-1}$	$\boldsymbol{B}^{-1}\boldsymbol{b}$

In the columns for the artificial variables, in the second row, we find (the transpose of) the solution $\bar{\boldsymbol{y}}$ to the dual problem.

SECOND PROOF OF THE DUALITY THEOREM. This proof relies on the following lemma, which is based on Farkas' theorem.

LEMMA 2.
a) Assume that $S \neq \emptyset$. *Then*

$$\boldsymbol{c}^T\boldsymbol{x} \geq r \text{ for all } \boldsymbol{x} \in S \quad \Longleftrightarrow \quad \text{there is a } \bar{\boldsymbol{y}} \in T \text{ such that } \boldsymbol{b}^T\bar{\boldsymbol{y}} \geq r.$$

b) Assume that $T \neq \emptyset$. *Then*

$$\boldsymbol{b}^T\boldsymbol{y} \leq s \text{ for all } \boldsymbol{y} \in T \quad \Longleftrightarrow \quad \text{there is an } \bar{\boldsymbol{x}} \in S \text{ such that } \boldsymbol{c}^T\bar{\boldsymbol{x}} \leq s.$$

We postpone the proof for a moment, and show how theorem 7 follows from the lemma.

Assuming that $S \neq \emptyset$, $T \neq \emptyset$ it follows from (13) that the minimum and maximum both exist (finitely) and that

$$\min_S \boldsymbol{c}^T\boldsymbol{x} \geq \max_T \boldsymbol{b}^T\boldsymbol{y}.$$

We may choose $r = \min_S \boldsymbol{c}^T\boldsymbol{x}$ in part a) of the lemma. Then we conclude that there is a $\bar{\boldsymbol{y}} \in T$ such that

$$\text{(16)} \qquad \boldsymbol{b}^T\bar{\boldsymbol{y}} \geq r = \min_S \boldsymbol{c}^T\boldsymbol{x} \geq \max_T \boldsymbol{b}^T\boldsymbol{y}.$$

Hence we have equality here, and $\bar{\boldsymbol{y}}$ is an optimal solution for problem D.

If $\bar{\boldsymbol{x}}$ is an optimal solution to the primal problem it follows from (16) that

$$\boldsymbol{b}^T\bar{\boldsymbol{y}} = \boldsymbol{c}^T\bar{\boldsymbol{x}}.$$

□

Remark. Part a) of the lemma also shows that if $S \neq \emptyset$ and $T = \emptyset$ then $\min_S \boldsymbol{c}^T\boldsymbol{x} = -\infty$, and similarly for part b).

It remains to give the proof of the lemma. We prove only part a), leaving b) to the reader. The implication to the left is an obvious consequence of inequality (13). Thus, it is the implication to the right that should be proved.

The assumption is that the system

$$\text{(17)} \qquad \boldsymbol{A}\boldsymbol{x} = \boldsymbol{b}, \quad \boldsymbol{x} \geq \boldsymbol{0}, \quad \boldsymbol{c}^T\boldsymbol{x} < r$$

has no solution. Homogenizing by putting $\boldsymbol{x} = \dfrac{1}{z_0}\boldsymbol{z}$, with $z_0 > 0$, (17) implies that the system

$$\text{(18)} \qquad \boldsymbol{A}\boldsymbol{z} - \boldsymbol{b}z_0 = \boldsymbol{0}, \quad \boldsymbol{z} \geq \boldsymbol{0}, \quad z_0 > 0, \quad \boldsymbol{c}^T\boldsymbol{z} - rz_0 < 0$$

has no solution. In fact this is true also when $z_0 = 0$. Otherwise there would exist a $\bar{\boldsymbol{z}}$ such that

$$\boldsymbol{A}\bar{\boldsymbol{z}} = \boldsymbol{0}, \quad \bar{\boldsymbol{z}} \geq \boldsymbol{0}, \quad \boldsymbol{c}^T\bar{\boldsymbol{z}} < 0.$$

Then for any $\boldsymbol{x} \in S$ we have $\boldsymbol{x} + \lambda\bar{\boldsymbol{z}} \in S$ if $\lambda > 0$, and

$$\boldsymbol{c}^T(\boldsymbol{x} + \lambda\bar{\boldsymbol{z}}) = \boldsymbol{c}^T\boldsymbol{x} + \lambda\boldsymbol{c}^T\bar{\boldsymbol{z}} \to -\infty \quad \text{as } \lambda \to +\infty,$$

contradicting the unsolvability of (17).

Now we have in (18) a simplified version of the first case of the corollary to Farkas' theorem proved in example 4.17, page 138:

$$\begin{cases} \begin{pmatrix} \boldsymbol{A} & -\boldsymbol{b} \end{pmatrix} \begin{pmatrix} \boldsymbol{z} \\ z_0 \end{pmatrix} = \boldsymbol{0} \\[2ex] \begin{pmatrix} \boldsymbol{z} \\ z_0 \end{pmatrix} \geq \begin{pmatrix} \boldsymbol{0} \\ 0 \end{pmatrix} \\[2ex] \begin{pmatrix} -\boldsymbol{c}^T & r \end{pmatrix} \begin{pmatrix} \boldsymbol{z} \\ z_0 \end{pmatrix} > 0 \end{cases}$$

has no solution. By this corollary we conclude that

$$\begin{pmatrix} \boldsymbol{A}^T \\ -\boldsymbol{b}^T \end{pmatrix} \boldsymbol{y} \geq \begin{pmatrix} -\boldsymbol{c} \\ r \end{pmatrix}, \qquad \boldsymbol{y} \text{ free}$$

has a solution. Thus there is a $\tilde{\boldsymbol{y}}$ such that

$$\boldsymbol{A}^T \tilde{\boldsymbol{y}} \geq -\boldsymbol{c}, \qquad -\boldsymbol{b}^T \tilde{\boldsymbol{y}} \geq r.$$

Putting $\bar{\boldsymbol{y}} = -\tilde{\boldsymbol{y}}$ the lemma is proved. □

Summary

The following picture summarizes the connections between the primal and dual problems.

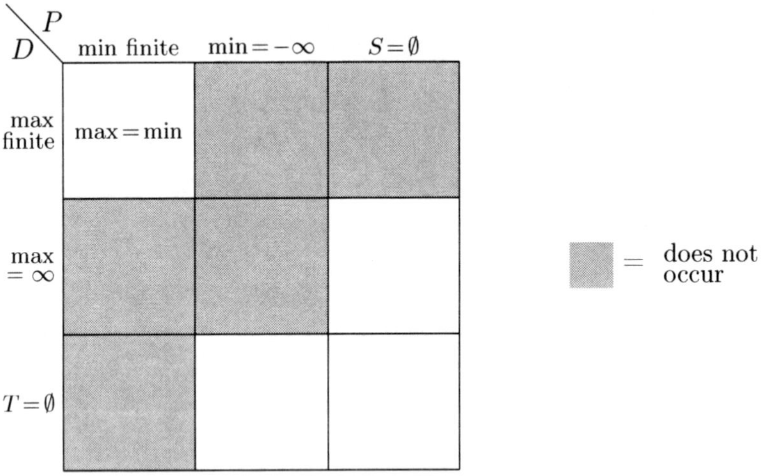

We should provide an example showing that the case $S = \emptyset$, $T = \emptyset$ really might occur. Here is one:

P: Minimize $x_1 + x_2$

subject to
$$\begin{aligned} -x_1 \qquad &\geq 1 \\ -x_2 &\geq 1 \\ x_2 &\geq 0 \end{aligned}$$

D: Maximize $y_1 + y_2$

subject to
$$\begin{aligned} -y_1 \qquad &= 1 \\ -y_2 &\leq 1 \\ y_1 \geq 0,\ y_2 &\geq 0. \end{aligned}$$

Exercises

5.13 a) Write down the dual problem to the following LP problem:

$$\begin{array}{ll} \text{Minimize} & 3x_1 - 2x_2 + x_3 + x_4 \\ \text{subject to} & 2x_1 - x_2 + x_3 - x_4 = 1 \\ & x_1 - 2x_2 - x_3 + 2x_4 \geq -3 \\ & \boldsymbol{x} \geq \boldsymbol{0}. \end{array}$$

b) Solve the dual problem graphically.

c) Use the solution to the dual problem to solve the primal problem.

S 5.14 Use the complementary slackness principle to ascertain if $\hat{\boldsymbol{y}} = (2, 1, 3)$ is an optimal solution to the problem

$$\begin{array}{ll} \text{Maximize} & 4y_1 + 29y_2 + 13y_3 \\ \text{subject to} & y_1 + 2y_2 - y_3 \leq 1 \\ & 3y_1 - y_2 - y_3 \leq 2 \\ & -2y_1 + 4y_2 + 3y_3 \leq 9 \\ & 5y_1 + y_2 - y_3 \leq 9. \end{array}$$

5.15 a) Find the dual of the problem

$$\begin{array}{lll} P: & \text{Maximize} & 4y_1 - 2y_2 - y_3 \\ & \text{subject to} & y_1 + y_2 + y_3 \leq 3 \\ & & 2y_1 + 2y_2 + y_3 \leq 4 \\ & & y_1 - y_2 \leq 0 \\ & & \boldsymbol{y} \geq \boldsymbol{0}. \end{array}$$

b) Show that $\hat{\boldsymbol{y}} = (1, 1, 0)$ is an optimal solution to problem P. Find an optimal solution to the dual problem.

c) Are these optimal solutions unique?

5.16 For what values of c is $\overline{\boldsymbol{x}} = (0, -1, 1, 0)$ an optimal point for the following LP problem?

$$\begin{array}{ll} \text{Minimize} & 2x_1 + x_2 + cx_3 + 4x_4 \\ \text{subject to} & x_1 + x_2 + x_3 + x_4 = 0 \\ & x_1 - x_2 \leq 2 \\ & x_2 - x_3 \leq 1 \\ & x_3 - x_4 \leq 1 \\ & x_1 \geq 0,\ x_2 \text{ free},\ x_3 \geq 0,\ x_4 \geq 0. \end{array}$$

(2010–04–09, 5)

5.4 Examples of some applications

The goal in this section is to give some flavour of LP problems in practice. For a fuller discussion of the topics mentioned the reader should consult other literature.

Example 18. A company makes several different products $P_1, \ldots, P_n$. The profit on product P_j is c_j (€/unit). All products need to be worked at some or all of machines $Q_1, \ldots, Q_m$, where the available capacity of machine Q_i is denoted by b_i (hours). Let a_{ij} (hours/unit) denote the working time required for product P_j at machine Q_i.

Suppose that the company decides to make x_j units of P_j, where the numbers $x_1, \ldots, x_n$ should be chosen so as to maximize the profit. Then the function to be maximized is

$$\sum_{j=1}^{n} c_j x_j.$$

The constraints are $x_j \geq 0$ (of course) and the limitations of machine capacity:

$$\sum_{j=1}^{n} a_{ij} x_j \leq b_i, \qquad i = 1, \ldots, m.$$

The dual of this LP problem is to minimize $\sum_{i=1}^{m} b_i y_i$ when all $y_i \geq 0$ and

$$\sum_{i=1}^{m} a_{ij} y_i \geq c_j, \qquad j = 1, \ldots, n. \tag{19}$$

The dimension of y_i is €/hour, and this variable is connected to machine Q_i. In economic literature, depending on the context, the y_i:s are given names as "fictive prices", "shadow prices", "fictive costs", "margin costs". To understand the meaning of these we consider the situation from the point of view of *the market*. Suppose that the company, instead of working the products P_j at their machines, considers to offer the machine time to the market to use for something else. What value should then be assigned to the machine time?

The objective of the market is to minimize its costs, which are

$$\sum_{i=1}^{m} b_i y_i$$

if y_i denotes the cost per hour of machine Q_i. The constraints (19) state that the y_i must be such that the profit of the company when offering the machines to the market is not less than the profit made when producing the goods P_j.

At optimum

$$\sum_{i=1}^{m} b_i\hat{y}_i = \sum_{j=1}^{n} c_j\hat{x}_j.$$

Note that if there is strict inequality at the optimal point for some j in (19) then by the CSP $\hat{x}_j = 0$, i.e., product P_j is not produced. Correspondingly, assume that at optimum for some i

$$\sum_{j=1}^{n} a_{ij}\hat{x}_j < b_i,$$

i.e., that the corresponding slack variable is strictly positive. Then by the CSP $\hat{y}_i = 0$. The interpretation is that if the capacity of machine Q_i is not fully used then the marginal value of its time is zero. Any price offered by the market for use of the slack time will mean profit to the company. □

Example 19. We are going to give a simple example of so called *sensitivity analysis*. What happens to the optimal solution and the optimal value if some of the coefficients or the right hand side quantities are changed a little? Questions of this kind are of course of great importance in practical problems, where there often is uncertainty in the input data.

We consider an LP problem P in canonical form: minimize $\boldsymbol{c}^T\boldsymbol{x}$ subject to $\boldsymbol{A}\boldsymbol{x} = \boldsymbol{b}$, $\boldsymbol{x} \geq \boldsymbol{0}$. (Notation as before.) The dual problem is to maximize $\boldsymbol{b}^T\boldsymbol{y}$ when $\boldsymbol{A}^T\boldsymbol{y} \leq \boldsymbol{c}$. Let $\hat{\boldsymbol{x}}$ and $\hat{\boldsymbol{y}}$ be optimal solutions. Then $\boldsymbol{c}^T\hat{\boldsymbol{x}} = \boldsymbol{b}^T\hat{\boldsymbol{y}}$ and the variables can be partitioned so that

$$\hat{\boldsymbol{x}} = \begin{pmatrix} \hat{\boldsymbol{x}}_B \\ \boldsymbol{0} \end{pmatrix} = \begin{pmatrix} \boldsymbol{B}^{-1}\boldsymbol{b} \\ \boldsymbol{0} \end{pmatrix}.$$

We assume that $\hat{\boldsymbol{x}}$ is a non-degenerate solution, $\hat{\boldsymbol{x}}_B > \boldsymbol{0}$. The particular question we want to address is: what happens if $\boldsymbol{b}$ is changed to $\boldsymbol{b}' = \boldsymbol{b} + \Delta\boldsymbol{b}$?

Let $\boldsymbol{x}' = (\boldsymbol{x}'_B, \boldsymbol{0})$, with the same partition as before. For $\boldsymbol{x}'$ to be a solution the constraints require that

$$\boldsymbol{B}\boldsymbol{x}'_B = \boldsymbol{b}' \quad \Longleftrightarrow \quad \boldsymbol{x}'_B = \boldsymbol{B}^{-1}(\boldsymbol{b} + \Delta\boldsymbol{b}) = \hat{\boldsymbol{x}}_B + \boldsymbol{B}^{-1}(\Delta\boldsymbol{b}).$$

For $\boldsymbol{x}'$ to be a feasible solution it should also be non-negative. This is equivalent to

$$\hat{\boldsymbol{x}}_B + \boldsymbol{B}^{-1}(\Delta\boldsymbol{b}) \geq \boldsymbol{0},$$

and is certainly satisfied if $\Delta\boldsymbol{b}$ is small enough.[7]

By the CSP, since we assume $\hat{\boldsymbol{x}}_B > \boldsymbol{0}$, we have

$$\boldsymbol{B}^T\hat{\boldsymbol{y}} = \boldsymbol{c}_B, \qquad \boldsymbol{N}^T\hat{\boldsymbol{y}} \leq \boldsymbol{c}_N.$$

If we assume that $\boldsymbol{x}'_B > \boldsymbol{0}$ we can invoke the CSP again and conclude that in fact $\boldsymbol{x}' = (\boldsymbol{x}'_B, \boldsymbol{0})$ and $\hat{\boldsymbol{y}}$ are optimal for the perturbed problem and its dual respectively.

In particular, the perturbation has produced no change in what variables are the basic ones at the minimum point.

Now, by the CSP once more, the optimal value of the perturbed problem is

$$\boldsymbol{c}^T\boldsymbol{x}' = (\boldsymbol{b} + \Delta\boldsymbol{b})^T\hat{\boldsymbol{y}} = \boldsymbol{b}^T\hat{\boldsymbol{y}} + (\Delta\boldsymbol{b})^T\hat{\boldsymbol{y}} = \boldsymbol{c}^T\hat{\boldsymbol{x}} + (\Delta\boldsymbol{b})^T\hat{\boldsymbol{y}}.$$

Thus, the change in the optimal value is

$$(\Delta\boldsymbol{b})^T\hat{\boldsymbol{y}} = \sum_{i=1}^{m} \Delta b_i\, \hat{y}_i.$$

(Compare this result with the interpretation of $\hat{y}_i$ in example 18.) □

Example 20. *(Line fitting).* Suppose some measurements of the values of an unknown function f are made, resulting in the table

x	0	2	5
$f(x)$	0	3	8

[7]The precise limits can be computed if we assume that only one b_i is changed, that is, $\Delta b_j = 0$ if $j \neq i$. Any serious implementation of the simplex method has facilities for such computations.

Using these measurements we wish to find a first degree polynomial $g(x) = kx + l$ which approximates the function f in the sense that it minimizes

$$\max_{x=0,2,5} |f(x) - (kx+l)|.$$

(Note: no squares; this is not the least squares method.)

We obtain linear inequalities by the following reformulation:

$$\max_{x=0,2,5} |f(x) - (kx+l)| = m \iff$$

$$\iff \quad |f(x) - (kx+l)| \le m, \quad x = 0,2,5 \iff$$

$$\iff \quad -m \le f(x) - (kx+l) \le m, \quad x = 0,2,5 \iff$$

$$\iff \quad \begin{cases} kx + l + m \ge f(x), & x = 0,2,5 \\ kx + l - m \le f(x), & x = 0,2,5. \end{cases}$$

We have thus arrived at the following LP problem in the variables k, l, m: find the minimum of m when

$$\begin{cases} l + m \ge 0 \\ l - m \le 0 \\ 2k + l + m \ge 3 \\ 2k + l - m \le 3 \\ 5k + l + m \ge 8 \\ 5k + l - m \le 8 \\ m \ge 0, \; k, l \text{ free.} \end{cases}$$

The results of the measurements have been put in here.

We will not pursue the computations, which are rather tedious and should be left to a computer. □

Exercise

5.17 A manufacturer has the option of producing one or more of four different products using manpower and two different raw materials. The requirements of these for the different products are given in the table below, which also shows the profit per unit for the different items and the total resources per week which are available.

Determine the production levels which maximize the total profit. (We assume that non-integer levels are allowed.)

	Product 1	Product 2	Product 3	Product 4	Amount available
Man-weeks	1	1	1	1	15
Kg of raw-material 1	7	5	3	2	120
Kg of raw-material 2	1	1	2	3	20
Unit profit €	6	5.5	9	8	
Production levels	x_1	x_2	x_3	x_4	

(2009–04–17, 6)

Miscellanous exercises

5.18 For what values of the number b does the following problem have an optimal solution?

$$\begin{array}{ll} \text{Maximize} & 2y_1 + by_2 + y_3 + y_4 \\ \text{subject to} & y_1 + 3y_2 + y_4 \leq 4 \\ & 2y_1 + 2y_2 \leq 3 \\ & 3y_2 + 4y_3 + y_4 \leq 3 \\ & (y_1, y_2, y_3, y_4 \text{ unrestricted}). \end{array}$$

5.19 Consider the problem

$$\begin{array}{ll} \text{Minimize} & kx_1 - 2x_2 + x_3 - x_4 - x_5 \\ \text{subject to} & 2x_1 + 2x_3 + x_5 = 1 \\ & x_1 - x_3 + x_4 = 2 \\ & 2x_1 + x_2 - 3x_3 = 1 \\ & x_i \geq 0 \text{ for all } i. \end{array}$$

a) Solve the problem in the case $k = -1$. (Use x_2, x_4, x_5 as initial basic variables.)

b) For what values of k is your solution optimal?

(1997–04–04, 5)

5.20 A producer of olive oil mixes three different makes of raw oil, A, B and C, in his high quality brand. Of these C has the finest aroma, while A and B are comparable. The mixture must contain at least 30% of C to be acceptable to the consumers. Now A contains 5 units of acid per litre, B contains 2 units and C 15 units per litre. The authorities allow not more that 6 units of acid per litre in olive oil for the consumers.

The producer buys A at the price of 1 € per litre, and B and C for 2 € per litre. The price to the retail dealers is 4 €. What is the maximal profit possible per litre olive oil, neglecting the work and packing cost?

5.21 Consider the LP problem to minimize $-x_1 - x_2 - x_3 - 3x_4 - 3x_5$ subject to

$$\begin{array}{l} x_1 + 2x_2 + 3x_3 + 3x_4 + 5x_5 \leq 7 \\ x_1 + 3x_2 + 4x_3 + 5x_4 + 6x_5 \geq 10 \\ x_2 + x_4 = 2 \\ x_i \geq 0 \text{ for all } i. \end{array}$$

a) Compute a basic feasible solution using x_1, x_2 and x_4 as basic varibles.

b) Solve the problem. Is the solution unique?

c) State the dual problem.

d) Find a solution to the dual problem.

(2008–10–21, 3)

5.22 Using the *Phase I* method, find a feasible point of the set

$$\begin{aligned} 10x_1 \qquad\quad - 5x_3 &\leq 1 \\ 2x_1 - 3x_2 + 2x_3 &\leq 3 \\ 2x_1 + x_2 - 5x_3 &\geq 6 \\ \text{all } \ x_k \geq 0. & \end{aligned}$$

(2006–04–25, 4a)

Chapter 6

CONVEX FUNCTIONS

The subject matter of this chapter is the theory of convex functions. The main result is the fact that a stationary point of a convex function is automatically a global minimum point.

We begin in section 6.1 with a discussion of convex functions of one variable. It will be seen there that convexity has a very clear geometrical meaning. It proves profitable even in the case of several variables to rely on this geometrical interpretation to clarify many of the arguments.

6.1 Convex functions of one variable

Let f be a real-valued function of one variable, defined on an interval I.

DEFINITION 1. *The function f is called* **convex** *if*

$$\begin{cases} a, b \in I \\ 0 < \lambda < 1 \end{cases} \implies f(\lambda a + (1-\lambda)b) \leq \lambda f(a) + (1-\lambda)f(b).$$

It is called **strictly convex** *if*

$$\begin{cases} a, b \in I \\ 0 < \lambda < 1 \end{cases} \implies f(\lambda a + (1-\lambda)b) < \lambda f(a) + (1-\lambda)f(b).$$

The definition has the following important geometrical interpretation. Denote by A the point $(a, f(a))$ and by B the point $(b, f(b))$ on the curve $y = f(x)$. The condition that f is (strictly) convex is equivalent to the condition that the curve between A and B lies (strictly) below the

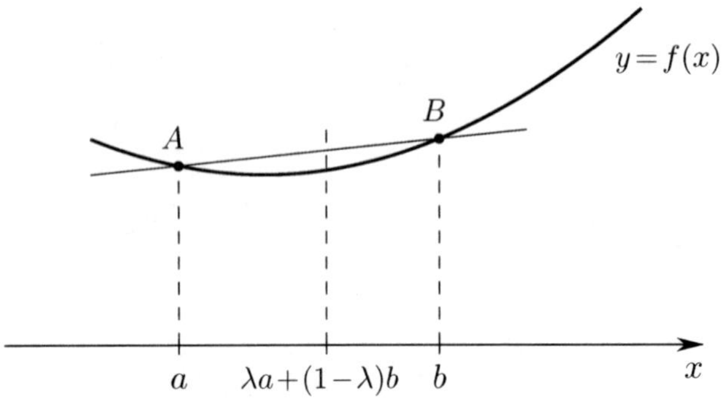

chord, i.e. the straight line, through A and B, for any points A and B on the curve. See the figure above.

Example 1. The function $f(x) = |x|$ is convex. This is a simple consequence of the triangle inequality. If $0 < \lambda < 1$ then

$$f(\lambda a + (1-\lambda)b) = |\lambda a + (1-\lambda)b| \leq \lambda|a| + (1-\lambda)|b| = \lambda f(a) + (1-\lambda)f(b).$$

It is not strictly convex.

It is obvious from the geometrical interpretation that any affine function $f(x) = kx+m$ is convex. □

We leave to the reader as an exercise to prove by induction that if f is convex then

$$f\Big(\sum_1^n \lambda_i a_i\Big) \leq \sum_1^n \lambda_i f(a_i) \tag{1}$$

for any points $a_1, \ldots, a_n$ in I and any numbers $\lambda_1, \ldots, \lambda_n$ that satisfy $\sum_i^n \lambda_i = 1$ and $\lambda_i \geq 0$ for all i.

DEFINITION 2. *The function f is called (strictly)* **concave** *if $-f$ is (strictly) convex.*

LEMMA 1. *Assume that f is convex on the interval I. Then, with notation as above, if $x < a$ or if $x > b$ the curve $y = f(x)$ lies above the chord AB.*

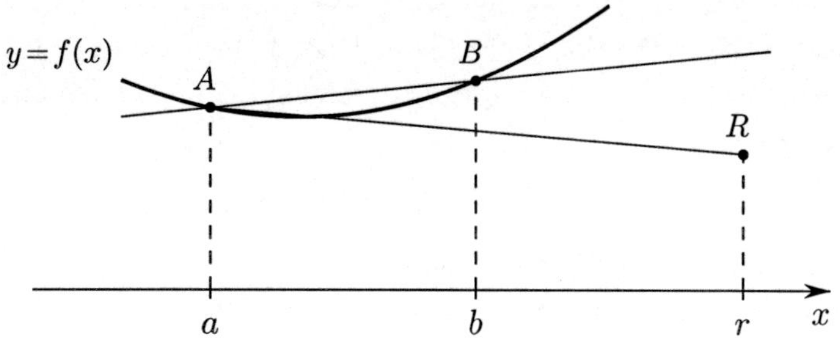

PROOF. Consider the case $x > b$. Assume that there is a point R: $(r, f(r))$ on the curve, $r > b$, which is strictly below the chord AB. Convexity then demands that the curve between a and r lies below the chord AR. However, the point B on the curve gives a contradiction to this. □

Remark. The reader should not believe that the proof uses the figure to draw conclusions. We merely state in geometrical language things that can be proved analytically. Assume that $b = \lambda a + (1-\lambda)r$ with $0 < \lambda < 1$. Computing the equation of the chord AB (a rather tedious but not difficult task) gives the analytic formulation of the assumption that R lies strictly below the chord:

$$f(r) < \frac{1}{1-\lambda} f(b) - \frac{\lambda}{1-\lambda} f(a).$$

The convexity of f implies in particular that

$$f(b) \leq \lambda f(a) + (1-\lambda) f(r).$$

From these two formulas we arrive at the contradiction $f(b) < f(b)$.

In the future we will mostly use geometrical arguments, thus avoiding technical, but in principle not difficult, computations.

COROLLARY 1. *If the convex function f has a local minimum at a point a in the domain I, then f has a global minimum there.*

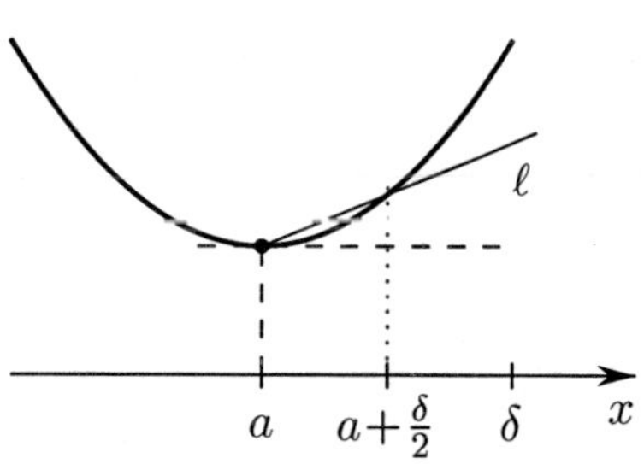

PROOF. First let a be an inner point of the interval. By assumption there is a neighbourhood $a-\delta < x < a+\delta$ such that $f(x) \geq f(a)$ for all x in this neighbourhood. Consider now the chord ℓ through the points on the curve with x-coordinates a and

$a+\frac{\delta}{2}$ respectively. We have $f(a+\frac{\delta}{2}) \geq f(a)$, Thus the slope of this chord is non-negative. By the lemma, the curve is above the chord for $x \geq a+\frac{\delta}{2}$. It follows that

$$f(x) \geq f(a) \quad \text{for all } x \in I \text{ with } x \geq a.$$

The case $x \leq a$ is treated analogously.

Of course this argument also covers the case when a is an endpoint of the interval. □

THEOREM 1. *If the function f is convex on the interval I it is continuous on the interior of I.*

PROOF. Let $a \in \operatorname{int}(I)$ and put $A = (a, f(a))$. Since a is an inner point of I we can consider two chords through A, one to the left and one to the right. The definition of convexity and the lemma imply that there is a right neighbourhood of a where the function curve lies between these two lines. Since they meet at A we have

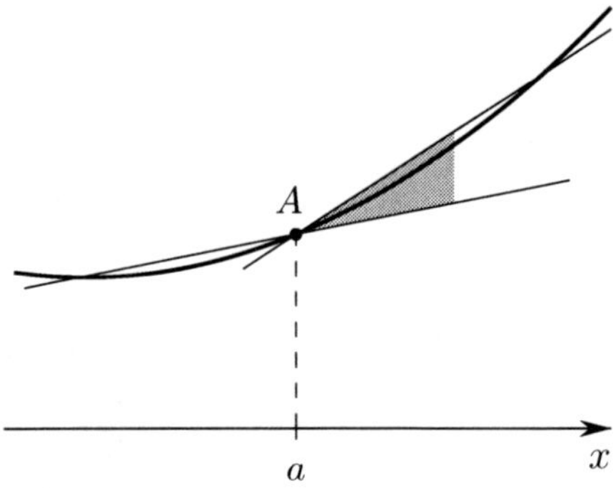

$$\lim_{x \to a^+} f(x) = f(a).$$

That $\lim_{x\to a^-} f(x) = f(a)$ is proved analogously. Thus f is continuous at a. □

Example 2. A convex function is not necessarily continuous at the endpoints of the domain I. The curve may look like the left figure below, however not like the right one.

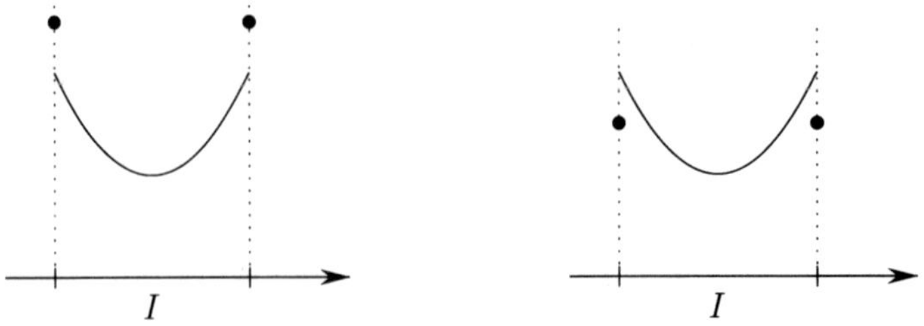

□

We now turn to the question of differentiability of convex functions. We should not expect convex functions to be differentiable everywhere in the domain; in fact $f(x) = |x|$ constitutes an example of a non-differentiable convex function. However, this function is differentiable everywhere except at the origin, where f' has a jump. The following theorem states that this is the worst singularity possible in f'. If the derivative of a convex function is discontinuous at some point then it has a jump there.

THEOREM 2. *Assume that f is convex on I and that a is an interior point of I. Then the following limits exist:*

$$\lim_{x\to a^+} \frac{f(x)-f(a)}{x-a} \stackrel{\text{def}}{=} f'_+(a),$$

$$\lim_{x\to a^-} \frac{f(x)-f(a)}{x-a} \stackrel{\text{def}}{=} f'_-(a).$$

Furthermore, $f'_-(a) \le f'_+(a)$.

PROOF. Consider chords through the point A as in the figure, where $q < a < r < s$.

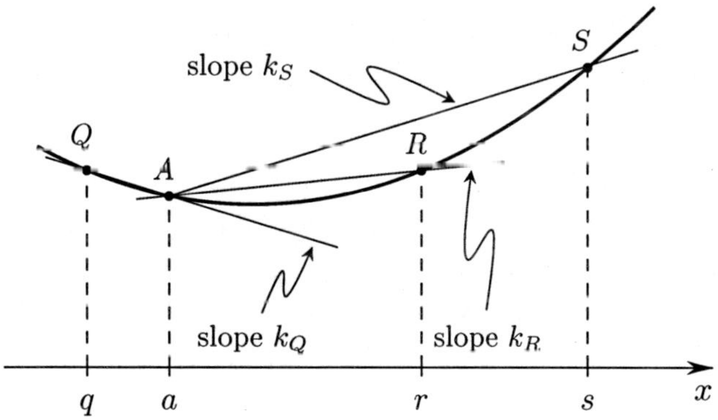

With the notation in the figure the convexity of f implies that the point R lies below the chord AS, which in turn implies the inequality

$$k_R \le k_S \tag{2}$$

for the slopes. Thus, the function

$$x \longmapsto \frac{f(x)-f(a)}{x-a}$$

is increasing for $x > a$. Furthermore it is bounded from below. This follows since by the lemma the curve for $x \geq a$ lies above the line through Q and A, implying that $k_R \geq k_Q$ for all R. These properties prove the existence of the right limit $f'_+(a)$. The existence of the left limit is proved in the same way. Obviously, $f'_-(a) \leq f'_+(a)$. □

Remark. If a is an endpoint of the interval I the limits might be $-\infty$ or $+\infty$. See the left figure of example 2.

THEOREM 3. *Assume that f is convex on I, that a is an interior point of I and that f is differentiable at a. Then the curve $y = f(x)$ lies above its tangent at a.*

PROOF. Consider the figure in the proof of theorem 2 above. As in (2) we have $k_R \leq k_S$. Taking the limit as $r \to a^+$ we get

$$f'(a) \leq k_S. \tag{3}$$

This inequality implies that the curve lies above the tangent for all $x \in I$ with $x > a$. The case $x < a$ is proved analogously. □

We leave as an exercise for the reader to prove the converse statement: *If f is differentiable on I and the function curve lies above all its tangents, then f is convex.*

COROLLARY 2. *Under the same assumptions as in the theorem, if f has a stationary point at a then f has a global minimum at a.*

PROOF. By assumption $f'(a) = 0$. Then the equation of the tangent is $y = f(a)$. Since the curve lies above the tangent

$$f(x) \geq f(a) \quad \text{for all } x \in I.$$

Thus f has a global minimum at a. □

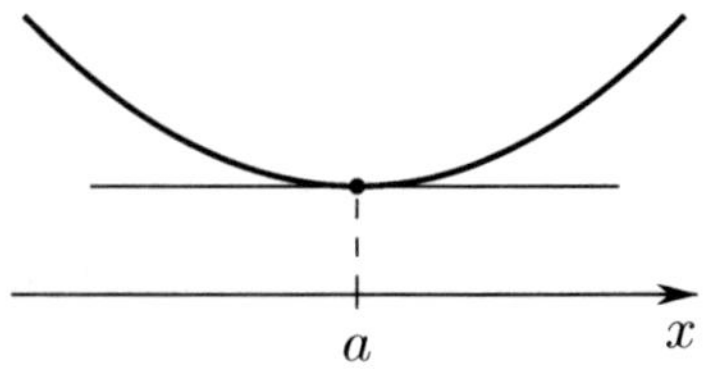

The class of k times differentiable functions on an open interval I, with continuous kth order derivative, is denoted by $C^k(I)$.

THEOREM 4. *Assume that* $f \in C^1(I)$. *Then*

1. f *is convex* $\iff$ f' *is increasing.*

2. f *is concave* $\iff$ f' *is decreasing.*

PROOF. Of course it is enough to prove the first statement.

$\Rightarrow$): Assume that $a < b$, and let C be a point between A and B on the curve. (See the figure.) The same reasoning that led to inequalites (2) and (3) above shows that

$$f'(a) \leq k_1 \leq k_2 \leq f'(b).$$

Thus f' is increasing.

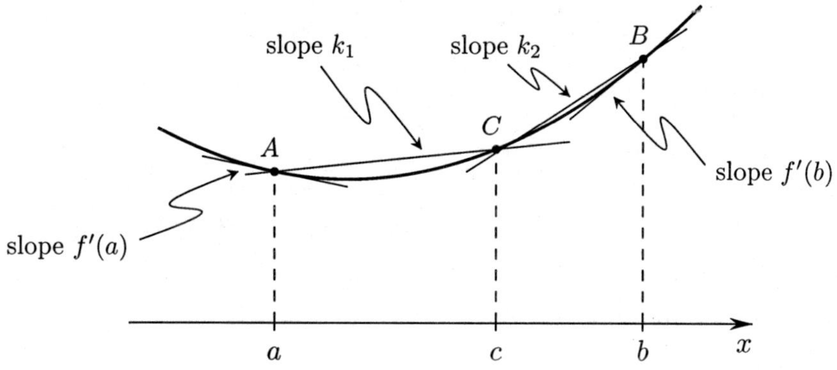

$\Leftarrow$): Now assume that f' is increasing. To prove the convexity we use the definition and the well known mean value theorem from differential calculus.

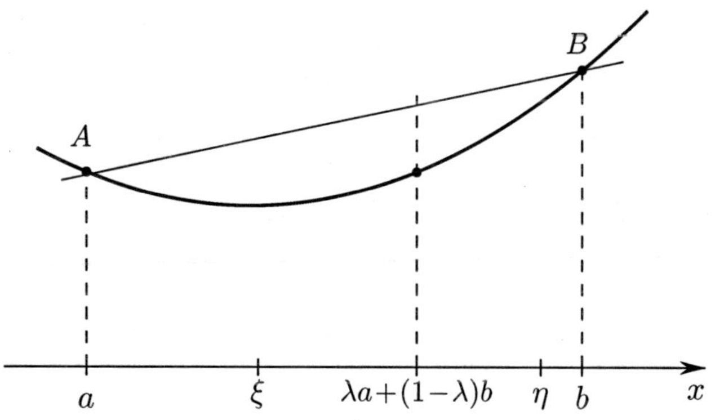

Let a and b be any points in I, with $a < b$, and let $0 < \lambda < 1$. Consider the difference

$$\begin{aligned}
&\lambda f(a) + (1-\lambda)f(b) - f(\lambda a + (1-\lambda)b) = \\
&= \lambda\Big(f(a) - f(\lambda a + (1-\lambda)b)\Big) + (1-\lambda)\Big(f(b) - f(\lambda a + (1-\lambda)b)\Big) = \\
&= \lambda\Big(-(1-\lambda)(b-a)f'(\xi)\Big) + (1-\lambda)\Big(\lambda(b-a)f'(\eta)\Big) = \\
&= \lambda(1-\lambda)(b-a)\Big(f'(\eta) - f'(\xi)\Big),
\end{aligned}$$

where by the mean value theorem ξ and η are points on the x-axis located as indicated in the figure. In particular $\xi < \eta$, which since f' is increasing implies that $f'(\eta) - f'(\xi) \geq 0$. Thus

$$f(\lambda a + (1-\lambda)b) \leq \lambda f(a) + (1-\lambda)f(b),$$

which was to be proven. □

It is left to the reader to check that corresponding statements are valid in the case of strictly convex f and strictly increasing f'.

We now state the theorem that connects convexity with the second order derivative.

THEOREM 5. *Assume that $f \in C^2(I)$. Then*

1. *f is convex on I $\iff$ $f''(x) \geq 0$ for all $x \in I$.*
2. *f is concave on I $\iff$ $f''(x) \leq 0$ for all $x \in I$.*

PROOF. This follows from theorem 4 by the well known correspondence between the monotonocity of a function (f') and the sign of its derivative (f''). □

Example 3. The functions $f(x) = e^x$ and $g(x) = x^4$ are convex on $\mathbf{R}$, since $f''(x) = e^x \geq 0$ and $g''(x) = 12x^2 \geq 0$.

The function $h(x) = \ln x$ is concave on $x > 0$, since $h''(x) = -\dfrac{1}{x^2} \leq 0$. □

Remark. Note that $f''(x) > 0$ implies that f is strictly convex, but that the converse statement is false; the function $g(x) = x^4$ is strictly convex, however $g''(0) = 0$.

DEFINITION 3. *Let $A : (a, f(a))$ be a point on the curve $y = f(x)$ such that f is convex in a left neighbourhood of a and concave in a right neighbourhood of a or vice versa. Then A is called a* **point of inflection** *for the curve, and a is called a point of inflection for the function f.*

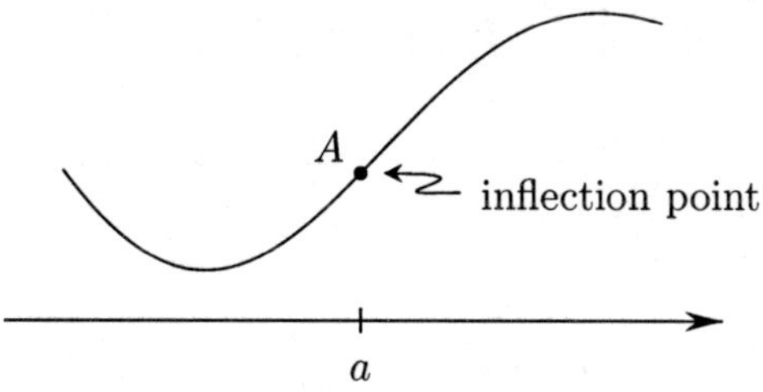

Example 4. The curve $y = x^3$ has an inflection point at the origin. □

Obviously, if a is a point of inflection then $f''(a) = 0$. However, the corresponding statement in the opposite direction is false. The function $f(x) = x^4$ is a counterexample.

We close this section with two examples which are somewhat off the main stream of the course. The first is an important application of the convexity of the exponential function.

Example 5. Let $a_1, \ldots, a_N$ be positive numbers, and put $b_i = \ln a_i$ for $i = 1, \ldots, N$. Using the convexity of the function e^x in the form (1) we get

$$\left(\prod_{i=1}^{N} a_i\right)^{1/N} = \left(\prod_{i=1}^{N} e^{b_i}\right)^{1/N} = \exp\left(\sum_{i=1}^{N} \frac{1}{N} b_i\right) \leq$$

$$\leq \sum_{i=1}^{N} \frac{1}{N} \exp b_i = \frac{1}{N} \sum_{i=1}^{N} a_i.$$

Put

$$A = \frac{1}{N} \sum_{i=1}^{N} a_i \qquad \text{(the } \textit{arithmetic mean} \text{ of } a_1, \ldots, a_N\text{)},$$

$$G = \left(\prod_{i=1}^{N} a_i\right)^{1/N} \qquad \text{(the } \textit{geometric mean} \text{ of } a_1, \ldots, a_N\text{)}.$$

We have proved the well known *arithmetic-geometric inequality*

$$G \leq A.$$

□

In the next example we state and prove the so called *Jensen's inequality*.

Example 6. Let h be an e.g. continuous function on the interval $[0,1]$, and f a convex function on $\mathbf{R}$. (The assumptions can be relaxed.) Prove the inequality

$$f\Big(\int_0^1 h(t)\,dt\Big) \leq \int_0^1 f(h(t))\,dt. \tag{4}$$

Solution: We consider Riemann sums. Let $0 = t_0 < t_1 < \ldots < t_n = 1$ be a subdivision of the interval $[0,1]$, let τ_i be a point in the subinterval $[t_{i-1}, t_i]$, and put $\Delta t_i = t_i - t_{i-1}$, $i = 1, \ldots, n$. Then $\sum_1^n \Delta t_i = 1$, and by the definition of convexity

$$f\Big(\sum\nolimits_1^n h(\tau_i)\,\Delta t_i\Big) \leq \sum_{i=1}^{n} f(h(\tau_i))\,\Delta t_i. \tag{5}$$

To the right is a Riemann sum corresponding to the integral to the right in (4). The argument of f on the left side of (5) is a Riemann sum of the integral $\int_0^1 h(t)\,dt$. Consequently, considering a sequence of subdivisions such that $\max_i \Delta t_i$ tends to zero, in the limit (5) becomes (4), since a convex function is continuous (theorem 1). □

Remark. In the special case when $f(x) = e^x$ (4) takes the form

$$\exp \int_0^1 h(t)\,dt \leq \int_0^1 e^{h(t)}\,dt.$$

Exercises

6.1 Let I be an interval in $\mathbf{R}$, and $f\colon I \to \mathbf{R}$ a function of one variable. Show that f is convex if and only if for any integer $k \geq 2$ the following holds true: $x_1, \ldots, x_k \in I$ implies that

$$f(\textstyle\sum_{j=1}^k \lambda_j x_j) \leq \displaystyle\sum_{j=1}^{k} \lambda_j f(x_j),$$

where $\sum_{j=1}^k \lambda_j = 1$ and $\lambda_j \geq 0$ for $j = 1, \ldots, k$.

6.2 Which of the following functions are convex on $x > 0$ (a is any real number)?

a) $f(x) = \ln x$,

b) $f(x) = -\ln x$,

c) $f(x) = \ln(1 + e^{ax})$,

d) $f(x) = e^{ax}$.

6.3 Over what domain is the function $f(x) = x^4 - 12x^2$ convex?

6.4 Let $f_1, \ldots, f_k \colon \mathbf{R} \to \mathbf{R}$ be convex functions. Consider the function f defined by

$$f(x) = \sum_{j=1}^{k} \alpha_j f_j(x),$$

where $\alpha_j > 0$ for $j = 1, \ldots, k$. Show that f is convex. — State and prove a similar result for concave functions.

6.5 Prove that if f is a differentiable function on an interval I with the property that the curve $y = f(x)$ is above all its tangents, then f is convex.

6.2 Convex functions of several variables

6.2.1 Definition and general observations

Consider a function f defined on a convex set $S \subseteq \mathbf{R}^n$.

DEFINITION 4. *The function f is called **convex** if*

$$\left\{ \begin{array}{l} \boldsymbol{a}, \boldsymbol{b} \in S \\ 0 < \lambda < 1 \end{array} \right. \quad \Longrightarrow \quad f(\lambda \boldsymbol{a} + (1-\lambda)\boldsymbol{b}) \leq \lambda f(\boldsymbol{a}) + (1-\lambda) f(\boldsymbol{b}).$$

We leave to the reader to formulate the definition of a strictly convex function. As in the one variable case a function f is called *concave* if $-f$ is convex.

Remark. The point $\lambda \boldsymbol{a} + (1-\lambda)\boldsymbol{b}$ is on the straight line between the points $\boldsymbol{a}$ and $\boldsymbol{b}$. This explains why it is necessary for f to be defined on a *convex* set S.

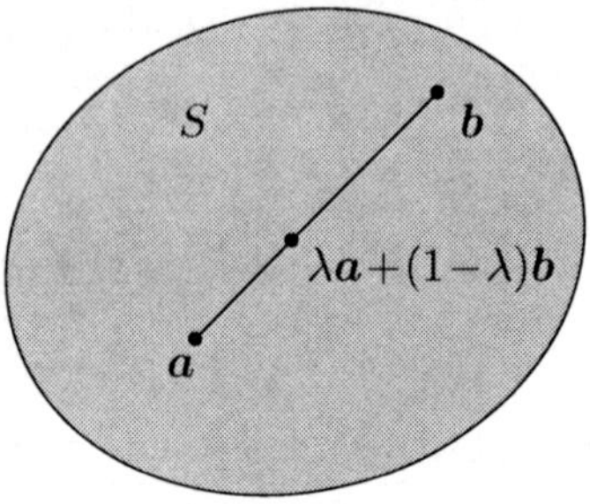

This observation also gives a geometrical interpretation of convexity. Consider the points $A : (\boldsymbol{a}, f(\boldsymbol{a}))$ and $B : (\boldsymbol{b}, f(\boldsymbol{b}))$ on the surface $y = f(\boldsymbol{x})$ in $\mathbf{R}^{n+1}$. The function f is convex if and only if for all $\boldsymbol{a}$ and $\boldsymbol{b}$ in S the surface above the straight line between $\boldsymbol{a}$ and $\boldsymbol{b}$ lies below the chord through A and B.

In fact, $t \mapsto t\boldsymbol{a} + (1-t)\boldsymbol{b} = \boldsymbol{b} + t(\boldsymbol{a}-\boldsymbol{b})$ is a parametrization of the line through $\boldsymbol{a}$ and $\boldsymbol{b}$. Thus, the following theorem is obvious from the definitions.

THEOREM 6. *The function f is convex if and only if the restriction of f to any straight line,*

$$t \longmapsto f(\boldsymbol{x} + t\boldsymbol{d}),$$

is convex.

This theorem will make it possible to deduce many results on convex functions of several variables from the one variable case.

Example 7. A quadratic function $f(\boldsymbol{x}) = \frac{1}{2}\boldsymbol{x}^T \boldsymbol{H}\boldsymbol{x}$ is convex if and only if the matrix $\boldsymbol{H}$ is positive semidefinite.

PROOF. Consider the restriction of f to a line:

$$F(t) = \frac{1}{2}(\boldsymbol{x} + t\boldsymbol{d})^T \boldsymbol{H}(\boldsymbol{x} + t\boldsymbol{d}) = \frac{1}{2}\boldsymbol{x}^T \boldsymbol{H}\boldsymbol{x} + t\boldsymbol{d}^T \boldsymbol{H}\boldsymbol{x} + \frac{1}{2}t^2\boldsymbol{d}^T \boldsymbol{H}\boldsymbol{d}.$$

Differentiating with respect to t gives

$$F'(t) = \boldsymbol{d}^T \boldsymbol{H}\boldsymbol{x} + t\boldsymbol{d}^T \boldsymbol{H}\boldsymbol{d}, \qquad F''(t) = \boldsymbol{d}^T \boldsymbol{H}\boldsymbol{d}.$$

Using theorems 5 and 6 we now have

$$
\begin{aligned}
f \text{ is convex} \quad &\iff \quad F \text{ is convex for all } \boldsymbol{d} \neq \boldsymbol{0} \text{ and all } \boldsymbol{x} \quad \iff \\
&\iff \quad F''(t) \geq 0 \text{ for all } \boldsymbol{d} \neq \boldsymbol{0} \text{ and all } \boldsymbol{x} \quad \iff \\
&\iff \quad \boldsymbol{d}^T \boldsymbol{H} \boldsymbol{d} \geq 0 \text{ for all } \boldsymbol{d} \neq \boldsymbol{0} \quad \iff \\
&\iff \quad \boldsymbol{H} \text{ is positive semidefinite.}
\end{aligned}
$$

Exercise: Prove that if $\boldsymbol{H}$ is positive definite then the quadratic function is strictly convex. What about the reverse statement? □

Example 8. For what values of the number a is the quadratic function defined by

$$
\boldsymbol{H} = \begin{pmatrix} 1 & 2 & 3 \\ 2 & 5 & 7 \\ 3 & 7 & a \end{pmatrix}
$$

convex?

Solution: By example 7 we have to decide those values of a for which $\boldsymbol{H}$ is positive semidefinite. We do that by writing out the quadratic form in full and completing squares. (The factor $\frac{1}{2}$ is skipped.)

$$
\begin{aligned}
\boldsymbol{x}^T \boldsymbol{H} \boldsymbol{x} &= x_1^2 + 5x_2^2 + ax_3^2 + 4x_1x_2 + 6x_1x_3 + 14x_2x_3 = \\
&= (x_1 + 2x_2 + 3x_3)^2 + x_2^2 + (a-9)x_3^2 + 2x_2x_3 = \\
&= (x_1 + 2x_2 + 3x_3)^2 + (x_2 + x_3)^2 + (a-10)x_3^2.
\end{aligned}
$$

Now we see that $\boldsymbol{H}$ is positive semidefinite if and only if $a - 10 \geq 0$, that is, if and only if $a \geq 10$. Hence this is the condition for the quadratic function to be convex. □

Remark. The completing av squares can also be achieved by Gauss elimination in the rows of $\boldsymbol{H}$. See the solution to exercise A.4a, page 403. — Another way of solving the problem is to use Sylvester's criterion in the form of theorem A.8, page 360.

THEOREM 7. *Assume that f is convex on a convex set S in $\mathbf{R}^n$. Then the sets*

$$S_\alpha = \{\, \boldsymbol{x} \in S\,;\, f(\boldsymbol{x}) < \alpha \,\}$$

are convex for all α.

PROOF. Let $\boldsymbol{a}, \boldsymbol{b} \in S_\alpha$, and let $0 < \lambda < 1$. Then $\lambda\boldsymbol{a} + (1-\lambda)\boldsymbol{b}$ is in S_α, since by the convexity of f

$$f(\lambda\boldsymbol{a} + (1-\lambda)\boldsymbol{b}) \le \lambda f(\boldsymbol{a}) + (1-\lambda)f(\boldsymbol{b}) < \lambda\alpha + (1-\lambda)\alpha = \alpha.$$

□

Remark. Note that convexity of all S_α does not imply that f is convex. Give a counterexample by drawing a suitable function curve in the one variable case! (Cf. exercise 6.33.)

The **graph** of a function f is, as is well known, another word for the surface $y = f(\boldsymbol{x})$ in $\mathbf{R}^{n+1}$. We now introduce the **epigraph** of f, which is the set

$$\operatorname{epi}(f) = \{\, (\boldsymbol{x}, y)\,;\, y \ge f(\boldsymbol{x}),\ \boldsymbol{x} \in S \,\}$$

in $\mathbf{R}^{n+1}$.

THEOREM 8. *Assume that S is a convex set in $\mathbf{R}^n$, and that f is a function from S to $\mathbf{R}^n$. Then f is convex if and only if* $\operatorname{epi}(f)$ *is a convex subset of $\mathbf{R}^{n+1}$.*

PROOF. By theorem 6 it is enough to prove the statement in the one variable case.

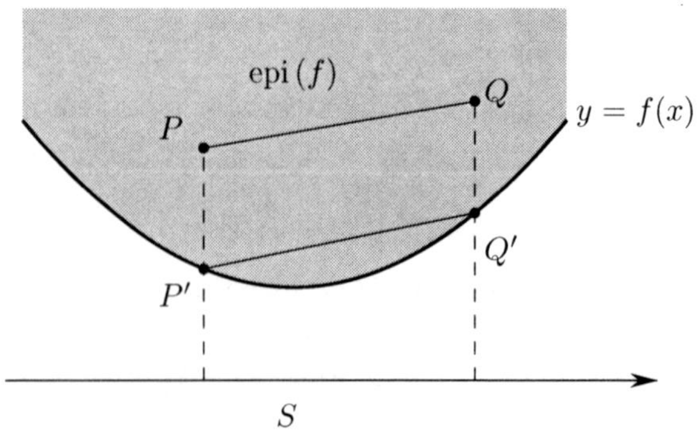

First assume that f is a convex function of one variable. If P and Q are points in epi(f) then (see the figure) let P' and Q' be the corresponding points on the graph. By assumption, the chord $P'Q'$ lies above the curve, hence in the epigraph. It follows that the line segment between P and Q is in the epigraph. By definition, this proves that the epigraph is a convex set.

Conversely, assume that epi(f) is convex. Let P' and Q' be two points on the graph. In particular they belong to the epigraph. Hence the chord $P'Q'$ is in the epigraph. This proves that the curve is below the chord. Then by definition f is convex. □

The following lemma shows how you can build complicated convex functions from simpler ones.

LEMMA 2.

1. *Assume that the functions f_i are convex, $i = 1, \dots, k$, and that $\alpha_i \geq 0$ are real numbers. Then $\sum_{i=1}^{k} \alpha_i f_i$ is a convex function.*

2. *Assume that the functions f_i are convex, $i = 1, \dots, k$. Then $\max_{1 \leq i \leq k} f_i$ is a convex function.*

3. *Assume that $g\colon \mathbf{R}^n \to \mathbf{R}$ is concave. Then $f(\boldsymbol{x}) = \dfrac{1}{g(\boldsymbol{x})}$ is a convex function on $S = \{\, \boldsymbol{x}\,;\, g(\boldsymbol{x}) > 0 \,\}$.*

4. *Assume that $g\colon \mathbf{R} \to \mathbf{R}$ is a nondecreasing convex function, and that the function $h\colon \mathbf{R}^n \to \mathbf{R}$ is convex. Then $f(\boldsymbol{x}) = g(h(\boldsymbol{x}))$ is a convex function.*

5. *Assume that $g\colon \mathbf{R}^m \to \mathbf{R}$ is a convex function and that $\boldsymbol{h}(\boldsymbol{x}) = \boldsymbol{A}\boldsymbol{x} + \boldsymbol{b}$ is affine, where $\boldsymbol{A}$ is of type $m \times n$. Then $f(\boldsymbol{x}) = g(\boldsymbol{h}(\boldsymbol{x}))$ is a convex function.*

PROOF. We prove only statement number 3, leaving the others to the reader as an exercise.

Let $\boldsymbol{x}, \boldsymbol{y} \in S$, and let $0 < \lambda < 1$. Then $\lambda\boldsymbol{x} + (1-\lambda)\boldsymbol{y} \in S$, since S is convex by theorem 7 (because $-g$ is convex). Furthermore

$$f(\lambda\boldsymbol{x} + (1-\lambda)\boldsymbol{y}) = \frac{1}{g(\lambda\boldsymbol{x} + (1-\lambda)\boldsymbol{y})} \leq \frac{1}{\lambda g(\boldsymbol{x}) + (1-\lambda) g(\boldsymbol{y})},$$

where the inequality follows from the fact that g is concave and positive. Using that the function $x \mapsto \dfrac{1}{x}$ is convex for $x > 0$ we now get

$$f(\lambda\boldsymbol{x} + (1-\lambda)\boldsymbol{y}) \le \lambda\frac{1}{g(\boldsymbol{x})} + (1-\lambda)\frac{1}{g(\boldsymbol{y})} = \lambda f(\boldsymbol{x}) + (1-\lambda)f(\boldsymbol{y}).$$

By definition, this proves that f is convex. □

Example 9. a) A general quadratic function $f(\boldsymbol{x}) = \frac{1}{2}\boldsymbol{x}^T\boldsymbol{H}\boldsymbol{x}+\boldsymbol{c}^T\boldsymbol{x}+d$ is convex if $\boldsymbol{H}$ is positive semidefinite, since it is a sum of a convex function (by example 7) and an affine (hence convex) function.

b) If $\boldsymbol{H}$ is positive semidefinite the function $f(\boldsymbol{x}) = e^{\frac{1}{2}\boldsymbol{x}^T\boldsymbol{H}\boldsymbol{x}}$ is convex. This follows from the 4th statement above, since e^x is a nondecreasing convex function on $\mathbf{R}$.

c) The function $f(\boldsymbol{x}) = \dfrac{1}{c_1x_1 + c_2x_2 + d}$ is convex on the half-plane

$$c_1x_1 + c_2x_2 + d > 0.$$

This follows from statement 3 above.

d) If $\boldsymbol{H}$ is positive semidefinite the function

$$f(\boldsymbol{x}) = \frac{1}{2}(\boldsymbol{A}\boldsymbol{x} + \boldsymbol{b})^T\boldsymbol{H}(\boldsymbol{A}\boldsymbol{x} + \boldsymbol{b})$$

is convex. This follows from the last statement in the lemma. □

Example 10. The function

$$f(\boldsymbol{x}) = \sum_{i=1}^{n} x_i \ln x_i,$$

which is connected to the *entropy* concept, is convex on the set S in $\mathbf{R}^n$ defined by $x_i > 0$ for all i. To prove this we first verify that the function of one variable $g(x) = x \ln x$ is convex on $x > 0$. This follows from the sign of its second derivative by theorem 5, since

$$g''(x) = \frac{1}{x} > 0 \qquad \text{when} \quad x > 0.$$

As a consequence the functions

$$S \ni \boldsymbol{x} \longmapsto x_i \ln x_i$$

are convex for $i = 1, \ldots, n$. Our function f is the sum of these functions, hence convex by the first part of the lemma above. □

6.2.2 Continuity

THEOREM 9. *Assume that f is convex on the convex set S in $\mathbf{R}^n$. Then f is continuous on the interior of S.*

The proof is much more difficult than in the one variable case. This can be explained by the fact that convexity is essentially a one dimensional concept (see the geometrical interpretation of the definition), while the notion of continuity of a function of more than one variable is not. The proof should be skipped at first reading.

PROOF. Let $\bar{\boldsymbol{x}}$ be an interior point of S. Since subtracting a constant does not affect neither convexity nor continuity, it is no restriction to assume that $f(\bar{\boldsymbol{x}}) = 0$. Furthermore, by a translation of coordinates we can assume that $\bar{\boldsymbol{x}} = \boldsymbol{0}$. After these simplifications of notation our objective is to prove that $f(\boldsymbol{x}) \to 0$ as $\boldsymbol{x} \to \boldsymbol{0}$.

Since $\boldsymbol{0}$ is supposed to be an interior point of S, there is a (spherical) neighbourhood of $\boldsymbol{0}$ which is contained in S. We assume from now on that $\boldsymbol{x} \neq \boldsymbol{0}$ is in this neighbourhood. Denote by $\boldsymbol{e}_i$, $i = 1, \ldots, n$, the basis vectors, and let $\boldsymbol{x} = \sum_{i=1}^{n} x_i \boldsymbol{e}_i$. We introduce the notation $|\boldsymbol{x}| = \sum_{i=1}^{n} |x_i|$. Consider for a moment the special case when all $x_i \geq 0$. Then

$$\boldsymbol{x} = \sum_{i=1}^{n} \frac{x_i}{|\boldsymbol{x}|} |\boldsymbol{x}| \boldsymbol{e}_i, \tag{6}$$

where $\boldsymbol{x}$ is written as a convex combination of the vectors $|\boldsymbol{x}|\boldsymbol{e}_i$. Thus by the convexity of f

$$f(\boldsymbol{x}) \leq \sum_{i=1}^{n} \frac{x_i}{|\boldsymbol{x}|} f(|\boldsymbol{x}|\boldsymbol{e}_i) \leq \sum_{i=1}^{n} f(|\boldsymbol{x}|\boldsymbol{e}_i). \tag{7}$$

In the general case, when one or more x_i may be negative, we have to introduce some minus signs in (6) to make it a convex combination. Then (7) is replaced by

$$f(\boldsymbol{x}) \leq \sum_{i=1}^{n} f(\delta_i |\boldsymbol{x}| \boldsymbol{e}_i) \tag{8}$$

where $\delta_i = \pm 1$ with the plus sign if $x_i \geq 0$ and the minus sign if $x_i < 0$.

In order to obtain an inequality in the opposite direction we use the convexity of f once more:

$$0 = f(\boldsymbol{0}) = f\Big(\frac{1}{2}\boldsymbol{x} + \frac{1}{2}(-\boldsymbol{x})\Big) \leq \frac{1}{2} f(\boldsymbol{x}) + \frac{1}{2} f(-\boldsymbol{x}).$$

From this inequality we obtain that $f(\boldsymbol{x}) \geq -f(-\boldsymbol{x})$, and by using (8) for $-\boldsymbol{x}$ we get

$$f(\boldsymbol{x}) \geq -\sum_{i=1}^{n} f(-\delta_i|\boldsymbol{x}|\boldsymbol{e}_i). \tag{9}$$

Now, all functions

$$t \longmapsto f(t\delta_i \boldsymbol{e}_i), \qquad i = 1, \ldots, n,$$

are restrictions of f to straight lines, hence convex and hence continuous by the one variable case. Since $|\boldsymbol{x}| \to 0$ as $\boldsymbol{x} \to \boldsymbol{0}$ it follows from (8) and (9) that

$$f(\boldsymbol{x}) \to 0 \quad \text{as } \boldsymbol{x} \to \boldsymbol{0}.$$

Thus the proof is finished. □

6.2.3 Differentiability

Let $\boldsymbol{a}$ be a point in $\mathbf{R}^n$ and let $\boldsymbol{d} \neq \boldsymbol{0}$ be a direction in $\mathbf{R}^n$. The **derivative** at $\boldsymbol{a}$ of a function f in **direction** $\boldsymbol{d}$ is denoted by $f'(\boldsymbol{a}; \boldsymbol{d})$ and is defined as

$$f'(\boldsymbol{a}; \boldsymbol{d}) = \lim_{t \to 0+} \frac{f(\boldsymbol{a} + t\boldsymbol{d}) - f(\boldsymbol{a})}{t}.$$

Putting $F(t) = f(\boldsymbol{a} + t\boldsymbol{d})$, $t \geq 0$, the function F describes the values of f along the ray from $\boldsymbol{a}$ in direction $\boldsymbol{d}$. Obviously

$$f'(\boldsymbol{a}; \boldsymbol{d}) = F'_+(0).$$

In the case of convex f the function F is also convex and $F'_+(0)$ exists by theorem 2, page 201. Thus the following theorem follows directly from the one dimensional case.

THEOREM 10. *Assume that f is convex on the convex set S and that $\boldsymbol{a}$ is an interior point of S. Then the directional derivative $f'(\boldsymbol{a}; \boldsymbol{d})$ exists for all $\boldsymbol{d} \neq \boldsymbol{0}$.*

Note that the existence of a directional derivative in all directions from $\boldsymbol{a}$ does not imply that f is differentiable at $\boldsymbol{a}$.

We are now going to study differentiable convex functions, and go on with the generalization to functions of several variables of the one variable case. The next theorem states that the surface $y = f(\boldsymbol{x})$ everywhere lies above any of its tangent planes.

THEOREM 11. *Assume that f is convex on S and that $\boldsymbol{a}$ is an interior point of S. If f is differentiable at $\boldsymbol{a}$ then*

$$f(\boldsymbol{x}) \geq f(\boldsymbol{a}) + \boldsymbol{\nabla} f(\boldsymbol{a})^T(\boldsymbol{x} - \boldsymbol{a}) \qquad \text{for all } \boldsymbol{x} \in S. \tag{10}$$

PROOF. Consider the restriction of f to the line through $\boldsymbol{x}$ and $\boldsymbol{a}$. Then the theorem follows from the corresponding statement for convex functions of one variable, theorem 3, page 202. □

Exercises:
1) Verify that if f is strictly convex then (10) is valid with strict inequality for $\boldsymbol{x} \neq \boldsymbol{a}$.
2) Formulate and prove a converse statement as in the one variable case.

COROLLARY 1. *Under the same assumptions as in the theorem, if $\boldsymbol{a}$ is a stationary point for f then f has a global minimum at $\boldsymbol{a}$.*

PROOF. If $\boldsymbol{a}$ is a stationary point then $\boldsymbol{\nabla} f(\boldsymbol{a}) = \boldsymbol{0}$ and (10) becomes

$$f(\boldsymbol{x}) \geq f(\boldsymbol{a}) \qquad \text{for all } \boldsymbol{x} \in S.$$

□

Let as before $F(t) = f(\boldsymbol{a} + t\boldsymbol{d})$ denote the restriction of f to the line through $\boldsymbol{a}$ with direction $\boldsymbol{d} \neq \boldsymbol{0}$. Then

$$\begin{aligned} f \text{ convex} \quad &\Longleftrightarrow \quad F \text{ convex (for all } \boldsymbol{a} \in S \text{ and all } \boldsymbol{d} \neq \boldsymbol{0}) \quad \Longleftrightarrow \\ &\Longleftrightarrow \quad F' \text{ increasing} \quad \Longleftrightarrow \\ &\Longleftrightarrow \quad F'(t_2) - F'(t_1) \geq 0 \quad \text{if } t_2 - t_1 \geq 0. \end{aligned}$$

But $F'(t) = \boldsymbol{\nabla} f(\boldsymbol{a} + t\boldsymbol{d})^T\boldsymbol{d}$ by the chain rule. Changing the notation to $\boldsymbol{a} = \boldsymbol{x}_1$, $\boldsymbol{d} = \boldsymbol{x}_2 - \boldsymbol{x}_1$, $t_1 = 0$, $t_2 = 1$ we have proved the following theorem.

THEOREM 12. *Let S be an open convex set and let $f \in \mathcal{C}^1(S)$. Then*

$$f \text{ is convex} \iff \Big(\nabla f(\boldsymbol{x}_2) - \nabla f(\boldsymbol{x}_1)\Big)^T (\boldsymbol{x}_2 - \boldsymbol{x}_1) \geq 0 \text{ for all } \boldsymbol{x}_1, \boldsymbol{x}_2 \in S.$$

Now assume that f is twice differentiable. Then

$$f \text{ convex} \iff F \text{ convex (for all } \boldsymbol{a} \in S \text{ and all } \boldsymbol{d} \neq \boldsymbol{0}) \iff$$

$$\iff F''(t) \geq 0 \quad \text{for all } t.$$

But by the chain rule

$$F''(t) = \boldsymbol{d}^T \boldsymbol{H}(\boldsymbol{a}+t\boldsymbol{d})\, \boldsymbol{d},$$

where $\boldsymbol{H}(\boldsymbol{x})$ denotes the Hessian of $f(\boldsymbol{x})$. Hence

$$f \text{ convex} \iff \boldsymbol{d}^T \boldsymbol{H}(\boldsymbol{a})\boldsymbol{d} \geq 0 \quad \text{for all } \boldsymbol{d} \neq \boldsymbol{0} \text{ and all } \boldsymbol{a} \in S \iff$$

$$\iff \boldsymbol{H}(\boldsymbol{a}) \text{ positive semidefinite for all } \boldsymbol{a} \in S.$$

We have proved

THEOREM 13. *Let S be an open convex set and let $f \in \mathcal{C}^2(S)$. Then f is convex if and only if its Hessian $\boldsymbol{H}(\boldsymbol{x})$ is positive semidefinite for all $\boldsymbol{x} \in S$.*

Example 11. Let

$$f(\boldsymbol{x}) = x_1^2 + 4x_1x_2 + 4x_2^2 + 2x_2^3.$$

Show that f is convex on the set $S : x_2 > 0$.

Solution: Differentiating gives

$$f_1'(\boldsymbol{x}) = 2x_1 + 4x_2, \qquad f_2'(\boldsymbol{x}) = 4x_1 + 8x_2 + 6x_2^2$$

and

$$\boldsymbol{H}(\boldsymbol{x}) = \begin{pmatrix} 2 & 4 \\ 4 & 8+12x_2 \end{pmatrix}.$$

The matrix $\boldsymbol{H}(\boldsymbol{x})$ is positive semidefinite if and only if

$$\boldsymbol{d}^T \boldsymbol{H}(\boldsymbol{x})\boldsymbol{d} \geq 0 \qquad \text{for all } \boldsymbol{d} = \begin{pmatrix} d_1 \\ d_2 \end{pmatrix}.$$

After some (easy) computing we find that this is equivalent to

$$2(d_1 + 2d_2)^2 + 12x_2\, d_2^2 \geq 0 \qquad \text{for all } \boldsymbol{d},$$

which is satisfied if and only if $x_2 \geq 0$.

As an alternative we can check for semidefiniteness using Sylvester's criterion. Since $2 > 0$, the Hessian is positive semidefinite if and only if

$$\begin{vmatrix} 2 & 4 \\ 4 & 8+12x_2 \end{vmatrix} \geq 0 \quad \Longleftrightarrow \quad 24x_2 \geq 0 \quad \Longleftrightarrow \quad x_2 \geq 0.$$

Here we have used theorem A.8, page 360 in the appendix. □

Example 12. Prove that the set

$$S = \{\, (x_1, x_2, x_3) \in \mathbf{R}^3\, ;\, x_1x_2x_3 \geq 1,\ x_1 \geq 0,\ x_2 \geq 0,\ x_3 \geq 0 \,\}$$

is convex.

Solution: S is the epigraph of the function $f(x_1, x_2) = \dfrac{1}{x_1x_2}$ of two variables. By theorem 8 it is enough to show that f is convex. The Hessian of f is

$$\boldsymbol{H}(\boldsymbol{x}) = \begin{pmatrix} \dfrac{2}{x_1^3x_2} & \dfrac{1}{x_1^2x_2^2} \\ \dfrac{1}{x_1^2x_2^2} & \dfrac{2}{x_1x_2^3} \end{pmatrix}.$$

This matrix is positive definite by Sylvester's criterion, since $\dfrac{2}{x_1^3x_2} > 0$ and

$$\begin{vmatrix} \dfrac{2}{x_1^3x_2} & \dfrac{1}{x_1^2x_2^2} \\ \dfrac{1}{x_1^2x_2^2} & \dfrac{2}{x_1x_2^3} \end{vmatrix} = \frac{3}{x_1^4x_2^4} > 0$$

for $x_1, x_2 > 0$. Thus f is convex by theorem 13. □

Remark. The result should not be interpreted as showing that the function of three variables $g(x_1, x_2, x_3) = x_1x_2x_3$ is concave (or convex). In fact it is neither. It is easy to find a line such that the restriction of g to it is convex, and another line where the restriction is concave.

Exercises

6.6 Let $f_1, \dots, f_k \colon \mathbf{R}^n \to \mathbf{R}$ be convex functions. Put

$$f(\boldsymbol{x}) = \max_{1 \le j \le k} f_j(\boldsymbol{x}).$$

Show that f is convex. — State and prove a similar result for concave functions.

S 6.7 Let $h \colon \mathbf{R}^n \to \mathbf{R}$ be a convex function, and let $g \colon \mathbf{R} \to \mathbf{R}$ be a nondecreasing convex function. Show that the composite function defined by $f(\boldsymbol{x}) = g\big(h(\boldsymbol{x})\big)$ is convex.

6.8 Let $g \colon \mathbf{R}^m \to \mathbf{R}$ be a convex function, and let $\boldsymbol{h} \colon \mathbf{R}^n \to \mathbf{R}^m$ be an affine function of the form $\boldsymbol{h}(\boldsymbol{x}) = \boldsymbol{A}\boldsymbol{x} + \boldsymbol{b}$, where $\boldsymbol{A}$ is an $m \times n$-matrix and $\boldsymbol{b}$ is an m-vector. Show that the composite function $f \colon \mathbf{R}^n \to \mathbf{R}$ defined by $f(\boldsymbol{x}) = g\big(\boldsymbol{h}(\boldsymbol{x})\big)$ is a convex function.

6.9 Which of the following functions are convex on $\mathbf{R}^2$? (*Hint:* Avoid differentiation if possible. Work with the definition and general properties of convex functions, e.g. those stated in lemma 2.)

a) $f(x_1, x_2) = x_1^2 + x_2^2$,

b) $f(x_1, x_2) = (x_1^2 + x_2^2)^{1/2}$,

c) $f(x_1, x_2) = \ln(e^{x_1} + e^{x_2})$,

d) $f(x_1, x_2) = x_1 e^{-(x_1 + x_2)}$,

e) $f(x_1, x_2) = -x_1^2 - 5x_2^2 + 2x_1 x_2 + 10x_1 - 10x_2$.

6.10 For what values of a is $f(\boldsymbol{x}) = \frac{1}{2}\boldsymbol{x}^T \boldsymbol{H} \boldsymbol{x}$ convex if

$$\boldsymbol{H} = \begin{pmatrix} 1 & 1 & -2 \\ 1 & 2 & -3 \\ -2 & -3 & a \end{pmatrix}?$$

(1998–12–14, 1c)

S 6.11 Which of the following functions are convex, concave or neither.

a) $f(x, y, z) = e^x + e^{x+2y} + e^{x+2y+3z}$,

b) $g(x, y, z) = xyz + x^2 + y^2 + z^2$.

(2008–01–10, 2a)

6.12 Show that the function of three variables

$$f(\boldsymbol{x}) = x_1^2 + x_2^2 + 3x_3^2 + x_1x_2 + 3x_1x_3 + 3x_2x_3 + x_3^3$$

is convex on the set $x_3 \geq 0$. (1996–12–16, 1b)

6.13 Show that the set

$$\{ (x_1, x_2) \in \mathbf{R}^2 ; \, x_1x_2 \geq 1, \ x_1 > 0 \}$$

is convex by considering it as the epigraph of a suitable function.

6.3 Subgradients

Up to now we have treated the case when the convex functions under consideration are differentiable. However, this is often not the case. In fact, the simple operation of taking the maximum of two differentiable convex functions results in a new convex function (by lemma 2), which in general is not differentiable. For functions of one variable this is illustrated by the figure. The maximum function is not differentiable at the point a. For functions of two variables there may be a whole curve of such points.

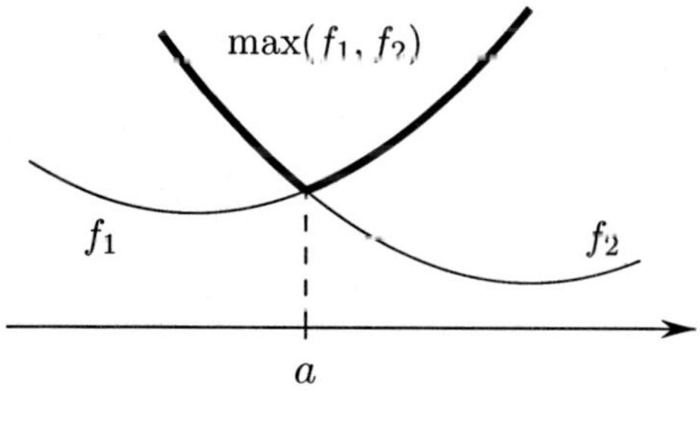

Of course, we know from theorem 9 that a convex function is always continuous.

DEFINITION 5. *Assume that the function f is convex on the convex set S in $\mathbf{R}^n$. The vector $\boldsymbol{\xi} \in \mathbf{R}^n$ is called a* **subgradient** *of f at $\boldsymbol{a} \in S$ if*

$$f(\boldsymbol{x}) \geq f(\boldsymbol{a}) + \boldsymbol{\xi}^T(\boldsymbol{x} - \boldsymbol{a}) \qquad \text{for all } \boldsymbol{x} \in S.$$

If f is differentiable at $\boldsymbol{a}$ then $\boldsymbol{\nabla} f(\boldsymbol{a})$ is a subgradient of f at $\boldsymbol{a}$ by theorem 11; in fact it can be shown to be the only one. We leave as an exercise to show that if $\boldsymbol{\xi} = \boldsymbol{0}$ is a subgradient at $\boldsymbol{a}$ then f has a global minimum at $\boldsymbol{a}$.

The geometrical interpretation of the subgradient is the following: the hyperplane in $\mathbf{R}^{n+1}$

$$y = f(\boldsymbol{a}) + \boldsymbol{\xi}^T(\boldsymbol{x} - \boldsymbol{a})$$

is a *support plane* to the epigraph of f at $(\boldsymbol{a}, f(\boldsymbol{a}))$. Note: This support plane has the normal vector $(\boldsymbol{\xi}, -1)$; $\boldsymbol{\xi}$ itself is a vector in $\mathbf{R}^n$. See the figure.

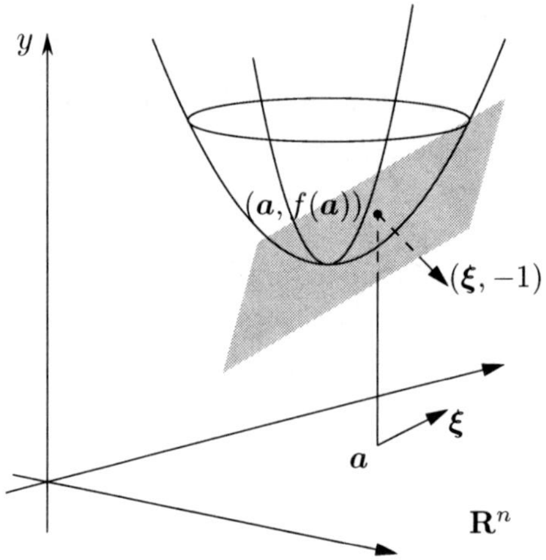

The figure below shows level curves of a nondifferentiable function of two variables. At the point $\boldsymbol{a}$ the set of subgradients might look as shown, or it might reduce to a mere line segment.

Exercise: Show that the set of subgradients at $\boldsymbol{a}$ is a *convex set*.

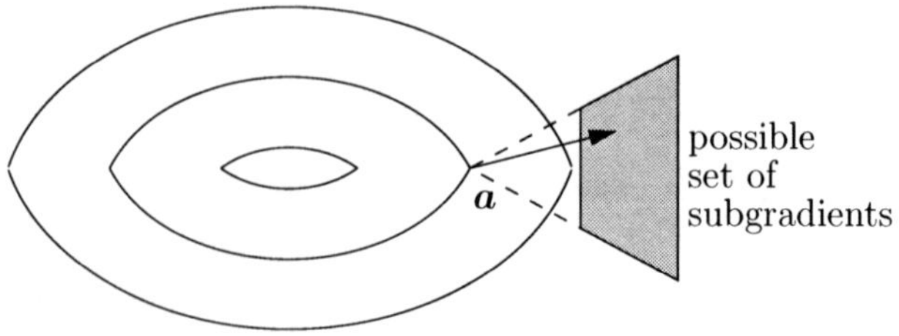

Of course the existence of a subgradient at every point should be proved. This we proceed to do now.

THEOREM 14. *Assume that f is a convex function on S. For any $\boldsymbol{a} \in$ int (S) there is a $\boldsymbol{\xi}$ such that*

$$y = f(\boldsymbol{a}) + \boldsymbol{\xi}^T(\boldsymbol{x} - \boldsymbol{a})$$

is a support plane of epi(f). *In particular, $\boldsymbol{\xi}$ is a subgradient of f at $\boldsymbol{a}$.*

PROOF. We know that epi(f) is a convex set with $(\boldsymbol{a}, f(\boldsymbol{a}))$ as a boundary point. Then by theorem 4.8, page 134, there is a support plane at this point, with equation say

$$\boldsymbol{\xi}_0^T(\boldsymbol{x} - \boldsymbol{a}) + \mu(y - f(\boldsymbol{a})) = 0, \qquad (\boldsymbol{\xi}_0, \mu) \neq (\boldsymbol{0}, 0).$$

We can choose the signs so that

$$\boldsymbol{\xi}_0^T(\boldsymbol{x} - \boldsymbol{a}) + \mu(y - f(\boldsymbol{a})) \geq 0 \qquad \text{for } (\boldsymbol{x}, y) \in \text{epi}(f). \tag{11}$$

What we need to prove is that $\mu > 0$, because then $\boldsymbol{\xi} = -\dfrac{1}{\mu}\boldsymbol{\xi}_0$ is a subgradient at $\boldsymbol{a}$.

That $\mu < 0$ is impossible can be seen at once from (11), since we get at contradiction by choosing y large enough. Suppose that $\mu = 0$. Then

$$\boldsymbol{\xi}_0^T(\boldsymbol{x} - \boldsymbol{a}) \geq 0 \qquad \text{for all } \boldsymbol{x} \in S.$$

However, since $\boldsymbol{a}$ is an inner point in S, $\boldsymbol{x}$ can be chosen a small step in direction $-\boldsymbol{\xi}_0$ from $\boldsymbol{a}$. This means making $\boldsymbol{x} - \boldsymbol{a}$ a positive multiple of $-\boldsymbol{\xi}_0$. Thus we get

$$-\boldsymbol{\xi}_0^T\boldsymbol{\xi}_0 \geq 0,$$

i.e., $\boldsymbol{\xi}_0 = \boldsymbol{0}$. Now $(\boldsymbol{\xi}_0, \mu) = (\boldsymbol{0}, 0)$, and we have a contradiction. — Consequently $\mu > 0$ is the only possibility, and the proof is finished. □

We also have a result in the opposite direction.

THEOREM 15. *Assume that S is a convex set in $\mathbf{R}^n$, and that f is a real-valued function on S. Suppose that for all $\boldsymbol{a} \in \operatorname{int}(S)$ there is a $\boldsymbol{\xi}$ such that*

$$f(\boldsymbol{x}) \geq f(\boldsymbol{a}) + \boldsymbol{\xi}^T(\boldsymbol{x} - \boldsymbol{a}) \qquad \text{for all } \boldsymbol{x} \in S. \tag{12}$$

Then f is convex on $\operatorname{int}(S)$.

PROOF. Let $\boldsymbol{b}, \boldsymbol{c} \in \operatorname{int}(S)$. As usual it is enough to consider the restriction of f to the line between $\boldsymbol{b}$ and $\boldsymbol{c}$. The points $B : (\boldsymbol{b}, f(\boldsymbol{b}))$ and $C : (\boldsymbol{c}, f(\boldsymbol{c}))$ lie on the graph. Let $0 < \lambda < 1$ and put $\boldsymbol{a} = \lambda \boldsymbol{b} + (1-\lambda)\boldsymbol{c}$. We want to prove that $(\boldsymbol{a}, f(\boldsymbol{a})$ is below the chord BC.

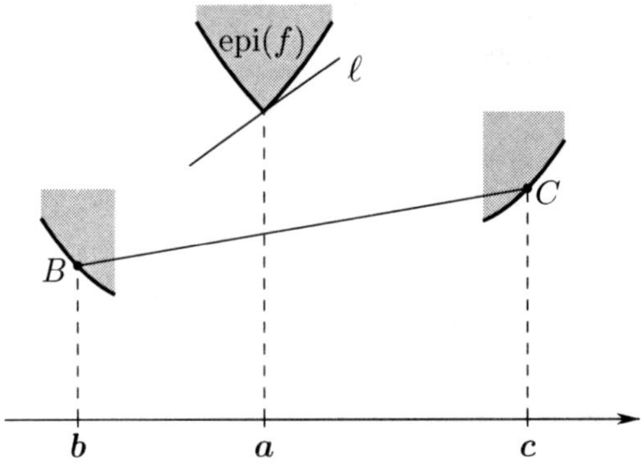

Suppose otherwise. (See the figure.) By assumption there is a support line ℓ of $\operatorname{epi}(f)$ at the point $(\boldsymbol{a}, f(\boldsymbol{a}))$. In particular, by (12) the points B and C are both above ℓ, which is clearly impossible. Thus the theorem is proved. □

Exercises

6.14 Let f be a convex function on $\mathbf{R}^n$. Show that the set of subgradients of f at a fixed point $\boldsymbol{a}$ is a closed convex set.

6.15 Let X be a polyhedral set, and define the function g by the following optimization problem:

$$g(\boldsymbol{u}) = \min_{x \in X} \; \boldsymbol{c}^T\boldsymbol{x} + \boldsymbol{u}^T(\boldsymbol{A}\boldsymbol{x} - \boldsymbol{b}).$$

a) Show that g is concave.

b) Give a description of the subgradients of g at any given $\bar{\boldsymbol{u}}$.

6.16 In exercise 6.15, find the function g explicitly and describe the set of subgradients at each point $\boldsymbol{u} \geq \boldsymbol{0}$ if

$$\boldsymbol{A} = \begin{pmatrix} 1 & 1 \\ -1 & 2 \end{pmatrix}, \quad \boldsymbol{b} = \begin{pmatrix} 3 \\ 4 \end{pmatrix}, \quad \boldsymbol{c} = \begin{pmatrix} -1 \\ -3 \end{pmatrix},$$

and

$$X = \{ (x_1, x_2) \,;\, 0 \leq x_1 \leq 2,\, 0 \leq x_2 \leq \tfrac{3}{2} \}.$$

6.4 Optimization of convex functions

6.4.1 More on minimization

For the convenience of the reader we review some definitions. Let f be a real-valued function on (the whole of) $\mathbf{R}^n$, and let S be a subset of[1] $\mathbf{R}^n$. Consider the problem

$$\min_{\boldsymbol{x} \in S} f(\boldsymbol{x}). \tag{13}$$

No assumption of convexity is made at the moment, neither of f nor S.

DEFINITION 6.
a) The function f has a **(global) minimum** *at $\boldsymbol{a} \in S$ if*

$$f(\boldsymbol{x}) \geq f(\boldsymbol{a}) \qquad \text{for all } \boldsymbol{x} \in S.$$

b) The function f has a **local minimum** *at $\boldsymbol{a} \in S$ if there exists a neighbourhood $B_\delta(\boldsymbol{a})$ of $\boldsymbol{a}$ such that*

$$f(\boldsymbol{x}) \geq f(\boldsymbol{a}) \qquad \text{for all } \boldsymbol{x} \in B_\delta(\boldsymbol{a}) \cap S.$$

Any point $\boldsymbol{a}$ in S is called a *feasible solution* to the problem (13). It is called a *global/local optimal solution* if f has a global/local minimum at $\boldsymbol{a}$.

We talk about a *strict* global/local minimum at $\boldsymbol{a}$ if there is strict inequality above when $\boldsymbol{x} \neq \boldsymbol{a}$.

In the case when f is convex on a convex set S, (13) is called a *convex programming problem*. It is to those that we now turn our attention.

[1] It is of course sufficient that f is defined on a set X such that $\operatorname{int}(X) \supseteq S$.

THEOREM 16. *Assume that $\boldsymbol{a}$ is a local optimal solution to (13), where $f: S \to \mathbf{R}^n$ is convex. Then*

1° $\boldsymbol{a}$ is a global optimal solution.

2° if $\boldsymbol{a}$ is a strict local minimum then the global optimal solution $\boldsymbol{a}$ is unique.

3° if f is strictly convex then the global optimal solution $\boldsymbol{a}$ is unique.

PROOF. 1°. Assume that there is a $\boldsymbol{b} \in S$ such that $f(\boldsymbol{b}) < f(\boldsymbol{a})$. Consider the values of f on the line between $\boldsymbol{a}$ and $\boldsymbol{b}$.

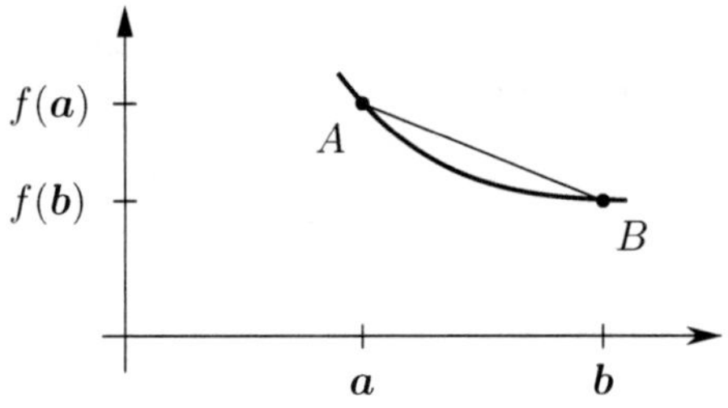

The convexity of f implies that the curve between $\boldsymbol{a}$ and $\boldsymbol{b}$ lies below the chord AB (see the figure). From this it follows that f has no local minimum at $\boldsymbol{a}$, and we have a contradiction.

2° and 3° are proved analogously to 1°. The details are left to the reader. □

Remark. In the case of f differentiable and $\boldsymbol{a}$ an inner point of S, the statements of the theorem are already well-known.

We now state a necessary and sufficient condition for a convex function to have a global minimum at a point.

THEOREM 17. *Assume that f is a convex real-valued function on $\mathbf{R}^n$, and that S is a (nonempty) convex subset of $\mathbf{R}^n$. Then $\boldsymbol{a} \in S$ is an optimal solution to the problem (13) if and only if there is a subgradient $\boldsymbol{\xi}$ at $\boldsymbol{a}$ such that*

(14) $$\boldsymbol{\xi}^T(\boldsymbol{x} - \boldsymbol{a}) \geq 0 \qquad \textit{for all } \boldsymbol{x} \in S.$$

PROOF. Assume that there is a subgradient $\boldsymbol{\xi}$ satisfying (14). By the definition of subgradient we then have

$$f(\boldsymbol{x}) \geq f(\boldsymbol{a}) + \boldsymbol{\xi}^T(\boldsymbol{x} - \boldsymbol{a}) \geq f(\boldsymbol{a}) \qquad \text{for all } \boldsymbol{x} \in S.$$

Thus f has a global minimum at $\boldsymbol{a}$.

In order to prove the converse we first simplify the notation by letting $\boldsymbol{a} = \boldsymbol{0}$ and $f(\boldsymbol{a}) = 0$. This is of course no restriction, since it only requires a translation of coordinates.

Now introduce the following subsets of $\mathbf{R}^{n+1}$:

$$S_1 = \{ (\boldsymbol{x}, y) \,;\, \boldsymbol{x} \in \mathbf{R}^n,\ y > f(\boldsymbol{x}) \},$$
$$S_2 = \{ (\boldsymbol{x}, y) \,;\, \boldsymbol{x} \in S,\ y \leq 0 \}.$$

The reader should convince himself that these sets are convex; furthermore they are disjoint since by assumption $f(\boldsymbol{0}) = 0$ is a global minimum of f in S. As a consequence they have a separating hyperplane[2]

$$\boldsymbol{\xi}_0^T \boldsymbol{x} + \mu y = \alpha,$$

where $(\boldsymbol{\xi}_0, \mu) \neq (\boldsymbol{0}, 0)$. We can choose signs so that

$$\boldsymbol{\xi}_0^T \boldsymbol{x} + \mu y \leq \alpha \qquad \text{for } (\boldsymbol{x}, y) \in S_1, \tag{15}$$

$$\boldsymbol{\xi}_0^T \boldsymbol{x} + \mu y \geq \alpha \qquad \text{for } (\boldsymbol{x}, y) \in S_2. \tag{16}$$

We are going to prove that $\mu < 0$ and $\alpha = 0$. Assuming this for a moment and letting $\boldsymbol{\xi} = -\dfrac{1}{\mu}\boldsymbol{\xi}_0$, (15) shows that

$$\boldsymbol{\xi}^T \boldsymbol{x} \leq y \qquad \text{whenever } \boldsymbol{x} \in \mathbf{R}^n \text{ and } y > f(\boldsymbol{x}).$$

This of course implies that

$$f(\boldsymbol{x}) \geq \boldsymbol{\xi}^T \boldsymbol{x} \qquad \text{for all } \boldsymbol{x} \in \mathbf{R}^n,$$

i.e., that $\boldsymbol{\xi}$ is a subgradient of f at $\boldsymbol{0}$. Also, putting $y = 0$ in (16) shows that $\boldsymbol{\xi}^T \boldsymbol{x} \geq 0$ for all $\boldsymbol{x} \in S$, so that (14) is satisfied.

We now proceed to the proof that $\mu < 0$ and $\alpha = 0$. Putting $\boldsymbol{x} = \boldsymbol{0}$ and $y = \eta$ in (15) yields

$$\mu\eta \leq \alpha \qquad \text{for all } \eta > 0.$$

---

[2]Here we must use a modification of our earlier theorems, since the set S_1 is not closed. We will not go into details, but note that there is weak inequality only in (15) and (16).

This implies that $\mu \leq 0$ and $\alpha \geq 0$. Putting $\boldsymbol{x} = \boldsymbol{0}$ and $y = 0$ in (16) yields $\alpha \leq 0$. Hence $\alpha = 0$. Suppose now that $\mu = 0$. Then (15) takes the form

$$\boldsymbol{\xi}_0^T \boldsymbol{x} \leq 0 \qquad \text{for all } \boldsymbol{x} \in \mathbf{R}^n.$$

In particular, letting $\boldsymbol{x} = \boldsymbol{\xi}_0$ leads to $\|\boldsymbol{\xi}_0\|^2 \leq 0$, i.e., to $\boldsymbol{\xi}_0 = \boldsymbol{0}$. Since $(\boldsymbol{\xi}_0, \mu) \neq (\boldsymbol{0}, 0)$ by assumption, we have arrived at a contradiction. Thus, $\mu < 0$ and $\alpha = 0$. The proof is finished. □

COROLLARY 1. *Assume beside the other conditions in the theorem that f is differentiable at $\boldsymbol{a}$. Then $\boldsymbol{a}$ is an optimal solution to the problem (13) if and only if*

$$\nabla f(\boldsymbol{a})^T (\boldsymbol{x} - \boldsymbol{a}) \geq 0 \qquad \textit{for all } \boldsymbol{x} \in S.$$

The corollary follows immediately from the theorem, since the only subgradient at a point $\boldsymbol{a}$ where f is differentiable is the gradient $\nabla f(\boldsymbol{a})$. However, it is easy to give a direct proof of the corollary without resorting to subgradients. Note that the directional derivative in direction $\boldsymbol{d}$ is given by

$$f'(\boldsymbol{a}; \boldsymbol{d}) = \nabla f(\boldsymbol{a})^T \boldsymbol{d}.$$

Thus what we need to prove is that $\boldsymbol{a}$ is optimal if and only if $f'(\boldsymbol{a}; \boldsymbol{d}) \geq 0$ for any direction that points into the set S. Then we are back at the one variable case; to prove that a is an optimal solution to the problem of minimizing f over an interval $[a, b]$ if and only if the right derivative at a is greater than or equal to 0. This last statement is almost trivial. The details are left to the reader.

The following figure gives a geometrical interpretation of the statement in the corollary.

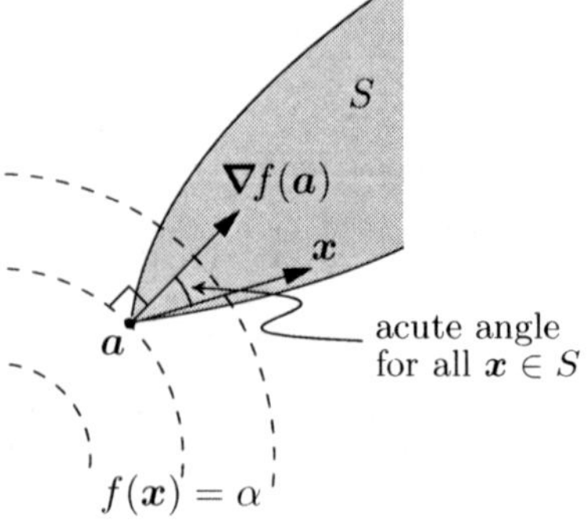

As a particular case of the corollary we obtain the corollary of theorem 11, page 215, that if $\boldsymbol{a}$ is an interior point of S then

$$\boldsymbol{a} \text{ is an optimal solution} \quad \Longleftrightarrow \quad \nabla f(\boldsymbol{a}) = \boldsymbol{0}.$$

Example 13. Consider the function $f(\boldsymbol{x}) = 4x_1 + 3x_2$ in the first quadrant $S\colon x_1 \geq 0, x_2 \geq 0$. Of course f is convex. At every point in S the only subgradient is

$$\nabla f(\boldsymbol{x}) = \begin{pmatrix} 4 \\ 3 \end{pmatrix}.$$

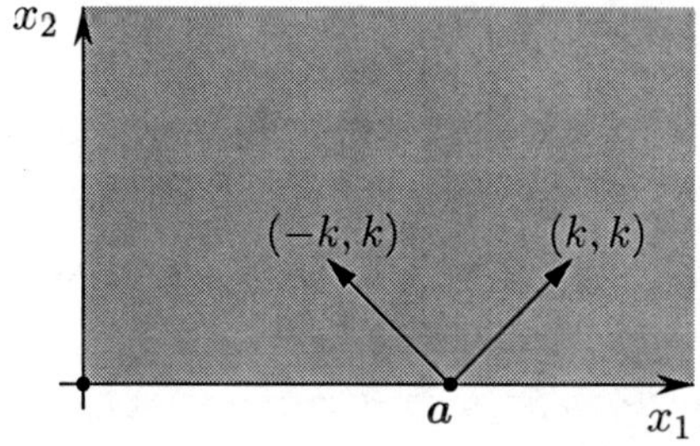

Assume first that $\boldsymbol{a} = (a_1, 0)$ is a point on the x_1-axis. We then have

$$\nabla f(\boldsymbol{a})^T(\boldsymbol{x} - \boldsymbol{a}) = \begin{pmatrix} 4 & 3 \end{pmatrix} \begin{pmatrix} x_1 - a_1 \\ x_2 - 0 \end{pmatrix} = 4(x_1 - a_1) + 3x_2.$$

This expression is positive e.g. if $x_1 - a_1 = k$, $x_2 = k > 0$, and is negative if $x_1 - a_1 = -k < 0$, $x_2 = k > 0$. The condition of the corollary is not satisfied, and f has no minimum at $\boldsymbol{a}$.

Now assume that $\boldsymbol{a} = \boldsymbol{0}$. Then

$$\nabla f(\boldsymbol{a})^T(\boldsymbol{x} - \boldsymbol{a}) = \begin{pmatrix} 4 & 3 \end{pmatrix} \begin{pmatrix} x_1 - 0 \\ x_2 - 0 \end{pmatrix} = 4x_1 + 3x_2 \geq 0$$

for all $x_1 \geq 0$, $x_2 \geq 0$.

Hence by the corollary f has a minimum at $\boldsymbol{0}$. □

6.4.2 Maximization

Maximizing a convex function is a much more difficult enterprise than minimizing it. Not even in the one variable case is it from the behaviour of the function near a local maximum possible to decide whether this point is in fact a *global* maximum. E.g. for the function in the figure you can not from the behaviour near a draw any conclusions as to whether the value of f at b is larger or smaller than those near a.

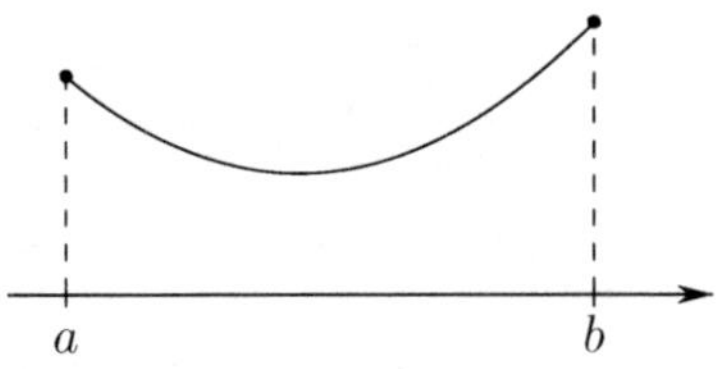

We now state a necessary (but not sufficient) condition for a convex function to have a *local* maximum at a point $\boldsymbol{a}$.

THEOREM 18. *Assume that f is a convex real-valued function on $\mathbf{R}^n$, and that S is a (nonempty) convex subset of $\mathbf{R}^n$. Consider the problem to maximize f subject to $\boldsymbol{x} \in S$. If $\boldsymbol{a}$ is a local optimal solution to this problem, then for any subgradient $\boldsymbol{\xi}$ at $\boldsymbol{a}$*

$$\boldsymbol{\xi}^T(\boldsymbol{x} - \boldsymbol{a}) \leq 0 \qquad \text{for all } \boldsymbol{x} \in S. \tag{17}$$

PROOF. Let $\boldsymbol{x} \in S$ and put $\boldsymbol{x}_\lambda = \lambda \boldsymbol{x} + (1-\lambda)\boldsymbol{a}$. By the convexity of S $\boldsymbol{x}_\lambda$ belongs to S for $0 < \lambda < 1$, and since $\boldsymbol{a}$ is a local point of maximum

$$f(\boldsymbol{x}_\lambda) \leq f(\boldsymbol{a}) \qquad \text{if } \lambda \text{ is sufficiently close to } 0.$$

By the definition of a subgradient

$$f(\boldsymbol{x}_\lambda) \geq f(\boldsymbol{a}) + \boldsymbol{\xi}^T(\boldsymbol{x}_\lambda - \boldsymbol{a}).$$

Combining these inequalities leads to

$$\boldsymbol{\xi}^T(\boldsymbol{x}_\lambda - \boldsymbol{a}) \leq 0 \quad \text{or} \quad \lambda\boldsymbol{\xi}^T(\boldsymbol{x} - \boldsymbol{a}) \leq 0.$$

Dividing by λ we arrive at (17). □

The reader should provide an example to show that the converse of the theorem is false.

COROLLARY 2. *Under the assumptions of theorem 18 and also assuming that f is differentiable at $\boldsymbol{a}$, then*

$$\nabla f(\boldsymbol{a})^T(\boldsymbol{x}-\boldsymbol{a}) \le 0 \qquad \text{for all } \boldsymbol{x} \in S.$$

We close this section with a result which generalizes to convex objective functions the usual LP problem of maximizing a linear function over a polyhedral set.

THEOREM 19. *Assume that f is a convex function on $\mathbf{R}^n$, and that S is a compact polyhedral set. Then there exists an extreme point of S at which f attains its maximum over S.*

PROOF. The convex function f is continuous on S by theorem 9, page 213. Thus, since S is compact, there exist one or several points at which f has its maximum. It remains to prove that at least one of these is an extreme point of S.

If a certain maximum point $\bar{\boldsymbol{x}}$ is not itself an extreme point, then by theorem 5.1, page 151, it can at least be written as a convex combination of the extreme points of S, $\boldsymbol{x}_1, \ldots, \boldsymbol{x}_N$:

$$\bar{\boldsymbol{x}} = \sum_{i=1}^{N} \lambda_i \boldsymbol{x}_i \qquad \text{where } \sum_{i=1}^{N} \lambda_i = 1 \quad \text{and all } \lambda_i > 0.$$

From this it follows that

$$f(\bar{\boldsymbol{x}}) = f\Big(\sum_{i=1}^{N} \lambda_i \boldsymbol{x}_i\Big) \le \sum_{i=1}^{N} \lambda_i f(\boldsymbol{x}_i) \le \sum_{i=1}^{N} \lambda_i f(\bar{\boldsymbol{x}}) = \Big(\sum_{i=1}^{N} \lambda_i\Big) f(\bar{\boldsymbol{x}}) = f(\bar{\boldsymbol{x}}),$$

where the first inequality is due to the convexity of f and the second one to the optimality of $\bar{\boldsymbol{x}}$. Obviously there must be equality everywhere above. In particular,

$$f(\boldsymbol{x}_i) = f(\bar{\boldsymbol{x}}) \qquad \text{for those } i \text{ for which } \lambda_i \neq 0,$$

implying that these extreme points $\boldsymbol{x}_i$ are optimal solutions. □

Exercises

S 6.17 Show that

$$f(x,y) = e^x + x^2 - 2xy + 2y^2 + x - 4y + 8$$

has its minimum at $(0,1)$.

6.18 For the problem

$$\begin{aligned} \text{Minimize} \quad & (x_1-5)^2 + (x_2-8)^2 \\ \text{subject to} \quad & x_2 \geq x_1^2 \\ & x_2 \leq 4, \end{aligned}$$

and the point $\boldsymbol{a} = (2,4)$, show that for all feasible points $\boldsymbol{x}$

$$\boldsymbol{\nabla} f(\boldsymbol{a})^T(\boldsymbol{x} - \boldsymbol{a}) \geq 0.$$

Hence conclude that $\boldsymbol{a}$ is an optimal solution.

6.19 Find an optimal solution to the following problem.

$$\begin{aligned} \text{Minimize} \quad & x^2 + y^2 - x - 3 \\ \text{subject to} \quad & 2x^2 + y^2 \leq 1. \end{aligned}$$

6.20 a) Let

$$f(x_1,x_2) = be^{ax_1^2+2x_1x_2+2x_2^2} + a^2x_1^2 + (13-20a)x_1 - 8x_2.$$

Prove that f is convex if $a \geq \frac{1}{2}$ and $b \geq 0$.

b) Find those values of $a \geq \frac{1}{2}$ and $b \geq 0$ for which f has a minimum at $(2,1)$.

(1999–04–09, 3bc)

Miscellanous exercises

S 6.21 Consider the linear least squares problem to minimize

$$f(\boldsymbol{x}) = \|\boldsymbol{A}\boldsymbol{x} - \boldsymbol{b}\|^2,$$

where $\boldsymbol{A}$ is an $m \times n$ matrix and $\boldsymbol{b}$ is an m-vector. (Cf. subsection 3.6.2, page 92.)

a) Find necessary and sufficient conditions for a minimum by first showing that f is convex.

b) Specify some additional assumption which makes the minimum point unique.

6.22 Show that the following set S is convex:

$$S = \{\, (x_1, \ldots, x_n) \in \mathbf{R}^n \,;\, x_i^2 \le x_i \text{ for all } i \,\}.$$

(1998–04–16, 2c)

6.23 Let f be a convex function defined on a compact polyhedral set K. Assume that $f(\boldsymbol{x}) \le a$ for all extreme points of K. Show that $f(\boldsymbol{x}) \le a$ for all $\boldsymbol{x} \in K$.

6.24 a) Find out if the following function f is convex:

$$f(x_1, x_2) = 2(x_2 - x_1^2)^2 - 10.$$

defined on the set $S = \{\, (x_1, x_2) \,;\, -1 \le x_1 \le 1,\ -1 \le x_2 \le 1 \,\}$.

b) Do the same for an arbitrary convex set S which is a subset of

$$\{\, (x_1, x_2) \,;\, x_1^2 \ge x_2 \,\}.$$

S 6.25 Suppose that f is convex. Show that a vector $\boldsymbol{d} \ne \boldsymbol{0}$ is a direction of descent at $\boldsymbol{a}$ if and only if $f'(\boldsymbol{a}; \boldsymbol{d}) < 0$ with strict inequality. Does the result remain true without the convexity assumption on f?

6.26 Let f be a convex function on $x > 0$. Show that

$$f(x_1) + f(x_2) \le f(0) + f(x_1 + x_2), \qquad x_1, x_2 \ge 0.$$

(1997–12–17, 6)

6.27 Determine whether the set

$$\{\, (x_1, x_2, x_3) \,;\, x_1^4 + 6x_1^2x_2^2 + x_2^4 + x_3^4 \le 1 \,\}$$

is convex or not.

6.28 Investigate whether the function

$$f(x_1, x_2) = x_1^2 + x_2^2 + x_1^2 x_2^2$$

is convex on $\mathbf{R}^2$. (1999–10–21, 2c)

6.29 Prove that if $g_1(t)$ and $g_2(t)$ are convex functions on $\mathbf{R}$, and if $f(x, y)$ is a convex function on $\mathbf{R}^2$ which satisfies the condition

$$x_1 \le x_2, \quad y_1 \le y_2 \quad \Longrightarrow \quad f(x_1, y_1) \le f(x_2, y_2),$$

then the function $\varphi(t) = f(g_1(t), g_2(t))$ is convex. (2007–04–10, 5a)

6.30 For what real values of a and b is

$$f(x, y) = x^a y^b$$

convex in $x > 0$, $y > 0$?

6.31 The *distance* to a non-empty set S from a point $\boldsymbol{x}$ in $\mathbf{R}^n$ is defined as

$$f(\boldsymbol{x}) = \inf\{\, \|\boldsymbol{x} - \boldsymbol{y}\| \,;\, \boldsymbol{y} \in S \,\}.$$

Prove that f is convex.

6.32 a) Let $\boldsymbol{A}$ be an $m \times n$-matrix and $\boldsymbol{b}$ a column vector. Prove that the function

$$f(\boldsymbol{x}) = \|\boldsymbol{A}\boldsymbol{x} - \boldsymbol{b}\|$$

is convex on $\mathbf{R}^n$.

b) Let

$$\boldsymbol{A} = \begin{pmatrix} 1 & -2 \\ 2 & 4 \\ 3 & 0 \\ 4 & 1 \end{pmatrix}, \qquad \boldsymbol{b} = \begin{pmatrix} 2 \\ 1 \\ 4 \\ -4 \end{pmatrix}$$

Find the maximum of $f(\boldsymbol{x})$ when $\boldsymbol{x}$ belongs to the rectangle

$$0 \le x_1 \le 1, \quad 0 \le x_2 \le 2.$$

(2010–04–09, 6)

6.33 A function $f\colon \mathbf{R}^n \to \mathbf{R}$ is called *quasi-convex* if it has the property

$$f(\lambda \boldsymbol{x} + (1-\lambda)\boldsymbol{y}) \le \max(f(\boldsymbol{x}), f(\boldsymbol{y}))$$

for all $\boldsymbol{x}$ and $\boldsymbol{y}$ in $\mathbf{R}^n$ and all numbers λ with $0 \le \lambda \le 1$.

Prove that the function f is quasi-convex if and only if the set

$$\{\, \boldsymbol{x} \in \mathbf{R}^n \,;\, f(\boldsymbol{x}) \le c \,\}$$

is convex for alla numbers c (unless it is empty). (2009–12–16, 6)

6.34 a) Let f be an convex function on $X \subseteq \mathbf{R}^n$, and let C be a number. Prove that the set $S = \{\, \boldsymbol{x} \in X \,;\, f(\boldsymbol{x}) \leq C \,\}$ is convex.

b) Is the function $f\colon (x_1, x_2, x_2) \mapsto \ln(e^{x_1+x_2} + 1)$ convex on $\mathbf{R}^3$?

c) Is the set

$$S = \{\, (x_1, x_2) \,;\, |x_1 + x_2 - 4| \leq 5,\ \ln(1 + x_1^2 + x_2^2) \leq 1 \,\}$$

convex in $\mathbf{R}^2$?

(2009–01–09, 3)

6.35 Let $a_1, \ldots, a_N$ be positive numbers, and introduce numbers $t_1 \geq 0$, $i = 1, \ldots, N$, such that $\sum_{i=1}^{N} t_i = 1$. Define the *weighted arithmetic mean* and *weighted geometric mean* by

$$A = \sum_{i=1}^{N} t_i a_i \qquad \text{and} \qquad G = \prod_{i=1}^{N} a_i^{t_i}$$

respectively. Prove the inequality

$$G \leq A.$$

(The ordinary arithmetic and geometric means are the special case when $t_i = \dfrac{1}{N}$ for all i.)

Chapter 7

OPTIMIZATION WITH CONSTRAINTS

It is a common occurence that the variables of a function are restricted by some explicit constraints. In particular this is the case in linear programming. The general theory of optimization of such functions, known as *Karush-Kuhn-Tucker theory*, is the subject matter of this chapter. We should state explictly at this point that the theory itself is concerned with *local* minima only. In fact, as is well known, a global minimum may not always exist. To prove the existence of such a minimum some further assumptions are usually needed (convexity, compactness, ...).

7.1 Introduction

As a preliminary we first review the problem of minimizing a function when no (explicit) constraints are present. Thus consider a differentiable function $f: \mathbf{R}^n \to \mathbf{R}$ and let S be a subset of $\mathbf{R}^n$. For a given point $\boldsymbol{a} \in S$ we call a direction $\boldsymbol{d} \neq \boldsymbol{0}$ **feasible** at $\boldsymbol{a}$ if $\boldsymbol{a} + t\boldsymbol{d} \in S$ for $0 < t < \delta$, if $\delta > 0$ is small enough.

LEMMA 1. *Assume that $\bar{\boldsymbol{x}}$ locally solves the optimization problem*

$$\min_{\boldsymbol{x} \in S} f(\boldsymbol{x}).$$

Then

(1) $$\nabla f(\bar{\boldsymbol{x}})^T \boldsymbol{d} \geq 0 \quad \textit{for all feasible } \boldsymbol{d}.$$

PROOF. That $\bar{\boldsymbol{x}}$ locally solves the minimum problem means that $\bar{\boldsymbol{x}} \in S$ and that

$$f(\bar{\boldsymbol{x}}) \leq f(\boldsymbol{x}) \quad \text{for all } \boldsymbol{x} \in S \text{ in a neighbourhood of } \bar{\boldsymbol{x}}.$$

Put

$$F(t) = f(\bar{\boldsymbol{x}} + t\boldsymbol{d}), \qquad t \geq 0.$$

For fixed $\boldsymbol{d} \in \mathbf{R}^n$ this function describes the values of f on the half-ray in direction $\boldsymbol{d}$ from $\bar{\boldsymbol{x}}$. If $\boldsymbol{d}$ is a feasible direction F has a local minimum at $t = 0$. Then $F'_+(0) \geq 0$. Thus (1) follows by the chain rule. □

Example 1. If $\bar{\boldsymbol{x}}$ is an inner point of S all $\boldsymbol{d}$ are feasible. Then (1) implies that

$$\nabla f(\bar{\boldsymbol{x}}) = \boldsymbol{0}.$$

This is the familiar result that if the function f has a local minimum at an inner point $\bar{\boldsymbol{x}}$ then $\bar{\boldsymbol{x}}$ is a stationary point for f. □

Example 2. If $\bar{\boldsymbol{x}}$ is a boundary point of S not all $\boldsymbol{d}$ may be feasible. The figures show (for $n = 2$) some examples of feasible directions. Note that in the third case no feasible direction exists. (Exercise: draw a situation where $\bar{\boldsymbol{x}} \in \partial S$ and *all* directions are feasible.)

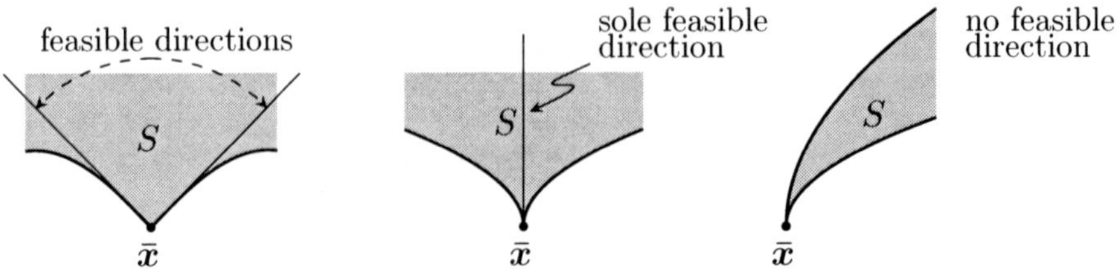

□

Example 3. Let S be a half-space in $\mathbf{R}^n$ defined by the condition $g(\boldsymbol{x}) \leq 0$, where $g(\boldsymbol{x}) = \boldsymbol{a}^T\boldsymbol{x} - c$. Assume that $\bar{\boldsymbol{x}}$ is a boundary point of S, so that $\boldsymbol{a}^T\bar{\boldsymbol{x}} - c = 0$. Then, if $t > 0$,

$$\bar{\boldsymbol{x}} + t\boldsymbol{d} \in S \iff g(\bar{\boldsymbol{x}} + t\boldsymbol{d}) \leq 0 \iff \boldsymbol{a}^T(\bar{\boldsymbol{x}} + t\boldsymbol{d}) - c \leq 0 \iff \boldsymbol{a}^T\boldsymbol{d} \leq 0.$$

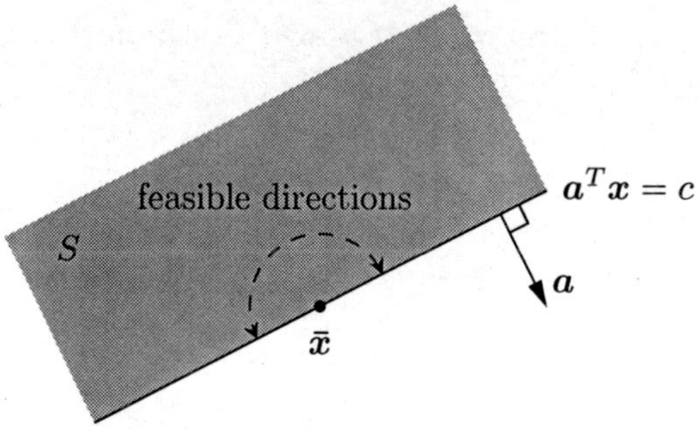

Hence a direction $\boldsymbol{d}$ is feasible if and only if $\boldsymbol{a}^T\boldsymbol{d} \leq 0$. It follows that condition (1) in this case can be stated as

$$\nabla f(\bar{\boldsymbol{x}})^T\boldsymbol{d} \geq 0 \qquad \text{whenever} \quad \boldsymbol{a}^T\boldsymbol{d} \leq 0.$$

This implies that the vectors $\nabla f(\bar{\boldsymbol{x}})$ and $\boldsymbol{a}$ are parallel, with $\nabla f(\bar{\boldsymbol{x}})$ pointing in the direction of $-\boldsymbol{a}$. In other words, there is a number $u \geq 0$ such that

$$\nabla f(\bar{\boldsymbol{x}}) = u(-\boldsymbol{a}) = -u\boldsymbol{a}.$$

In theorem 1 we will generalize this result to the case of more than one constraint. □

7.2 Affine constraints

7.2.1 Constraints in the form of inequalities

Let f be a differentiable function on an open set $X \subseteq \mathbf{R}^n$. Our first aim is to give necessary conditions for f to have a local minimum at a point $\bar{\boldsymbol{x}}$ under constraints

$$g_i(\boldsymbol{x}) \leq 0, \qquad i = 1, \ldots, m,$$

where the functions g_i are *affine*, i.e. of the form

$$g_i(\boldsymbol{x}) = \boldsymbol{a}_i^T\boldsymbol{x} - c_i.$$

We call the constraint $g_i(\boldsymbol{x}) \leq 0$ *active* at $\bar{\boldsymbol{x}}$ if $g_i(\bar{\boldsymbol{x}}) = 0$.

THEOREM 1. *Let f and g_i be as above. Assume that $\bar{\boldsymbol{x}}$ is a local solution to the problem*[1]

(2) $\min f(\boldsymbol{x})$ *when* $\boldsymbol{x} \in X$ *and* $g_i(\boldsymbol{x}) \leq 0, \quad i = 1, \dots, m.$

Then there exist real numbers $u_1, u_2, \dots, u_m \geq 0$ such that

$$\nabla f(\bar{\boldsymbol{x}}) + \sum_{i=1}^{m} u_i \boldsymbol{a}_i = \boldsymbol{0}. \tag{3}$$

For those constraints that are non-active at $\bar{\boldsymbol{x}}$ the corresponding u_i are equal to 0.

PROOF. If $\boldsymbol{a}_i^T \bar{\boldsymbol{x}} - c_i < 0$ for some i then obviously $\boldsymbol{a}_i^T \boldsymbol{x} - c_i < 0$ for all $\boldsymbol{x}$ that are sufficiently close to $\bar{\boldsymbol{x}}$. Thus we can disregard all non-active constraints when discussing the local behaviour of f at $\bar{\boldsymbol{x}}$.

Assume then that all constraints are active at $\bar{\boldsymbol{x}}$. From lemma 1 we know that

$$\nabla f(\bar{\boldsymbol{x}})^T \boldsymbol{d} \geq 0 \quad \text{for all feasible } \boldsymbol{d}.$$

As in example 3 (since X is open) a direction $\boldsymbol{d}$ is feasible if and only if

$$\boldsymbol{a}_i^T \boldsymbol{d} \leq 0, \qquad i = 1, \dots, m.$$

Let $\boldsymbol{A}$ denote the matrix whose rows are $\boldsymbol{a}_i^T$, $i = 1, \dots, m$. The assumption implies that the system of inequalities

$$\boldsymbol{A}\boldsymbol{d} \leq 0, \qquad (-\nabla f(\bar{\boldsymbol{x}}))^T \boldsymbol{d} > 0$$

has no solution. Consequently, by Farkas' theorem (page 136), the system

$$\boldsymbol{A}^T \boldsymbol{u} = -\nabla f(\bar{\boldsymbol{x}}), \qquad \boldsymbol{u} \geq \boldsymbol{0}$$

is solvable. This is the statement of the theorem, since the columns of $\boldsymbol{A}^T$ are precisely the vectors $\boldsymbol{a}_i$. □

[1] The purpose of the set X is to take care of possible side conditions that are not explicitly given by the constraints.

Consider again condition (3). It states that the vector $-\nabla f(\bar{\boldsymbol{x}})$ is a positive linear combination of the vectors $\boldsymbol{a}_1, \ldots, \boldsymbol{a}_m$. The set of all such combinations is as we know (example 4.22, page 144) a closed convex cone.[2] Thus the theorem has the following geometrical interpretation:

if $\bar{\boldsymbol{x}}$ is a local solution to the minimum problem (2) then $-\nabla f(\bar{\boldsymbol{x}})$ belongs to the cone of positive linear combinations of the vectors $\boldsymbol{a}_1, \ldots, \boldsymbol{a}_m$.

Example 4. Let $n = 2$, $m = 2$ and

$$g_1(\boldsymbol{x}) = 3x_1 - 2x_2 + 1, \qquad g_2(\boldsymbol{x}) = x_1 + 2x_2 - 5.$$

Then $\boldsymbol{a}_1 = (3, -2)$ and $\boldsymbol{a}_2 = (1, 2)$. The domain over which the objective function f is minimized is defined by the constraints

$$\begin{cases} 3x_1 - 2x_2 + 1 \leq 0 \\ \quad x_1 + 2x_2 - 5 \leq 0, \end{cases}$$

and is a sector in the plane bounded by two lines meeting at $\boldsymbol{z} = (1, 2)$. If f has a local minimum at this point $\boldsymbol{z}$ then $-\nabla f(\boldsymbol{z})$ is a positive linear combination of $\boldsymbol{a}_1$ and $\boldsymbol{a}_2$. See the figure on the next page. (Exercise: check the statements above by sketching some level curves of f in different cases.)

At e.g. $\boldsymbol{y} = (-1, -1)$ the first constraint is active but not the second one. Thus if f has minimum at $\boldsymbol{y}$ there is a positive number u such that

$$-\nabla f(\boldsymbol{y}) = u\boldsymbol{a}_1.$$

(Cf. example 3.)

Of course, if f has a local minimum at a point $\boldsymbol{x}$ where no constraint is active then $\nabla f(\boldsymbol{x}) = \boldsymbol{0}$.

---

[2]Actually it is the dual of the the cone of all feasible directions.

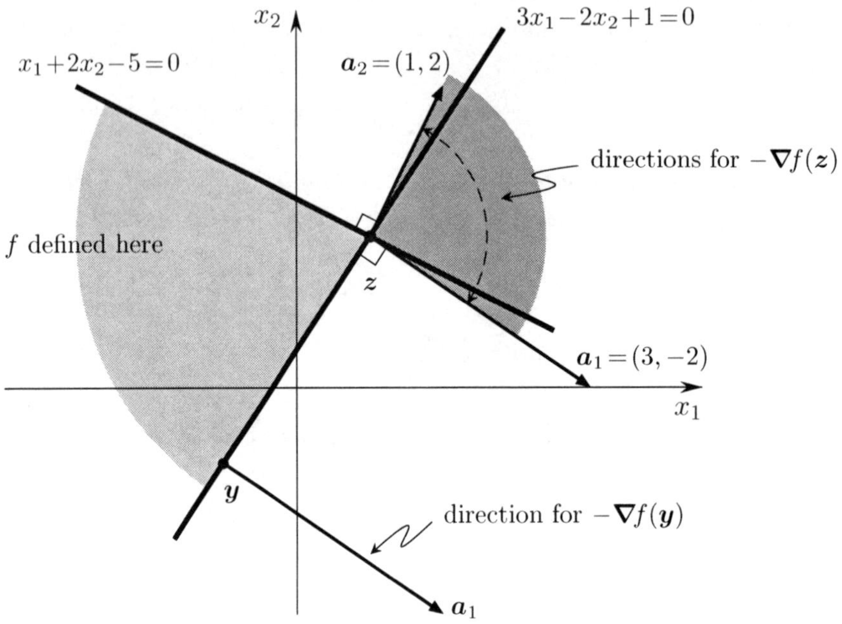

□

7.2.2 Equality constraints

The goal for the future is to successively improve on and reformulate theorem 1 to include constraints defined by non-affine functions, and to allow equalities in the constraints. The ultimate result will be the *Karush-Kuhn-Tucker theorem*, theorem 4, but we proceed stepwise, at first extending theorem 1 to include affine equality constraints.

Thus assume that f is constrained by some equalites as well, given by linear equations

$$h_j(\boldsymbol{x}) = \boldsymbol{b}_j^T\boldsymbol{x} - d_j = 0, \qquad j = 1, 2, \ldots, l.$$

Such an equality is equivalent to a pair of inequalities

$$\boldsymbol{b}_j^T\boldsymbol{x} - d_j \leq 0, \qquad (-\boldsymbol{b}_j)^T\boldsymbol{x} + d_j \leq 0.$$

Using theorem 1, in (3) these give rise to the two terms

$$u_j'\boldsymbol{b}_j + u_j''(-\boldsymbol{b}_j) = (u_j' - u_j'')\boldsymbol{b}_j, \qquad u_j', u_j'' \geq 0.$$

Putting $v_j = u'_j - u''_j$ we have

$$(u'_j - u''_j)\boldsymbol{b}_j = v_j \boldsymbol{b}_j, \qquad v_j \text{ unrestricted.}$$

Thus we now have obtained a generalization of theorem 1 to the case when some constraints are given by equalities. Observing that

$$\nabla g_i = \boldsymbol{a}_i, \qquad \nabla h_j = \boldsymbol{b}_j,$$

the theorem can be formulated in the following way, suitable for further generalizations later on.

THEOREM 2. *Let f be a differentiable function defined on an open set $X \subseteq \mathbf{R}^n$. Consider the problem of minimizing f subject to conditions*

$$\begin{aligned} &\boldsymbol{x} \in X \\ &g_i(\boldsymbol{x}) \leq 0, \quad i = 1, \ldots, m, \\ &h_j(\boldsymbol{x}) = 0, \quad j = 1, \ldots, l, \end{aligned}$$

where the constraints are affine:

$$g_i(\boldsymbol{x}) = \boldsymbol{a}_i^T \boldsymbol{x} - c_i, \qquad h_j(\boldsymbol{x}) = \boldsymbol{b}_j^T \boldsymbol{x} - d_j.$$

Assume that $\bar{\boldsymbol{x}}$ is a local point of minimum. Then there exist numbers $u_1, \ldots, u_m \geq 0$ and $v_1, \ldots, v_l$ (free) such that

$$\nabla f(\bar{\boldsymbol{x}}) + \sum_{i=1}^{m} u_i \nabla g_i(\bar{\boldsymbol{x}}) + \sum_{j=1}^{l} v_j \nabla h_j(\bar{\boldsymbol{x}}) = \boldsymbol{0}. \tag{4}$$

For those constraints that are non-active at $\bar{\boldsymbol{x}}$ the corresponding u_i are equal to 0.

Remark. In the case when all constraints are equalities (4) is the familiar necessary condition of Lagrange for a constrained minimum.[3] The numbers u_i, v_j in (4) are often called *Lagrange multipliers*.

Later on we will find it convenient in connection with the minimization problem above to introduce the function

$$L(\boldsymbol{x}; \boldsymbol{u}, \boldsymbol{v}) = f(\boldsymbol{x}) + \sum_{i=1}^{m} u_i g_i(\boldsymbol{x}) + \sum_{j=1}^{l} v_j h_j(\boldsymbol{x}), \qquad \boldsymbol{x} \in X, \quad \boldsymbol{u} \geq \boldsymbol{0}.$$

[3] At present for affine constraints, but that restriction will be removed later on.

We call L the *Lagrange function* associated with the minimum problem. Obviously $L(\boldsymbol{x};\boldsymbol{u},\boldsymbol{v}) \le f(\boldsymbol{x})$ whenever $\boldsymbol{x}$ satisfies the constraints, and

$$L(\boldsymbol{x};\boldsymbol{u},\boldsymbol{v}) = f(\boldsymbol{x}) \quad \Longleftrightarrow \quad \begin{cases} u_i g_i(\boldsymbol{x}) = 0, & i = 1,\dots,m \\ v_j h_j(\boldsymbol{x}) = 0, & j = 1,\dots,l. \end{cases}$$

Theorem 2 can now be reformulated as follows:

> *A necessary condition for $\bar{\boldsymbol{x}}$ to be a local minimum point is that there exist $\boldsymbol{u} \ge 0$ and $\boldsymbol{v}$ (free) such that*
>
> $$\nabla_{\boldsymbol{x}} L(\bar{\boldsymbol{x}};\boldsymbol{u},\boldsymbol{v}) = \boldsymbol{0}, \qquad L(\bar{\boldsymbol{x}};\boldsymbol{u},\boldsymbol{v}) = f(\bar{\boldsymbol{x}}).$$

Example 5. We are now going to apply the theorem to the following minimization problem (which admittedly is so simple that it can be solved without any theory at all).

Find the minimum of $f(\boldsymbol{x}) = 2x_1^2 + x_2^2$ subject to the constraints

$$\begin{aligned} x_1 + 3x_2 &\le 6, \\ x_1 + 2x_2 &\ge 2, \\ x_2 &\ge 0. \end{aligned}$$

Solution: The constraints define a compact triangle in the plane. Thus, since f is continuous, a global minimum certainly exists. Our strategy is to use the conditions of theorem 2 to compile a list of points which with certainty contains the minimum point, and then from this list select the point with the least value.

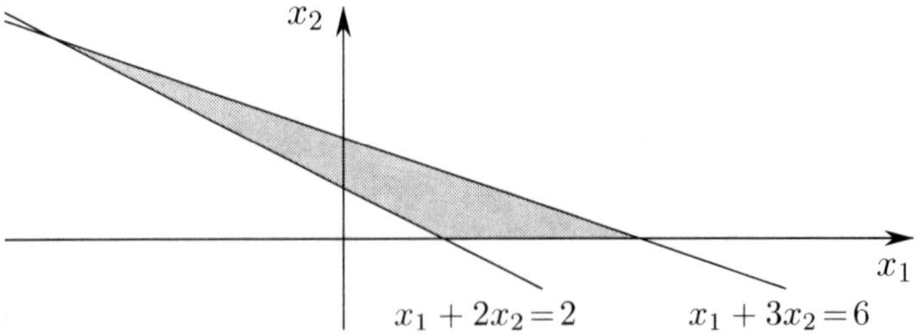

The set X which appears in the theorem is in this case (and often) the whole of $\mathbf{R}^2$. The Lagrange function is

$$L(\boldsymbol{x};\boldsymbol{u}) = 2x_1^2 + x_2^2 + u_1(x_1+3x_2-6) + u_2(-x_1-2x_2+2) + u_3(-x_2),$$

$$u_1, u_2, u_3 \ge 0.$$

Differentiating we get

$$\frac{\partial L}{\partial x_1} = 4x_1 + u_1 - u_2$$
$$\frac{\partial L}{\partial x_2} = 2x_2 + 3u_1 - 2u_2 - u_3.$$

We now divide the problem into 8 cases.

I. $u_1 = 0,\ u_2 = 0,\ u_3 = 0.$
II. $u_1 > 0,\ u_2 = 0,\ u_3 = 0.$
III. $u_1 = 0,\ u_2 > 0,\ u_3 = 0.$
IV. $u_1 = 0,\ u_2 = 0,\ u_3 > 0.$
V. $u_1 = 0,\ u_2 > 0,\ u_3 > 0.$
VI. $u_1 > 0,\ u_2 = 0,\ u_3 > 0.$
VII. $u_1 > 0,\ u_2 > 0,\ u_3 = 0.$
VIII. $u_1 > 0,\ u_2 > 0,\ u_3 > 0.$

If a constraint is not active at a point the corresponding Lagrange multiplier is zero. Hence (although it is not logically so) we can interpret case I as searching for a point where no constraints are active, i.e. a stationary point of f. Certainly the condition $u_1 > 0$ in case II implies that the first constraint is active, that is, $x_1 + 3x_2 = 6$. In case III the second condition is active, and so on.

Working through these cases using the equations

$$\frac{\partial L}{\partial x_1} = 0, \qquad \frac{\partial L}{\partial x_2} = 0$$

and the appropriate constraints we get respectively

I. $\begin{cases} x_1 = 0 \\ x_2 = 0 \end{cases}$ $(0, 0)$ does not satisfy the constraints.

II. $\begin{cases} 4x_1 + u_1 = 0 \\ 2x_2 + 3u_1 = 0 \\ x_1 + 3x_2 = 6 \end{cases} \iff \begin{cases} x_1 = -\frac{u_1}{4} \\ x_2 = -\frac{3u_1}{2} \\ u_1 = -\frac{24}{19} < 0 \end{cases}$

$$\text{III.}\quad \begin{cases} 4x_1 - u_2 = 0 \\ 2x_2 - 2u_2 = 0 \\ x_1 + 2x_2 = 2 \end{cases} \iff \begin{cases} x_1 = \dfrac{u_2}{4} \\ x_2 = u_2 \\ u_2 = \dfrac{8}{9} \end{cases} \iff \begin{cases} x_1 = \dfrac{2}{9} \\ x_2 = \dfrac{8}{9} \\ u_2 = \dfrac{8}{9} > 0 \end{cases}$$

$\left(\dfrac{2}{9}, \dfrac{8}{9}\right)$ *does* satisfy the constraints.

$$\text{IV.}\quad \begin{cases} 4x_1 = 0 \\ 2x_2 - u_3 = 0 \\ x_2 = 0 \end{cases} \iff \begin{cases} x_1 = 0 \\ x_2 = 0 \\ u_3 = 0 \end{cases}$$

$(0,0)$ does not satisfy the constraints.

$$\text{V.}\quad \begin{cases} 4x_1 - u_2 = 0 \\ 2x_2 - 2u_2 - u_3 = 0 \\ x_1 + 2x_2 = 2 \\ x_2 = 0 \end{cases} \iff \begin{cases} x_1 = 2 \\ x_2 = 0 \\ u_2 = 8 \\ u_3 = -16 < 0 \end{cases}$$

$$\text{VI.}\quad \begin{cases} 4x_1 + u_1 = 0 \\ 2x_2 + 3u_1 - u_3 = 0 \\ x_1 + 3x_2 = 6 \\ x_2 = 0 \end{cases} \iff \begin{cases} x_1 = 6 \\ x_2 = 0 \\ u_1 = -24 < 0 \\ u_3 = -72 < 0 \end{cases}$$

$$\text{VII.}\quad \begin{cases} 4x_1 + u_1 - u_2 = 0 \\ 2x_2 + 3u_1 - 2u_2 = 0 \\ x_1 + 3x_2 = 6 \\ x_1 + 2x_2 = 2 \end{cases} \iff \begin{cases} x_1 = -6 \\ x_2 = 4 \\ u_1 = -56 < 0 \\ u_2 = -80 < 0 \end{cases}$$

VIII. This case requires equality in all constraints, which is impossible.

As we now see, only one point, namely $(\frac{2}{9}, \frac{8}{9})$ (case III) satisfies the necessary conditions for a local minimum in theorem 2. Since a (global) minimum point exists, $(\frac{2}{9}, \frac{8}{9})$ must be that point. The minimum value is $\frac{8}{9}$. □

Remark. Suppose that we are searching for the maximum instead of the minimum of f. The maximum occurs where $-f$ has its minimum. Thus we can use the calculations above, requiring all $u_i \leq 0$. We find that the maximum occurs in one of the points $(\frac{6}{19}, \frac{36}{19})$, $(6,0)$, $(-6,4)$. A comparison between the values of f in these points gives as result that the maximum of f is $f(-6,4) = 88$.

Exercises

S 7.1 Suppose that some C^1-function f of two variables, defined for

$$-x_1 + 2x_2 \leq 0, \quad -3x_1 + x_2 \leq 0,$$

has a local minimum at the origin. Draw a figure indicating the possible directions of $\nabla f(0,0)$.

7.2 Consider the LP problem

$$\begin{aligned} \text{Minimize} \quad & -3x_1 - 3x_2 \\ \text{subject to} \quad & 3x_1 + 2x_2 \leq 18 \\ & -x_1 + 4x_2 \leq 8 \\ & x_1, x_2 \geq 0. \end{aligned}$$

For each extreme point of the domain, state the conditions of theorem 1 and determine whether they are true or not. Find the optimal solution.

7.3 General constraints; necessary conditions for a minimum

7.3.1 Inequality constraints only

To begin with we are going to study the problem to minimize $f(\boldsymbol{x})$ when $\boldsymbol{x}$ belongs to an open set $X \subseteq \mathbf{R}^n$ and satisfies general inequality constraints

$$g_i(\boldsymbol{x}) \leq 0, \qquad i = 1, \ldots, m.$$

The functions g_i are assumed to be differentiable in X.

Assume that f has a local minimum at $\bar{\boldsymbol{x}}$ under these constraints. We are going to prove that, under some extra conditions to be stated later, there exist positive real numbers $u_1, \ldots, u_m$ such that

(5) $$\nabla f(\bar{\boldsymbol{x}}) + \sum_{i=1}^{m} u_i \nabla g_i(\bar{\boldsymbol{x}}) = \boldsymbol{0}.$$

It is obvious that (5) becomes (3) when the functions g_i are affine.

It is no restriction to assume that all constraints are active at $\bar{\boldsymbol{x}}$, since if for some i we have $g_i(\bar{\boldsymbol{x}}) < 0$ then by continuity $g_i(\boldsymbol{x}) < 0$ for all $\boldsymbol{x}$ in a neighbourhood of $\bar{\boldsymbol{x}}$. Obviously such a constraint does not have to be considered when discussing conditions for a local minimum at $\bar{\boldsymbol{x}}$.

Using lemma 1 as before we obtain that

$$\nabla f(\bar{\boldsymbol{x}})^T \boldsymbol{d} \geq 0 \quad \text{for all feasible } \boldsymbol{d}.$$

Since the g_i are differentiable at $\bar{\boldsymbol{x}}$ we have

$$g_i(\bar{\boldsymbol{x}} + t\boldsymbol{d}) = g_i(\bar{\boldsymbol{x}}) + \nabla g_i(\bar{\boldsymbol{x}})^T (t\boldsymbol{d}) + o(t) = t\, \nabla g_i(\bar{\boldsymbol{x}})^T \boldsymbol{d} + o(t),$$

where $o(t)$ denotes a function with the property that $\dfrac{o(t)}{t} \to 0$ as $t \to 0$. For $\boldsymbol{d}$ to be feasible it is required that $g_i(\bar{\boldsymbol{x}} + t\boldsymbol{d}) \leq 0$ for $t > 0$ small enough. Hence, a *sufficient* condition for $\boldsymbol{d}$ to be feasible is that

$$(6) \qquad \nabla g_i(\bar{\boldsymbol{x}})^T \boldsymbol{d} < 0, \qquad i = 1, \ldots, m.$$

(Note the strict inequality.)

Denote by $\boldsymbol{A}$ the $m \times n$-matrix whose rows are $\nabla g_i(\bar{\boldsymbol{x}})^T$. Then condition (6) can be rewritten as $\boldsymbol{Ad} < \boldsymbol{0}$. Thus, a necessary condition for $\bar{\boldsymbol{x}}$ to be a local point of minimum is that

$$\boldsymbol{Ad} < \boldsymbol{0} \quad \Longrightarrow \quad \nabla f(\bar{\boldsymbol{x}})^T \boldsymbol{d} \geq 0,$$

or in other words, that the system of inequalities

$$(7) \qquad \boldsymbol{Ad} < \boldsymbol{0}, \quad (-\nabla f(\bar{\boldsymbol{x}}))^T \boldsymbol{d} > 0$$

has no solution.

To proceed further we need the following extension of Farkas' theorem. Here strict inequalities appear, and the conclusion is in one direction only.

LEMMA 2. *Assume that* $\boldsymbol{A}$ *is an* $m \times n$ *matrix with linearly independent rows. If the system of inequalities*

$$\boldsymbol{Ax} < \boldsymbol{0}, \qquad \boldsymbol{c}^T \boldsymbol{x} > 0$$

has no solution, then the system

$$\boldsymbol{A}^T \boldsymbol{y} = \boldsymbol{c}, \qquad \boldsymbol{y} \geq \boldsymbol{0}$$

is solvable.

PROOF. The proof proceeds in two stages. First, we show that some vector $\hat{\boldsymbol{x}}$ exists such that $\boldsymbol{A}\hat{\boldsymbol{x}} < \boldsymbol{0}$, i.e., such that

$$\boldsymbol{a}_i^T \hat{\boldsymbol{x}} < \boldsymbol{0} \qquad \text{for } i = 1, \ldots, m, \tag{8}$$

where $\boldsymbol{a}_i^T$ are the rows of $\boldsymbol{A}$. For this it is enough to prove that for each i there is an $\hat{\boldsymbol{x}}_i$ such that

$$\boldsymbol{a}_i^T \hat{\boldsymbol{x}}_i < 0, \qquad \boldsymbol{a}_j^T \hat{\boldsymbol{x}}_i \leq 0 \;\text{ if }\; j \neq i, \tag{9}$$

since putting $\hat{\boldsymbol{x}} = \hat{\boldsymbol{x}}_1 + \ldots + \hat{\boldsymbol{x}}_m$ makes $\hat{\boldsymbol{x}}$ satisfy (8).

Assume then that (9) is not true for some i. Then the system of inequalities

$$\boldsymbol{a}_j^T \boldsymbol{x} \leq 0, \qquad j \neq i,$$

$$(-\boldsymbol{a}_i)^T \boldsymbol{x} > 0$$

is not solvable. According to Farkas' theorem this implies that the system

$$\sum_{j \neq i} y_j \boldsymbol{a}_j = -\boldsymbol{a}_i, \qquad \text{all } \; y_j \geq 0 \tag{10}$$

is solvable. This is a contradiction to the assumption of linear independence of the rows $\boldsymbol{a}_j^T$ of $\boldsymbol{A}$. Thus the existence of $\hat{\boldsymbol{x}}$ is proved.

Now consider the system $\boldsymbol{A}^T \boldsymbol{y} = \boldsymbol{c}$, $\boldsymbol{y} \geq \boldsymbol{0}$ in the statement of the lemma. If it has no solution, Farkas' theorem implies that

$$\boldsymbol{A}\boldsymbol{x} \leq \boldsymbol{0}, \qquad \boldsymbol{c}^T \boldsymbol{x} > 0$$

is solvable. Let $\bar{\boldsymbol{x}}$ denote a solution. We perturb $\bar{\boldsymbol{x}}$ a little by considering, for small positive ε, the vector $\bar{\boldsymbol{x}} + \varepsilon \hat{\boldsymbol{x}}$. If ε is small enough we have

$$\boldsymbol{A}(\bar{\boldsymbol{x}} + \varepsilon \hat{\boldsymbol{x}}) = \boldsymbol{A}\bar{\boldsymbol{x}} + \varepsilon \boldsymbol{A}\hat{\boldsymbol{x}} < \boldsymbol{0}$$

and

$$\boldsymbol{c}^T(\bar{\boldsymbol{x}} + \varepsilon \hat{\boldsymbol{x}}) = \boldsymbol{c}^T \bar{\boldsymbol{x}} + \varepsilon \boldsymbol{c}^T \hat{\boldsymbol{x}} > 0.$$

Hence we have found a solution to $\boldsymbol{A}\boldsymbol{x} < \boldsymbol{0}$, $\boldsymbol{c}^T \boldsymbol{x} > 0$, contradicting the assumption. The lemma is proved. □

The lemma provides us with an extra condition which makes it possible to prove (5). If the gradients $\nabla g_i(\bar{\boldsymbol{x}})$ at the local minimum point $\bar{\boldsymbol{x}}$ are linearly independent, then the non-solvability of (7) implies (5). In conclusion we have proved the following theorem.

THEOREM 3. *Let f and g_i, $i = 1, \ldots, m$, be differentiable on an open set $X \subseteq \mathbf{R}^n$. Assume that $\bar{\boldsymbol{x}}$ is a local solution to the problem*

(11) $\min f(\boldsymbol{x}) \qquad \text{when} \quad \boldsymbol{x} \in X \quad \text{and} \quad g_i(\boldsymbol{x}) \leq 0, \quad i = 1, \ldots, m.$

Further assume that the gradients $\nabla g_i(\bar{\boldsymbol{x}})$ of those constraints that are active at $\bar{\boldsymbol{x}}$ are linearly independent. Then there exist real numbers $u_1, u_2, \ldots, u_m \geq 0$ such that

$$\nabla f(\bar{\boldsymbol{x}}) + \sum_{i=1}^{m} u_i \nabla g_i(\bar{\boldsymbol{x}}) = \boldsymbol{0}. \tag{12}$$

For those constraints that are non-active at $\bar{\boldsymbol{x}}$ the corresponding u_i are equal to 0.

Remark. Actually the conditions on the gradients of g_i can be somewhat relaxed. Refering back to (10) in the proof of the lemma, all that is needed to reach a contradiction is that the rows of $\boldsymbol{A}$ satisfy the following condition:

$$\left.\begin{array}{l} \sum_{i=1}^{m} \lambda_i \boldsymbol{a}_k = \boldsymbol{0} \\ \lambda_i \geq 0 \quad \text{for all } i \end{array}\right\} \quad \Longrightarrow \quad \lambda_i = 0 \quad \text{for all } i.$$

This property, which is weaker than linear independence, is called *positive linear independence.*

It is thus enough in the theorem to require that

(13) *the gradients $\nabla g_i(\bar{\boldsymbol{x}})$ are positive linearly independent.*

This side condition (or others close to it) in the theorem are usually called a *constraint qualification.*

Of course it is enough if (13) is fulfilled for those constraints that are active at $\bar{\boldsymbol{x}}$.

Another relaxation of the assumptions in the theorem is that only those g_i which correspond to active constraints at $\bar{\boldsymbol{x}}$ need to be differentiable. For non-active constraints continuity is sufficient.

Example 6. The reader should note that condition (12) is merely *necessary* for $\bar{\boldsymbol{x}}$ to be local minimum point; it is not a sufficient condition. The figure shows level curves (dashed) of a function f ($f(\boldsymbol{x}) = c$ with c growing upwards in the figure), and *one* constraint $g(\boldsymbol{x}) \leq 0$. Obviously, f has no local minimum at the point $\bar{\boldsymbol{x}}$, although (12) is satisfied there.

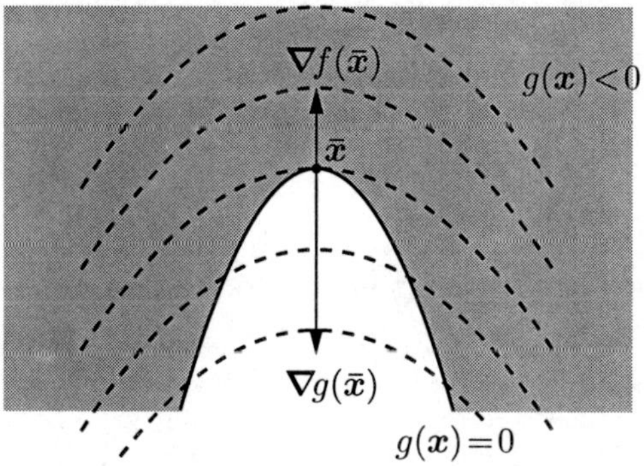

□

Example 7. Consider the problem in three variables of minimizing

$$f(\boldsymbol{x}) = 8x_1x_2 + 7x_3$$

under the constraints

$$g_1(\boldsymbol{x}) = x_1^2 + x_2^2 + x_3^3 - 2 \leq 0, \qquad g_2(\boldsymbol{x}) = -x_3 \leq 0.$$

The constraints define the upper half of a ball at the origin with radius $\sqrt{2}$. The point $\bar{\boldsymbol{x}} = (1, -1, 0)$ satisfies these conditions. The gradients at $\bar{\boldsymbol{x}}$ are

$$\begin{aligned}
\nabla f(\bar{\boldsymbol{x}}) &= (8x_2, 8x_1, 7) = (-8, 8, 7), \\
\nabla g_1(\bar{\boldsymbol{x}}) &= (2x_1, 2x_2, 3x_3^2) = (2, -2, 0), \\
\nabla g_2(\bar{\boldsymbol{x}}) &= (0,\ 0,\ -1).
\end{aligned}$$

Thus

$$\nabla f(\bar{\boldsymbol{x}}) + u_1 \nabla g_1(\bar{\boldsymbol{x}}) + u_2 \nabla g_2(\boldsymbol{x}) = \boldsymbol{0}$$

when $u_1 = 4$ and $u_2 = 7$, both numbers positive. Condition (12) is fulfilled at $\bar{\boldsymbol{x}}$. □

Remark. Note however that this fact does not show that $(1,-1,0)$ is a local minimum. (See example 6.) In fact, however, this is the case, and $(1,-1,0)$ is actually the global minimum of f under the given constraints.

To prove this we could follow the same strategy as in example 5. The constraints define a compact set. We need to find all points in this set which satisfy (12). These are $\pm(1,-1,0)$ and $(0,0,0)$. Also, we need to find all points (if any) which violate the constraint qualification (13). In this case there are no such points.

We leave the details of these computations to the reader.

7.3.2 The general case

We are now going to extend the previous discussion to the case when some constraints are given by equalities. Let h_j be continuously differentiable functions on X defining constraints

$$h_j(\boldsymbol{x}) = 0, \qquad j = 1, \ldots, l.$$

The problem before us is to minimize $f(\boldsymbol{x})$ under these constraints together with the previous conditions $\boldsymbol{x} \in X$ and

$$g_i(\boldsymbol{x}) \leq 0, \qquad i = 1, \ldots, m,$$

where as before f and the g_i are assumed differentiable.

The constraint qualification at $\bar{\boldsymbol{x}}$ now takes the form

$$\text{(CQ)} \quad \begin{cases} \sum_{i=1}^{m} \lambda_i \nabla g_i(\bar{\boldsymbol{x}}) + \sum_{j=1}^{l} \mu_j \nabla h_j(\bar{\boldsymbol{x}}) = \boldsymbol{0} \\ \lambda_i \geq 0 \quad \text{for all } i \end{cases} \implies \begin{cases} \lambda_i = 0 & \text{for all } i \\ \mu_j = 0 & \text{for all } j. \end{cases}$$

In particular we require that the gradients $\nabla g_i(\bar{\boldsymbol{x}})$ are positive linearly independent and that the gradients $\nabla h_j(\bar{\boldsymbol{x}})$ are linearly independent. Note that only constraints that are active att $\bar{\boldsymbol{x}}$ need to be taken into consideration.

The *Lagrange function* for this minimization problem is defined as

$$L(\boldsymbol{x}; \boldsymbol{u}, \boldsymbol{v}) = f(\boldsymbol{x}) + \sum_{i=1}^{m} u_i g_i(\boldsymbol{x}) + \sum_{j=1}^{l} v_j h_j(\boldsymbol{x}),$$

$$\boldsymbol{x} \in X, \quad \boldsymbol{u} \geq \boldsymbol{0}, \quad \boldsymbol{v} \text{ free.}$$

Then

$$(14) \qquad \nabla_{\boldsymbol{x}} L(\boldsymbol{x}; \boldsymbol{u}, \boldsymbol{v}) = \nabla f(\boldsymbol{x}) + \sum_{i=1}^{m} u_i \nabla g_i(\boldsymbol{x}) + \sum_{j=1}^{l} v_j \nabla h_j(\boldsymbol{x}).$$

As in the affine case $L(\boldsymbol{x}; \boldsymbol{u}, \boldsymbol{v}) \leq f(\boldsymbol{x})$ when $\boldsymbol{x}$ satisfies the constraints.

We use L to state the theorem giving necessary conditions for a local minimum at $\bar{\boldsymbol{x}}$.[4]

THEOREM 4. *Let f be a differentiable function defined on an open set $X \subseteq \mathbf{R}^n$. Consider the problem of minimizing f subject to the conditions*

$$\begin{aligned} &\boldsymbol{x} \in X \\ &g_i(\boldsymbol{x}) \leq 0, \quad i = 1, \ldots, m \\ &h_j(\boldsymbol{x}) = 0, \quad j = 1, \ldots, l. \end{aligned}$$

Assume that $\bar{\boldsymbol{x}}$ is a local point of minimum (in particular, that $\bar{\boldsymbol{x}}$ satisfies the constraints). Suppose that the functions g_i are differentiable and the functions h_j continuously differentiable at $\bar{\boldsymbol{x}}$, and that their gradients at $\bar{\boldsymbol{x}}$ satisfy the constraint qualification (CQ) above. Then there exist scalars u_i and v_j such that

$$(15) \qquad \nabla_{\boldsymbol{x}} L(\bar{\boldsymbol{x}}; \boldsymbol{u}, \boldsymbol{v}) = \boldsymbol{0},$$

$$(16) \qquad \boldsymbol{u} \geq \boldsymbol{0},$$

$$(17) \qquad \boldsymbol{u}^T \boldsymbol{g}(\bar{\boldsymbol{x}}) = 0, \quad \text{i.e.,} \quad u_i g_i(\bar{\boldsymbol{x}}) = 0 \quad \textit{for all } i.$$

In the case with no equality constraints the theorem becomes theorem 3. When all constraint functions are affine we have theorem 2. Note that condition (17) means exactly that $u_i = 0$ if the corresponding constraint $g_i(\boldsymbol{x}) \leq 0$ is non-active at $\bar{\boldsymbol{x}}$.

Conditions (17) are called the *complementary slackness conditions* (as in linear programming). The conditions implied by $\bar{\boldsymbol{x}}$ satisfying the

[4]The theorem is named after Kuhn and Tucker, who proved it in 1951. Later it was discovered that the theorem was in fact proved already in 1939 by Karush. Some authors now name it the KARUSH-KUHN-TUCKER THEOREM. This is the origin of the name KKT point for a point obeying the conditions in the theorem.

constraints are called *feasibility conditions* and (15) and sometimes (16) are called the *KKT conditions*.

A point $\bar{\boldsymbol{x}}$ satisfying the feasibility conditions and (15)–(17) is called a *KKT point* for the minimization problem.

The interpretation of theorem 4 is the following. Assuming that a feasible point $\bar{\boldsymbol{x}}$ is a local minimum point, there are two possibilities: $\bar{\boldsymbol{x}}$ is a KKT point or (CQ) is not satisfied at $\bar{\boldsymbol{x}}$. The last case should be thought of as an exceptional one. Neglecting this possibility the theorem states that $\bar{\boldsymbol{x}}$ being a KKT point is a *necessary* condition for a local minimum at $\bar{\boldsymbol{x}}$. Note the logic: there are other possibilities for a KKT point than a local minimum. See e.g. example 6 above. — Sufficient conditions will be discussed in section 7.4.

The theorem suggests the following strategy for problem solving. Suppose that you e.g. for compactness reasons know that a certain global minimum exists. Then this is of course in particular a local minimum. Draw up a list consisting of all KKT points. Add to this list the exceptional points where (CQ) is not satisfied, if any. Now the list with certainty contains the required global minimum point, and all you have to do to find it is to compare function values at all points on the list, and pick out the minimal one. See example 9 below and example 5 above.

Remark. Of course the constraint qualification should be applied only to gradients connected to constraints being active at $\bar{\boldsymbol{x}}$. Furthermore, for a non-active constraint we do not need the assumption of differentiability of g_i; continuity is sufficient.

The proof of theorem 4 is somewhat complicated and will be postponed until later.

Example 8. Find all KKT points for the problem of minimizing

$$f(x, y, z) = -\ln x - \ln y + z, \qquad x > 0, \quad y > 0$$

subject to the constraints

$$g_1(x, y, z) = x^2 + y^2 + z^2 - 1 \leq 0, \qquad g_2(x, y, z) = x^2 + y^2 - \frac{1}{2} \leq 0.$$

Solution: The set X is here defined by the conditions $x > 0$, $y > 0$ in $\mathbf{R}^3$. The Lagrange function is

$$L(x, y, z; u_1, u_2) =$$
$$= -\ln x - \ln y + z + u_1\Big(x^2 + y^2 + z^2 - 1\Big) + u_2\Big(x^2 + y^2 - \frac{1}{2}\Big),$$
$$(x, y, z) \in X, \quad u_1 \geq 0, \quad u_2 \geq 0.$$

We have

$$\nabla_{\boldsymbol{x}} L(x, y, z; u_1, u_2) = \begin{pmatrix} -1/x \\ -1/y \\ 1 \end{pmatrix} + u_1 \begin{pmatrix} 2x \\ 2y \\ 2z \end{pmatrix} + u_2 \begin{pmatrix} 2x \\ 2y \\ 0 \end{pmatrix},$$

and condition (15) for (x, y, z) to be a KKT point is

$$\text{(18)} \qquad \begin{cases} -\dfrac{1}{x} + 2u_1 x + 2u_2 x = 0 \\ -\dfrac{1}{y} + 2u_1 y + 2u_2 y = 0 \\ 1 + 2u_1 z = 0. \end{cases}$$

The complementary slackness conditions are

$$u_1(x^2 + y^2 + z^2 - 1) = 0,$$
$$u_2(x^2 + y^2 - \tfrac{1}{2}) = 0,$$

and the feasibility conditions are $x > 0$, $y > 0$ and

$$x^2 + y^2 + z^2 \leq 1, \qquad x^2 + y^2 \leq \frac{1}{2}.$$

It is obvious from the third equation in (18) that $u_1 \neq 0$, that is, the first constraint is active at any KKT point. Multiplying the first equation in (18) by x, the second by y and subtracting results in

$$2(u_1 + u_2)(x^2 - y^2) = 0.$$

Since we know that $u_1 > 0$ and $u_2 \geq 0$ this leads to $x^2 - y^2 = 0$, i.e. $x = y$, since both numbers are positive.

Consider first the case when $u_2 = 0$. Then (18) becomes

$$\begin{cases} -\dfrac{1}{x} + 2u_1 x = 0 \\ -\dfrac{1}{y} + 2u_1 y = 0 \\ 1 + 2u_1 z = 0. \end{cases}$$

This gives $x^2 = y^2 = -z$ $(= \dfrac{1}{2u_1})$. Using the equation $x^2 + y^2 + z^2 = 1$ we now obtain $z = 1 \pm \sqrt{2}$, where the plus sign gives a negative u_1 and must be dismissed. Thus we obtain the point

$$\left(\sqrt{\sqrt{2}-1},\ \sqrt{\sqrt{2}-1},\ 1-\sqrt{2}\right).$$

However, this point is not feasible, since $x^2 + y^2 = 2(\sqrt{2}-1) > \dfrac{1}{2}$.

Now consider the case when $u_2 > 0$. Then both constraints are active. We already know from (18) that $x = y$. Thus we have to solve the system

$$\begin{cases} x^2 + y^2 + z^2 = 1 \\ x^2 + y^2 = \dfrac{1}{2} \\ x = y. \end{cases}$$

We obtain $x = y = \dfrac{1}{2}$ and $z = \pm\dfrac{1}{\sqrt{2}}$. The minus sign gives a positive $u_1 = \dfrac{1}{\sqrt{2}}$. Finally, $u_2 = 2 - \dfrac{1}{\sqrt{2}} > 0$. — We have now found a point which is obviously feasible; it gives equality in both constraints.

Thus there is exactly one KKT point, namely $\bar{\boldsymbol{x}} = (\dfrac{1}{2}, \dfrac{1}{2}, -\dfrac{1}{\sqrt{2}})$. □

Remark. Let us check the constraint qualification (CQ). It is easy to see that the gradients ∇g_1 and ∇g_2 are linearly independent when $z \neq 0$. For $z = 0$ they are equal, hence positive linearly independent except at the origin, which is outside X. Thus the condition (CQ) is satisfied everywhere.

Note that the investigation above does not prove that $\bar{\boldsymbol{x}}$ *is* a local point of minimum; in fact there might not be one. We will return to this question in chapter 7.4, example 14, page 265.

Example 9. Solve the problem to minimize $f(\boldsymbol{x}) = x_1^3 + x_2^2$ under the conditions

$$x_1^2 + x_2^2 = 9, \qquad x_2 \geq 0.$$

Solution: Put

$$g(\boldsymbol{x}) = -x_2, \qquad h(\boldsymbol{x}) = x_1^2 + x_2^2 - 9.$$

The set in the plane defined by the constraints is compact, and f is continuous. Thus the minimum exists. In particular it is a local minimum. We now examine the (CQ) condition.

For the gradients

$$\nabla g(\boldsymbol{x}) = (0, -1), \qquad \nabla h(\boldsymbol{x}) = (2x_1, 2x_2)$$

to be parallel we must have $x_1 = 0$. But then $x_2 = \pm 3$ by the equality constraint. Thus the inequality constraint is not active and should be disregarded. The vector $\nabla h(\boldsymbol{x})$ is not $\boldsymbol{0}$. It follows that condition (CQ) is satisfied everywhere. Hence by theorem 4 the minimizer is to be found among the KKT points.

We now proceed to find all such points. The Lagrange function for this problem is

$$L(x_1, x_2; u, v) = x_1^3 + x_2^2 + u(-x_2) + v(x_1^2 + x_2^2 - 9), \qquad u \geq 0,$$

with

$$\nabla_{\boldsymbol{x}} L(x_1, x_2; u, v) = \begin{pmatrix} 3x_1^2 \\ 2x_2 \end{pmatrix} + u \begin{pmatrix} 0 \\ -1 \end{pmatrix} + v \begin{pmatrix} 2x_1 \\ 2x_2 \end{pmatrix}.$$

The KKT conditions are

$$\begin{cases} 3x_1^2 \qquad + 2vx_1 = 0 & \text{(a)} \\ 2x_2 - u + 2vx_2 = 0. & \text{(b)} \end{cases}$$

We also have the conditions

$$\begin{cases} x_1^2 + x_2^2 = 9 & \text{(c)} \\ ux_2 = 0 & \text{(d)} \\ x_2 \geq 0. \end{cases}$$

Assume first that $u > 0$. Then $x_2 = 0$ by (d), and we get a contradiction in (b). We conclude that $u = 0$ is the only possibility.

When $u = 0$ equation (b) takes the form

$$(1 + v)x_2 = 0.$$

Assuming first that $v \neq -1$ we get $x_2 = 0$. From (c) we obtain $x_1 = \pm 3$, and then (a) gives the result $v = \mp 9/2$. We thus have the two KKT points

$$(\pm 3, 0).$$

Now letting $v = -1$ we can solve (a) for x_1 and then (c) for x_2, obtaining two further KKT points

$$(0, 3) \quad \text{and} \quad \left(\frac{2}{3}, \frac{\sqrt{77}}{3}\right).$$

All these points are feasible.

Comparing the values of the objective function f at these four points (± 27, 9 and $\frac{239}{27}$ respectively) we finally find the minimizer; the minimum occurs att $(-3, 0)$ and is equal to -27. □

7.3.3 Proof of the Karush-Kuhn-Tucker theorem

It might be tempting to try to prove theorem 4 by converting every equality into two inequalities and use the earlier result of theorem 3, in the same way as in the affine case. However, an equality $h(\boldsymbol{x}) = 0$ is equivalent to the inequalities

$$h(\boldsymbol{x}) \leq 0, \qquad -h(\boldsymbol{x}) \leq 0,$$

and here the gradients are

$$\nabla h(\boldsymbol{x}) \quad \text{and} \quad -\nabla h(\boldsymbol{x}),$$

which are anti-parallel. Thus the constraint qualification (CQ) in theorem 3 is not fulfilled, so this idea is no good.

We must use another strategy for the proof. The idea in the proof below is to introduce new coordinates making the equality constraints confine $\boldsymbol{x}$ to some coordinate planes, i.e., making some variables fixed, and then to use theorem 3 on a problem in the remaining variables, with all constraints given by inequalities.

Now for the details (which should be skipped at first reading and maybe altogether).

PROOF. We begin by showing that the theorem is invariant under a change of coordinates $\boldsymbol{x} \mapsto \boldsymbol{y}$. Assume that

$$\begin{cases} x_1 = x_1(\boldsymbol{y}) = x_1(y_1, \ldots, y_n) \\ \ldots \\ x_n = x_n(\boldsymbol{y}) = x_n(y_1, \ldots, y_n) \end{cases}$$

is a bijective differentiable function from some region in $\mathbf{R}^n$ to a neighbourhood of $\bar{\boldsymbol{x}}$ in $\mathbf{R}^n$. Suppose that $\bar{\boldsymbol{x}}$ corresponds to $\bar{\boldsymbol{y}}$. Then the Jacobian matrix $\boldsymbol{x}'(\bar{\boldsymbol{y}})$ is invertible. By the chain rule

$$\nabla_{\boldsymbol{y}} F(\boldsymbol{x}(\bar{\boldsymbol{y}})) = \boldsymbol{x}'(\bar{\boldsymbol{y}})\, \nabla_{\boldsymbol{x}} F(\bar{\boldsymbol{x}})$$

for any differentiable function F. Hence, changing coordinates from $\boldsymbol{x}$ to $\boldsymbol{y}$ means multiplying all gradients in (CQ) and in (14) by the same invertible matrix $\boldsymbol{x}'(\bar{\boldsymbol{y}})$. Thus, none of the conditions in the theorem are affected.

The idea now is to change coordinates in such a way that all constraint surfaces $h_j(\boldsymbol{x}) = 0$ become coordinate planes, and then prove the theorem in the new variables. For simplicity of notation we assume that there are only two inequality constraints and only one equality constraint

$$g_1(\boldsymbol{x}) \le 0, \quad g_2(\boldsymbol{x}) \le 0, \quad h_1(\boldsymbol{x}) = 0.$$

Due to the constraint qualification (CQ) $\nabla h_1(\bar{\boldsymbol{x}}) \neq \boldsymbol{0}$. We may assume the variables to be numbered so that $\dfrac{\partial h_1}{\partial x_n}(\bar{\boldsymbol{x}}) \neq 0$. Then

$$\begin{cases} y_1 \quad\ = x_1 - \bar{x}_1 \\ \ldots \\ y_{n-1} = x_{n-1} - \bar{x}_{n-1} \\ y_n \quad\ = h_1(\boldsymbol{x}) \end{cases}$$

defines a change of coordinates by the inverse function theorem, since the Jacobian determinant at $\bar{x}$ is $\dfrac{\partial h_1}{\partial x_n}(\bar{\boldsymbol{x}})$, which is non-zero. We have

$$\boldsymbol{x} = \bar{\boldsymbol{x}} \quad \Longleftrightarrow \quad \boldsymbol{y} = \boldsymbol{0}$$

and

$$h_1(\boldsymbol{x}) = 0 \quad \Longleftrightarrow \quad y_n = 0.$$

Henceforth we work only in the $\boldsymbol{y}$-space. Let $F(\boldsymbol{y}) = f(\boldsymbol{x}(\boldsymbol{y}))$, and introduce similar notation for g_1, g_2 and h_1. Furthermore, put $\boldsymbol{y} = (\boldsymbol{y}', y_n)$ with $\boldsymbol{y}' \in \mathbf{R}^{n-1}$ and $\boldsymbol{0} = (\boldsymbol{0}', 0)$. By assumption the function F has a local minimum at $\boldsymbol{0}$ under the constraints

$$G_1(\boldsymbol{y}', y_n) \leq 0, \quad G_2(\boldsymbol{y}', y_n) \leq 0, \quad H_1(\boldsymbol{y}', y_n) = y_n = 0.$$

Putting $\widetilde{g}_i(\boldsymbol{y}') = G_i(\boldsymbol{y}', 0)$, $i = 1, 2$, this implies that the function of $n-1$ variables

$$\widetilde{f}(\boldsymbol{y}') = F(\boldsymbol{y}', 0)$$

has a local minimum at $\boldsymbol{0}'$ under the constraints

$$\widetilde{g}_1(\boldsymbol{y}') \leq 0, \quad \widetilde{g}_2(\boldsymbol{y}') \leq 0.$$

Now we use theorem 3. We then need to verify the constraint qualification in this theorem that the gradients $\boldsymbol{\nabla}_{\boldsymbol{y}'}\widetilde{g}_i(\boldsymbol{0}')$ are positive linearly independent. Assume that

$$\text{(19)} \qquad \lambda_1 \boldsymbol{\nabla}_{\boldsymbol{y}'}\widetilde{g}_1(\boldsymbol{0}') + \lambda_2 \boldsymbol{\nabla}_{\boldsymbol{y}'}\widetilde{g}_2(\boldsymbol{0}') = \boldsymbol{0}', \qquad \lambda_1, \lambda_2 \geq 0.$$

Choose μ_1 so that

$$\text{(20)} \qquad \lambda_1 \frac{\partial G_1}{\partial y_n}(\boldsymbol{0}) + \lambda_2 \frac{\partial G_2}{\partial y_n}(\boldsymbol{0}) + \mu_1 = 0.$$

Since $\boldsymbol{\nabla} H_1(\boldsymbol{0}) = \begin{pmatrix} 0 \\ \vdots \\ 1 \end{pmatrix}$ we can summarize (19) and (20) in

$$\lambda_1 \boldsymbol{\nabla} G_1(\boldsymbol{0}) + \lambda_2 \boldsymbol{\nabla} G_2(\boldsymbol{0}) + \mu_1 \boldsymbol{\nabla} H_1(\boldsymbol{0}) = \boldsymbol{0}.$$

By the assumption (CQ) this implies that $\lambda_1 = \lambda_2 \ (= \mu_1) = 0$.

The assumptions of theorem 3 are now verified. The conclusion is that there exist numbers $u_1, u_2 \geq 0$ such that

$$\boldsymbol{\nabla}_{\boldsymbol{y}'}\widetilde{f}(\boldsymbol{0}') + u_1 \boldsymbol{\nabla}_{\boldsymbol{y}'}\widetilde{g}_1(\boldsymbol{0}') + u_2 \boldsymbol{\nabla}_{\boldsymbol{y}'}\widetilde{g}_2(\boldsymbol{0}') = \boldsymbol{0}'.$$

Putting

$$v_1 = -\frac{\partial F}{\partial y_n}(\boldsymbol{0}) - u_1 \frac{\partial G_1}{\partial y_n}(\boldsymbol{0}) - u_2 \frac{\partial G_2}{\partial y_n}(\boldsymbol{0})$$

the result is that

$$\boldsymbol{\nabla} F(\boldsymbol{0}) + u_1 \boldsymbol{\nabla} G_1(\boldsymbol{0}) + u_2 \boldsymbol{\nabla} G_2(\boldsymbol{0}) + v_1 \begin{pmatrix} 0 \\ \vdots \\ 1 \end{pmatrix} = \boldsymbol{0}.$$

The theorem is proved. □

7.3.4 Some further comments on the constraint qualification

Example 10. Consider the problem of minimizing $f(x_1, x_2) = x_1$ under the constraints

$$g_1(x_1, x_2) = x_2 - x_1^3 \le 0, \qquad g_2(x_1, x_2) = -x_2 \le 0.$$

Obviously the minimum occurs at $(0,0)$. However, this is not a KKT point, since

$$\nabla f(0,0) = \begin{pmatrix} 1 \\ 0 \end{pmatrix}, \qquad \nabla g_1(0,0) = \begin{pmatrix} 0 \\ 1 \end{pmatrix}, \qquad \nabla g_2(0,0) = \begin{pmatrix} 0 \\ -1 \end{pmatrix}.$$

The explanation is that the constraint qualification is not satisfied at the origin; the vectors $\nabla g_1(0,0)$ and $\nabla g_2(0,0)$ are not positive linearly independent since their sum equals zero.

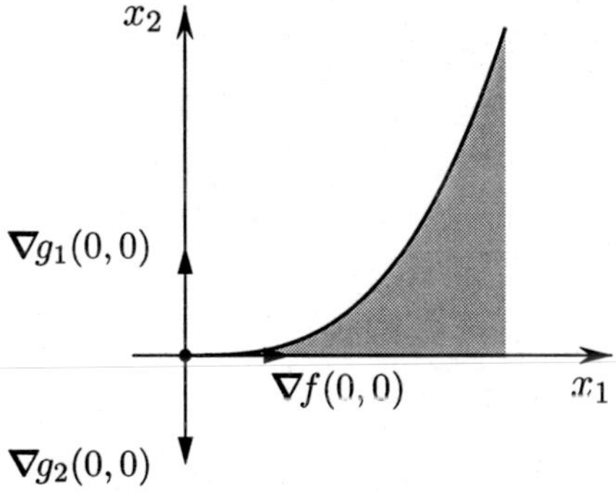

□

This example shows that without the condition (CQ) on the constraints theorem 4 is not true. Our goal in this subsection is to discuss in an informal way why some such side condition is required in the non-affine case, and how this is connected to some complicated behaviour of the boundary of the domain. At the end we make some other comments on the handling of constraints.

Denote by S the set defined by X and the constraints. In section 7.1 we defined a direction $\boldsymbol{d}$ at $\boldsymbol{a}$ as *feasible* if there is some small $\delta > 0$ such that $\boldsymbol{a} + t\boldsymbol{d} \in S$ for $0 \le t < \delta$. As shown in example 2, page 236, in the non-affine case there may be no feasible direction. Even with so simple a constraint as $x_1^2 + x_2^2 = 1$ *no* point has a feasible direction. On the other hand, in the affine case there are always feasible directions unless S consists of one point only.

To relax the concept of feasible direction, we now call a direction $\boldsymbol{d} \neq \boldsymbol{0}$ **attainable** at $\boldsymbol{a}$ if there exists in S a differentiable curve $\boldsymbol{x}(t)$, $0 \leq t < \delta$, with $\boldsymbol{x}(0) = \boldsymbol{a}$ and $\boldsymbol{x}'(0) = \boldsymbol{d}$. In other words, there should be a curve in S which has the tangent vector $\boldsymbol{d}$ at $\boldsymbol{a}$. Denote by $C(\boldsymbol{a})$ the cone of attainable directions at $\boldsymbol{a}$. (Exercise: prove that it really is a cone.)

Of course any feasible direction is attainable. For affine constraints also the reverse statement is true.

Example 11. The figures show some examples of $C(\boldsymbol{a})$ in different situations.

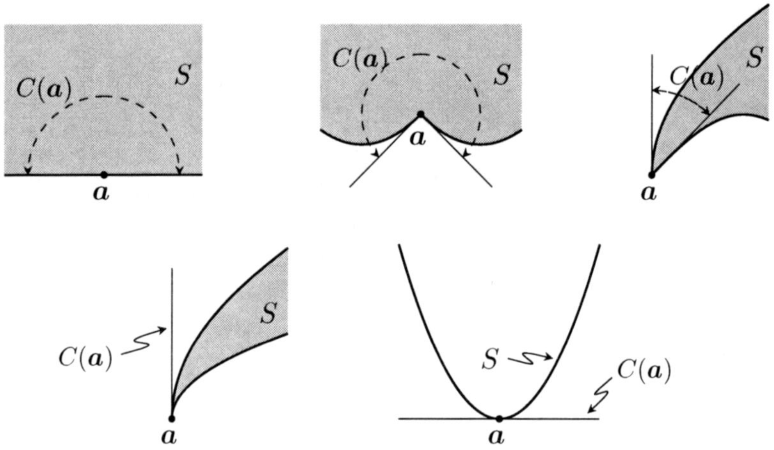

□

LEMMA 3. *Suppose that $\bar{\boldsymbol{x}}$ is a local solution to the problem* $\min_{\boldsymbol{x} \in S} f(\boldsymbol{x})$. *Then*

1. $\nabla f(\bar{\boldsymbol{x}})^T \boldsymbol{d} \geq 0$ *for all* $\boldsymbol{d} \in C(\bar{\boldsymbol{x}})$.

2. $\nabla f(\bar{\boldsymbol{x}})^T \boldsymbol{d} \geq 0$ *for all* $\boldsymbol{d} \in \mathrm{cl}(C(\bar{\boldsymbol{x}}))$.

PROOF. Let $\boldsymbol{d} \in C(\bar{\boldsymbol{x}})$ and $\boldsymbol{x}(t)$ be a curve in S with $\boldsymbol{x}(0) = \bar{\boldsymbol{x}}$ and tangent vector $\boldsymbol{d}$ at $\bar{\boldsymbol{x}}$. Put $F(t) = f(\boldsymbol{x}(t))$, $0 \leq t < \delta$. By assumption this function has a minimum at $t = 0$, consequently $F'_+(0) \geq 0$. Using the chain rule we obtain

$$\nabla f(\bar{\boldsymbol{x}})^T \boldsymbol{x}'(0) \geq 0.$$

Since $\boldsymbol{x}'(0) = \boldsymbol{d}$ this proves the first part of the lemma.

The second part follows from the first by continuity, choosing a sequence $\{\boldsymbol{d}_k\}_1^\infty$ in $C(\bar{\boldsymbol{x}})$ with limit $\boldsymbol{d}$. □

The lemma shows that we could have used attainable directions instead of feasible directions in the development of the theory in section 7.2.

Now assume for simplicity that there are only two constraints active at $\bar{\boldsymbol{x}}$:

$$g(\boldsymbol{x}) \leq 0, \qquad h(\boldsymbol{x}) = 0.$$

Put

$$\widetilde{C}(\bar{\boldsymbol{x}}) = \{\, \boldsymbol{d};\ \nabla g(\bar{\boldsymbol{x}})^T \boldsymbol{d} \leq 0,\ \nabla h(\bar{\boldsymbol{x}})^T \boldsymbol{d} = 0 \,\}$$

(exercise: this is a cone). In the case of affine constraints $C(\bar{\boldsymbol{x}}) = \widetilde{C}(\bar{\boldsymbol{x}})$, in general it can be proved that

$$C(\bar{\boldsymbol{x}}) \subseteq \widetilde{C}(\bar{\boldsymbol{x}}).$$

$\widetilde{C}(\bar{\boldsymbol{x}})$ is the cone of attainable (feasible) directions for the *linearized* problem.

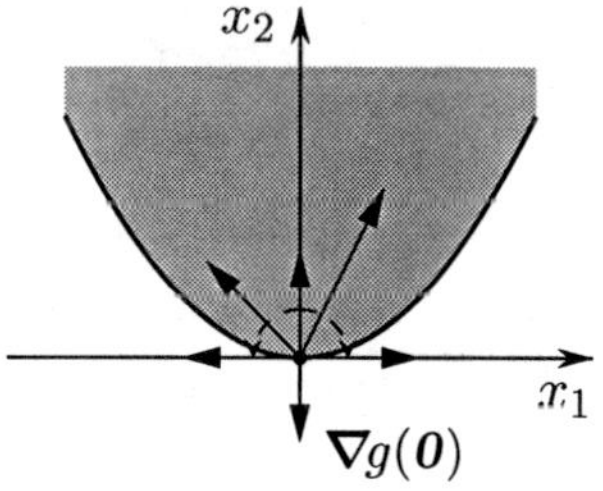

Example 12. Let S be defined by

$$g(\boldsymbol{x}) = x_1^2 - x_2 \leq 0$$

and consider $\bar{\boldsymbol{x}} = \boldsymbol{0}$. We have

$$C(\boldsymbol{0}) = \widetilde{C}(\boldsymbol{0}) = \{\, \boldsymbol{d} = (d_1, d_2)\,;\ d_2 \geq 0 \,\}.$$

□

Example 13. Consider again example 10. Here

$$C(\boldsymbol{0}) = \{\, (t, 0)\,;\ t \geq 0 \,\}, \qquad \widetilde{C}(\boldsymbol{0}) = \{\, (t, 0)\,;\ t \in \mathbf{R} \,\}.$$

The two cones are different! □

The point is that *some* condition on the constraints is required to avoid "pathological" cases like example 10. One such condition, in fact the original one considered by Kuhn and Tucker, is

$$\mathrm{cl}(C(\bar{\boldsymbol{x}})) = \widetilde{C}(\bar{\boldsymbol{x}}).$$

We will not discuss this here. We have chosen to formulate theorem 4 using the weaker condition (CQ), which is easier to verify in practice.

Some further comments. Suppose you happen to substitute two inequality constraints $h(\boldsymbol{x}) \leq 0$ and $-h(\boldsymbol{x}) \leq 0$ for one equality constraint. This does of course not affect the set S of feasible points. However, you have now introduced the two gradients $\nabla h(\bar{\boldsymbol{x}})$ and $-\nabla h(\bar{\boldsymbol{x}})$ in (CQ). Then this condition is *not* satisfied!

Another comment is this: Suppose you add the condition $G(\boldsymbol{x}) = \|\boldsymbol{x} - \bar{\boldsymbol{x}}\|^2 \geq 0$ to the others in a minimization problem. Since this condition is always satisfied, the problem is the same. But

$$\nabla G(\bar{\boldsymbol{x}}) = 2(\bar{\boldsymbol{x}} - \bar{\boldsymbol{x}}) = \boldsymbol{0},$$

and the constraint qualification (CQ) is not satisfied.

The way you handle your constraints is important for the solution of the problem!

Exercises

S 7.3 Consider the problem to minimize $6ax_1 - 4x_2$ subject to

$$\begin{aligned} x_1x_2 - x_2 &\leq 10 \\ 6x_1 + 4x_2 &\leq 38 \\ x_1 \geq 0,\ x_2 &\geq 0. \end{aligned}$$

For what values of a is $(3, 5)$ a KKT point? (1999–04–09, 1)

7.4 Consider the problem to minimize x^2y in the set S defined by

$$\begin{aligned} x^2 + y^2 &\leq 9 \\ x^2 + y &\geq 3. \end{aligned}$$

a) Show that the global minimum exists.

b) Calculate all (CQ) points in S.

c) Calculate all KKT points and solve the problem.

(2006–01–12, 5)

7.5 Consider the problem

$$\begin{aligned} \text{Minimize} \quad & xy + 3x^2 + cy \\ \text{subject to} \quad & y - x^2 \le 1 \\ & x^2 + y^2 \le 5 \\ & y - x \le 2. \end{aligned}$$

For what values of c is $(1, 2)$ a KKT point? (2009–12–16, 4)

7.6 Solve the problem

$$\begin{aligned} \text{Minimize} \quad & x_1 x_2 \\ \text{subject to} \quad & x_1^2 + x_2^2 \le 4 \\ & x_1^2 - x_2^2 \le 1. \end{aligned}$$

S 7.7 The numbers a, b and c are all positive. Find the least value of the sum of three positive numbers x, y and z subject to the constraint

$$\frac{a}{x} + \frac{b}{y} + \frac{c}{z} = 1$$

by the method of Lagrange multipliers.

Hint: Note that the positivity conditions are not active at a minimum point.

7.8 Find the maximum and minimum of $x + 8y$ when $x^4 + y^4 = 1$.

7.9 Solve the problem

$$\begin{aligned} \text{Minimize} \quad & x + y + z \\ \text{subject to} \quad & x^2 + y^2 = 1 \\ & x - y - z = 1. \end{aligned}$$

7.10 From literature in economy we have the following problem.

Let a_j, c_j, $j = 1, \ldots, n$, and b be positive constants. Consider the problem

$$\begin{aligned} \text{Minimize} \quad & \sum_{j=1}^{n} \frac{c_j}{x_j} \\ \text{subject to} \quad & \sum_{j=1}^{n} a_j x_j = b \\ & x_j > 0 \text{ for } j = 1, \ldots, n. \end{aligned}$$

Write down the KKT conditions, and solve the resulting equations.

7.4 Sufficient conditions for a minimum

7.4.1 First order sufficient conditions

We remind the reader that the problem we are considering is to minimize a function f subject to conditions $\boldsymbol{x} \in X$ and constraints

$$\begin{aligned} g_i(\boldsymbol{x}) &\leq 0, \quad i = 1, \ldots, m, \\ h_j(\boldsymbol{x}) &= 0, \quad j = 1, \ldots, l. \end{aligned}$$

Let as usual S denote the set of those $\boldsymbol{x} \in X$ which satisfy the constraints.

Theorem 4 states that if f has a local minimum at $\bar{\boldsymbol{x}} \in S$ and if the constraint qualification is satisfied at $\bar{\boldsymbol{x}}$ then $\bar{\boldsymbol{x}}$ is a KKT point. In other words, assuming (CQ), being a KKT point is a *necessary* condition for $\bar{\boldsymbol{x}}$ to be a local minimum point. We now turn our attention to the question of finding *sufficient* conditions for a local minimum. It turns out that a crucial concept in this context is *convexity*.

To be able to state the following theorems comfortably we need some more notation. For a given point $\bar{\boldsymbol{x}} \in S$ we let

$$I = \{\, i\,;\ g_i(\bar{\boldsymbol{x}}) = 0 \,\},$$

i.e., I contains the indices of those inequality constraints that are active at $\bar{\boldsymbol{x}}$.

THEOREM 5. *Assume that $\bar{\boldsymbol{x}} \in S$ is a KKT point for the minimum problem of theorem 4, page 251, i.e., that $\bar{\boldsymbol{x}}$ is a feasible point satisfying conditions (15)–(17) in that theorem for some $\boldsymbol{u}$, $\boldsymbol{v}$. Also assume that the functions f and g_i, $i \in I$, are convex in a neighbourhood of $\bar{\boldsymbol{x}}$, and that all the h_j are affine there. Then f has a local minimum at $\bar{\boldsymbol{x}}$.*

PROOF. The assumptions imply that the function

$$\boldsymbol{x} \longmapsto L(\boldsymbol{x}; \boldsymbol{u}, \boldsymbol{v}) = f(\boldsymbol{x}) + \sum_{i \in I} u_i g_i(\boldsymbol{x}) + \sum_{j=1}^{l} v_j h_j(\boldsymbol{x})$$

is convex in a neighbourhood of $\bar{\boldsymbol{x}}$. Furthermore, $\bar{\boldsymbol{x}}$ is a stationary point for this function. Consequently $\bar{\boldsymbol{x}}$ is a minimum point for L, by the corollary to theorem 6.11, page 215. Then for any feasible $\boldsymbol{x}$ close to $\bar{\boldsymbol{x}}$

$$f(\bar{\boldsymbol{x}}) = L(\bar{\boldsymbol{x}}; \boldsymbol{u}, \boldsymbol{v}) \leq L(\boldsymbol{x}; \boldsymbol{u}, \boldsymbol{v}) =$$

$$= f(\boldsymbol{x}) + \sum_{i \in I} u_i g_i(\boldsymbol{x}) + \sum_{j=0}^{l} v_j h_j(\boldsymbol{x}) \leq f(\boldsymbol{x}).$$

Thus $\bar{\boldsymbol{x}}$ is a local minimum point for f. □

If the set X is convex and *all* g_i, $i = 1, \ldots, m$, are convex in X, the function f is defined on a convex set S. Then, if f is convex on S a local minimum is also a global minimum. Thus

COROLLARY 1. *If we strengthen the assumptions of the theorem to convexity of the set X, convexity of f and all g_i in all of X and affinity of the h_j in X, then f has a global minimum at $\bar{\boldsymbol{x}}$.*

Example 14. Consider again example 8 on page 252. Here the function f is convex, the set X is convex and the constraint functions g_1 and g_2 are convex. So, once we have found the feasible KKT point $\bar{\boldsymbol{x}} = (\frac{1}{2}, \frac{1}{2}, -\frac{1}{\sqrt{2}})$, we know by the corollary that it is *the* minimum. We do not have to look any further. In particular, we need not worry about the possible existence of points not satisfying the condition (CQ). □

7.4.2 Second order sufficient conditions

In this section all functions are assumed twice continuously differentiable.

We denote by $\boldsymbol{\nabla}^2 f(\boldsymbol{x})$ the **Hessian** of the function f, i.e., the matrix

$$\left(\frac{\partial^2 f}{\partial x_j \partial x_k}(\boldsymbol{x}) \right)$$

of second order derivatives of f at $\boldsymbol{x}$. This matrix is symmetric. We remind the reader of the well known result that for the unconstrained problem of minimizing f a sufficient condition for the point $\bar{\boldsymbol{x}}$ to yield a local minimum is that it is a stationary point and that the Hessian is positive definite, that is,

$$\boldsymbol{d}^T \boldsymbol{\nabla}^2 f(\bar{\boldsymbol{x}}) \boldsymbol{d} > 0 \qquad \text{for all } \boldsymbol{d} \neq \boldsymbol{0}.$$

Note that this condition is not necessary; even for functions of one variable the reader can find examples of functions with a minimum at the origin but the Hessian (i.e., the second derivative) equal to zero there.

That the Hessian is positive *semi*definite is a necessary condition for a local minimum. However, this condition is not sufficient. One example is the following, over which the reader might ponder some minutes. Let

$$f(x_1, x_2) = (x_1^2 - x_2)(3x_1^2 - x_2).$$

At $(0,0)$ the Hessian is

$$\begin{pmatrix} 0 & 0 \\ 0 & 2 \end{pmatrix},$$

which is positive semidefinite. The function has neither a minimum nor a maximum at $(0,0)$, as can be seen from a contemplation of its sign. See the figure.

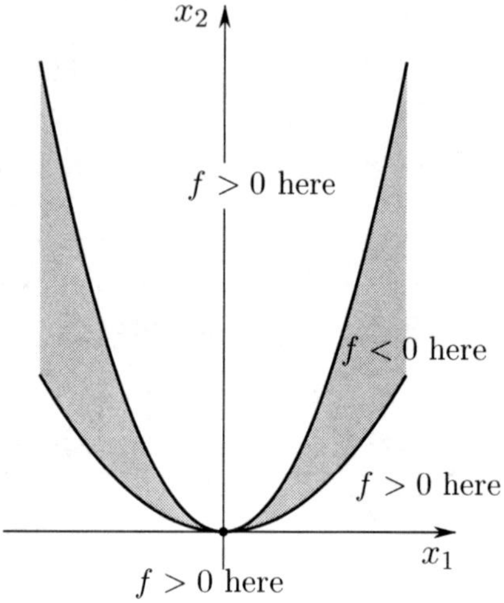

We now focus on the constrained problem. As usual $L(\boldsymbol{x}; \boldsymbol{u}, \boldsymbol{v})$ denotes the Lagrange function. We denote by $\boldsymbol{\nabla}^2_{\boldsymbol{xx}} L(\boldsymbol{x}; \boldsymbol{u}, \boldsymbol{v})$ the Hessian of L with respect to the x-variables. The following theorem states a sufficient condition for a KKT point $\bar{\boldsymbol{x}}$ to be a minimum point. It is not the ultimate result; we will improve somewhat on it later.

THEOREM 6. *Assume that $\bar{\boldsymbol{x}} \in S$, $\bar{\boldsymbol{u}}$ and $\bar{\boldsymbol{v}}$ satisfy the conditions*

1. $\boldsymbol{\nabla}_{\boldsymbol{x}} L(\bar{\boldsymbol{x}}; \bar{\boldsymbol{u}}, \bar{\boldsymbol{v}}) = \boldsymbol{0}$,

   $\bar{\boldsymbol{u}} \geq \boldsymbol{0}$,

   $\bar{u}_i g_i(\bar{\boldsymbol{x}}) = 0, \quad i = 1, \ldots, m.$

2. $\boldsymbol{d}^T \boldsymbol{\nabla}^2_{\boldsymbol{xx}} L(\bar{\boldsymbol{x}}; \bar{\boldsymbol{u}}, \bar{\boldsymbol{v}}) \boldsymbol{d} > 0$ *for all* $\boldsymbol{d} \neq \boldsymbol{0}$ *satisfying*

(21) $$\boldsymbol{\nabla} g_i(\bar{\boldsymbol{x}})^T \boldsymbol{d} \leq 0, \quad i \in I$$

and

(22) $$\boldsymbol{\nabla} h_j(\bar{\boldsymbol{x}})^T \boldsymbol{d} = 0, \quad j = 1, \ldots, l.$$

Then $\bar{\boldsymbol{x}}$ is a strict local minimum for the constrained function f.

The first condition means of course exactly that $\bar{\boldsymbol{x}}$ is a KKT point.

Remark. It might happen that conditions (21)–(22) are satisfied by $\boldsymbol{d} = \boldsymbol{0}$ only. Then the second condition is void. In this case the KKT point $\bar{\boldsymbol{x}}$ is certainly a minimum.

PROOF. Assume that $\bar{\boldsymbol{x}}$ is not a strict local minimum. Then there exists a sequence $\{\boldsymbol{x}_k\}_1^\infty$ in S, all $\boldsymbol{x}_k \neq \bar{\boldsymbol{x}}$, such that $\boldsymbol{x}_k \to \bar{\boldsymbol{x}}$ as $k \to \infty$ and $f(\boldsymbol{x}_k) \leq f(\bar{\boldsymbol{x}})$. Put

$$\lambda_k = ||\boldsymbol{x}_k - \bar{\boldsymbol{x}}||, \qquad \boldsymbol{d}_k = \frac{\boldsymbol{x}_k - \bar{\boldsymbol{x}}}{||\boldsymbol{x}_k - \bar{\boldsymbol{x}}||}.$$

Since $||\boldsymbol{d}_k|| = 1$, by Bolzano-Weierstrass' theorem, page 117, the sequence $\{\boldsymbol{d}_k\}_1^\infty$ has a convergent subsequence, with limit $\boldsymbol{d}$ say. After a change of notation we may assume that $\{\boldsymbol{d}_k\}_1^\infty$ itself is this subsequence. Then

$$\boldsymbol{x}_k = \bar{\boldsymbol{x}} + \lambda_k \boldsymbol{d}_k$$

where as $k \to \infty$

$$\boldsymbol{x}_k \to \bar{\boldsymbol{x}}, \qquad \lambda_k \to 0, \qquad \boldsymbol{d}_k \to \boldsymbol{d}$$

and $||\boldsymbol{d}|| = 1$.

We first show that $\boldsymbol{d}$ satisfies (21) and (22). If $i \in I$ we have by Taylor's theorem that

$$\begin{aligned} 0 \geq g_i(\boldsymbol{x}_k) &= g_i(\bar{\boldsymbol{x}}) + \nabla g_i(\bar{\boldsymbol{x}})^T \lambda_k \boldsymbol{d}_k + \lambda_k^2 O(1) = \\ &= 0 + \lambda_k \nabla g_i(\bar{\boldsymbol{x}})^T \boldsymbol{d}_k + \lambda_k^2 O(1). \end{aligned}$$

Here $O(1)$ denotes a sequence that is bounded for all k. Since $\lambda_k > 0$ this leads to

$$0 \geq \nabla g_i(\bar{\boldsymbol{x}})^T \boldsymbol{d}_k + \lambda_k O(1).$$

Letting $k \to \infty$ we obtain (21). Similarly, for $j = 1, \ldots, l$,

$$\begin{aligned} 0 = h_j(\boldsymbol{x}_k) &= h_j(\bar{\boldsymbol{x}}) + \nabla h_j(\bar{\boldsymbol{x}})^T \lambda_k \boldsymbol{d}_k + \lambda_k^2 O(1) = \\ &= 0 + \lambda_k \nabla h_j(\bar{\boldsymbol{x}})^T \boldsymbol{d}_k + \lambda_k^2 O(1). \end{aligned}$$

Dividing by λ_k and letting $k \to \infty$ proves (22).

We are now going to show that in fact

$$d^T \nabla^2_{xx} L(\bar{x}; \bar{u}, \bar{v}) d \leq 0. \tag{23}$$

This leads to a contradiction to the second assumption in the theorem. Consequently, the theorem will be proved.

Since $x_k \in S$ we have

$$L(x_k; \bar{u}, \bar{v}) \leq f(x_k) \leq f(\bar{x}) = L(\bar{x}; \bar{u}, \bar{v}).$$

Thus by Taylor's formula

$$0 \geq L(x_k; \bar{u}, \bar{v}) - L(\bar{x}; \bar{u}, \bar{v}) =$$

$$= \lambda_k \nabla_x L(\bar{x}; \bar{u}, \bar{v})^T d_k + \lambda_k^2 d_k^T \nabla^2_{xx} L(\bar{x}; \bar{u}, \bar{v}) d_k + \lambda_k^3 O(1).$$

The first term on the right is zero by the first assumption, and since $\lambda_k \neq 0$ we get

$$0 \geq d_k^T \nabla^2_{xx} L(\bar{x}; \bar{u}, \bar{v}) d_k + \lambda_k O(1).$$

Letting $k \to \infty$ we obtain (23), and the proof is finished. □

Example 15. Consider again example 9, page 255, and forget that we already know for compactness reasons that $(-3, 0)$ is a global minimum. At this point, with $u = 0$ and $v = \frac{9}{2}$ we have

$$\nabla^2_{xx} L(x_1, x_2; u, v) = \begin{pmatrix} 6x_1+2v & 0 \\ 0 & 2+2v \end{pmatrix} = \begin{pmatrix} -9 & 0 \\ 0 & 11 \end{pmatrix}.$$

This matrix is if course indefinite. However, we should only consider its effect on those d which satisfy (21) and (22). In this case

$$\nabla g(-3, 0)^T d \leq 0 \iff \begin{pmatrix} 0 & -1 \end{pmatrix} \begin{pmatrix} d_1 \\ d_2 \end{pmatrix} \leq 0 \iff -d_2 \leq 0$$

and

$$\nabla h(-3, 0)^T d = 0 \iff \begin{pmatrix} -6 & 0 \end{pmatrix} \begin{pmatrix} d_1 \\ d_2 \end{pmatrix} = 0 \iff d_1 = 0.$$

Thus $d = (0, t)$ with $t > 0$, and for such a d we have

$$d^T \nabla^2_{xx} L(-3, 0;\ 0, \tfrac{9}{2}) d = 11t^2 > 0.$$

By the theorem this shows that $(-3, 0)$ is a local minimum point. □

Example 16. Let us go back to example 7, page 249. The task was to minimize $f(\boldsymbol{x}) = 8x_1x_2 + 7x_3$ under constraints

$$g_1(\boldsymbol{x}) = x_1^2 + x_2^2 + x_3^3 - 2 \le 0, \qquad g_2(\boldsymbol{x}) = -x_3 \le 0.$$

In example 7 we confirmed that $\bar{\boldsymbol{x}} = (1, -1, 0)$ is a KKT point with $\bar{\boldsymbol{u}} = (4, 7)$.

The Lagrange function is

$$L(\boldsymbol{x}; \boldsymbol{u}) = 8x_1x_2 + 7x_3 + u_1(x_1^2 + x_2^2 + x_3^3 - 2) + u_2(-x_3), \qquad \boldsymbol{u} \ge \boldsymbol{0}.$$

The Hessian of L at $(\bar{\boldsymbol{x}}, \bar{\boldsymbol{u}})$ is

$$\nabla^2_{\boldsymbol{xx}} L(\bar{\boldsymbol{x}}; \bar{\boldsymbol{u}}) = \begin{pmatrix} 2\bar{u}_1 & 8 & 0 \\ 8 & 2\bar{u}_1 & 0 \\ 0 & 0 & 6\bar{u}_1\bar{x}_3 \end{pmatrix} = \begin{pmatrix} 8 & 8 & 0 \\ 8 & 8 & 0 \\ 0 & 0 & 0 \end{pmatrix}.$$

Since $\nabla g_1(\bar{\boldsymbol{x}}) = (2, -2, 0)$ and $\nabla g_2(\bar{\boldsymbol{x}}) = (0, 0, -1)$ conditions (21)–(22) become

$$2d_1 - 2d_2 \le 0, \qquad -d_3 \le 0.$$

We have

$$\boldsymbol{d}^T \nabla^2_{\boldsymbol{xx}} L(\bar{\boldsymbol{x}}; \bar{\boldsymbol{u}}) \boldsymbol{d} = 8d_1^2 + 16d_1d_2 + 8d_2^2 = 8(d_1 + d_2)^2,$$

which is not strictly positive for all these $\boldsymbol{d}$. The theorem produces no result in this case. However, a slight refinement is possible, which we will now discuss. □

In order to improve on theorem 6 we divide the index set I into two parts:

$$I^+ = \{\, i \in I\,;\, \bar{u}_i > 0 \,\}, \qquad I^0 = \{\, i \in I\,;\, u_i = 0 \,\}.$$

The in some sense normal case for an active constraint is that $\bar{u}_i > 0$; hence the set I^0 should be thought of as representing exceptional — but possible — cases. Referring back to the proof of the theorem we are going to show that the vector $\boldsymbol{d}$ appearing there in fact satisfies a stronger condition than (21), namely

$$(21') \qquad \begin{cases} \nabla g_i(\bar{\boldsymbol{x}})^T \boldsymbol{d} = 0, & i \in I^+ \\ \nabla g_i(\bar{\boldsymbol{x}})^T \boldsymbol{d} \le 0, & i \in I^0. \end{cases}$$

To do this we use the fact that $\nabla_{\boldsymbol{x}} L(\bar{\boldsymbol{x}}; \bar{\boldsymbol{u}}, \bar{\boldsymbol{v}}) = \boldsymbol{0}$. Taking the scalar product with $\boldsymbol{d}$ results in

$$\nabla f(\bar{\boldsymbol{x}})^T \boldsymbol{d} + \sum_{i \in I} \bar{u}_i \nabla g_i(\bar{\boldsymbol{x}})^T \boldsymbol{d} + \sum_{j=1}^{l} \bar{v}_j \nabla h_j(\bar{\boldsymbol{x}})^T \boldsymbol{d} = 0. \tag{24}$$

Here the last sum is equal to zero and all other terms are less than or equal to zero. This is proved for the g_i and h_j in the proof of theorem 6, and the proof for f is done similarly. Now it follows that all terms appearing in (24) must be equal to zero. In particular, $\bar{u}_i \nabla g_i(\bar{\boldsymbol{x}})^T \boldsymbol{d} = 0$ for all $i \in I$, which implies that

$$\nabla g_i(\bar{\boldsymbol{x}})^T \boldsymbol{d} = 0 \quad \text{when } i \in I^+.$$

With this we have obtained a proof of the following generalization of theorem 6.

THEOREM 7. *Assume that $\bar{\boldsymbol{x}} \in S$, $\bar{\boldsymbol{u}}$ and $\bar{\boldsymbol{v}}$ satisfy the conditions*

1. $\nabla_{\boldsymbol{x}} L(\bar{\boldsymbol{x}}; \bar{\boldsymbol{u}}, \bar{\boldsymbol{v}}) = \boldsymbol{0}$,

   $\bar{\boldsymbol{u}} \geq \boldsymbol{0}$,

   $\bar{u}_i g_i(\bar{\boldsymbol{x}}) = 0, \quad i = 1, \dots, m.$

2. $\boldsymbol{d}^T \nabla^2_{\boldsymbol{xx}} L(\bar{\boldsymbol{x}}; \bar{\boldsymbol{u}}, \bar{\boldsymbol{v}}) \boldsymbol{d} > 0$ *for all* $\boldsymbol{d} \neq 0$ *satisfying*

$$\begin{cases} \nabla g_i(\bar{\boldsymbol{x}})^T \boldsymbol{d} = 0, & i \in I^+ \\ \nabla g_i(\bar{\boldsymbol{x}})^T \boldsymbol{d} \leq 0, & i \in I^0 \end{cases} \tag{21'}$$

   and

$$\nabla h_j(\bar{\boldsymbol{x}})^T \boldsymbol{d} = 0, \quad j = 1, \dots, l. \tag{22}$$

Then $\bar{\boldsymbol{x}}$ is a strict local minimum for the constrained function f.

Since (21′) is a stronger condition than (21), the set of directions defined by (21′) is smaller. Hence condition 2 in theorem 7 is weaker than condition 2 in theorem 6.

Example 17. Using (21′) instead of (21) in example 16 we obtain the following equations for $\boldsymbol{d}$:

$$\begin{cases} 2d_1 - 2d_2 = 0 \\ -d_3 = 0 \end{cases} \iff \begin{cases} d_1 = t \\ d_2 = t \\ d_3 = 0. \end{cases}$$

With $\boldsymbol{d} = (t, t, 0)$, $t \neq 0$, we have

$$\boldsymbol{d}^T \nabla^2_{\boldsymbol{xx}} L(\bar{\boldsymbol{x}}; \bar{\boldsymbol{u}})\boldsymbol{d} = 32t^2 > 0.$$

Hence there is a local minimum at $\bar{\boldsymbol{x}}$ by theorem 7. □

Exercises

S 7.11 Solve the problem

$$\begin{array}{ll} \text{Minimize} & f(x,y) = x^2 + 2xy + 2y^2 - 5x - 5y \\ \text{subject to} & x + y \leq 1 \\ & y \leq 1 - x^2. \end{array}$$

7.12 Solve the problem

$$\begin{array}{ll} \text{Maximize} & f(x,y) = 2x + y - x^2 \\ \text{subject to} & 2x + 3y \leq 6 \\ & 2x + y \leq 4. \end{array}$$

7.13 Solve the problem

$$\begin{array}{ll} \text{Minimize} & x^2 + y^2 + xy + z^2 \\ \text{subject to} & x + y \geq z + 1 \\ & y^2 \leq x \\ & z \leq 7. \end{array}$$

S 7.14 Show that at the global minimizer $\boldsymbol{x}^*$ of the problem

$$\begin{array}{ll} \text{Minimize} & x_1 x_2 \\ \text{subject to} & x_2 - (x_1 - 1)^3 = 0, \end{array}$$

the matrix $\nabla^2_{\boldsymbol{xx}} L(\boldsymbol{x}^*; v^*)$ is indefinite, but the second order sufficient conditions are nevertheless satisfied.

7.15 Consider the optimization problem

$$\begin{aligned} \text{Minimize} \quad & x^3 - x^2 y \\ \text{subject to} \quad & x^4 + 2y^4 \le \frac{32}{9}, \\ & 5x^4 + y^4 \le 6. \end{aligned}$$

a) Prove that a solution exists.

b) Check which of the following four points that are KKT points: $(0,0)$, $(-1,1)$, $(1,-1)$ and $(-\frac{4}{3},\frac{2}{3})$.

c) Test the KKT points found in b) by the second order sufficient conditions for a local minimum. — Can you conclude, without further calculation, that one of these KKT points is a solution to the problem, i.e. a global minimum?

(2006–04–25, 5)

7.16 Consider the following problem:

$$\begin{aligned} \text{Minimize} \quad & -4x_1 + 2x_2 - 3x_3^2 \\ \text{subject to} \quad & x_1 + x_2 + 9x_3 \le 0 \\ & -x_1 + 2x_2 + 3x_3^2 = 0. \end{aligned}$$

a) Find all KKT points, and test for the second order optimality conditions.

b) Show that in fact no minimum exists.

7.5 Quadratic programming

In this section we specialize to the case when the objective function is quadratic,

$$f(\boldsymbol{x}) = \frac{1}{2}\boldsymbol{x}^T \boldsymbol{H}\boldsymbol{x} + \boldsymbol{c}^T\boldsymbol{x} + d,$$

($\boldsymbol{H}$ symmetric) and the constraints are affine, so called *quadratic programming*.

This is an important subject in its own right. Many methods and algorithms have been devised for this particular application, as in the case of linear programming. We will only touch upon the subject here.

Consider first the case of equality constraints, i.e., the problem of minimizing $f(\boldsymbol{x})$ subject to

$$\boldsymbol{A}\boldsymbol{x} = \boldsymbol{b},$$

where $\boldsymbol{A}$ is an $m \times n$-matrix and $\boldsymbol{b}$ an m-vector. We assume as usual that the set S defined by these restrictions is non-empty, and that the rank of $\boldsymbol{A}$ is m.

If $\boldsymbol{H}$ is positive definite f is strictly convex. Since the constraints are affine there is a unique minimum. The Lagrange function is

$$L(\boldsymbol{x}; \boldsymbol{v}) = f(\boldsymbol{x}) + \sum_{i=1}^{m} v_i(\boldsymbol{A}\boldsymbol{x} - \boldsymbol{b})_i,$$

if the ith constraint is written $(\boldsymbol{A}\boldsymbol{x} - \boldsymbol{b})_i = 0$. We prefer to use vector notation here, writing

$$L(\boldsymbol{x}; \boldsymbol{v}) = \tfrac{1}{2}\boldsymbol{x}^T \boldsymbol{H}\boldsymbol{x} + \boldsymbol{c}^T\boldsymbol{x} + d + \boldsymbol{v}^T(\boldsymbol{A}\boldsymbol{x} - \boldsymbol{b}).$$

Then

$$\nabla_{\boldsymbol{x}} L(\boldsymbol{x}; \boldsymbol{v}) = \boldsymbol{H}\boldsymbol{x} + \boldsymbol{c} + \boldsymbol{A}^T\boldsymbol{v},$$

and the conditions for a local minimum are

$$(25) \qquad \begin{cases} \boldsymbol{H}\boldsymbol{x} + \boldsymbol{c} + \boldsymbol{A}^T\boldsymbol{v} = \boldsymbol{0} \\ \boldsymbol{A}\boldsymbol{x} - \boldsymbol{b} \qquad\quad = \boldsymbol{0}. \end{cases}$$

Using block matrices this system of linear equations can be written

$$\begin{pmatrix} \boldsymbol{H} & \boldsymbol{A}^T \\ \boldsymbol{A} & \boldsymbol{O} \end{pmatrix} \begin{pmatrix} \boldsymbol{x} \\ \boldsymbol{v} \end{pmatrix} = \begin{pmatrix} -\boldsymbol{c} \\ \boldsymbol{b} \end{pmatrix}.$$

If $\boldsymbol{H}$ is invertible we have

$$\begin{pmatrix} \boldsymbol{H} & \boldsymbol{A}^T \\ \boldsymbol{A} & \boldsymbol{O} \end{pmatrix}^{-1} = \begin{pmatrix} \boldsymbol{H}^{-1} - \boldsymbol{H}^{-1}\boldsymbol{A}^T\widetilde{\boldsymbol{H}}\boldsymbol{A}\boldsymbol{H}^{-1} & \boldsymbol{H}^{-1}\boldsymbol{A}^T\widetilde{\boldsymbol{H}} \\ \widetilde{\boldsymbol{H}}\boldsymbol{A}\boldsymbol{H}^{-1} & -\widetilde{\boldsymbol{H}} \end{pmatrix},$$

where we have introduced the notation $\widetilde{\boldsymbol{H}} = (\boldsymbol{A}\boldsymbol{H}^{-1}\boldsymbol{A}^T)^{-1}$. (Exercise: check this by multiplying out the matrix and its inverse.) Thus

$$\begin{cases} \boldsymbol{x} = -(\boldsymbol{H}^{-1} - \boldsymbol{H}^{-1}\boldsymbol{A}^T\widetilde{\boldsymbol{H}}\boldsymbol{A}\boldsymbol{H}^{-1})\boldsymbol{c} + \boldsymbol{H}^{-1}\boldsymbol{A}^T\widetilde{\boldsymbol{H}}\boldsymbol{b} \\ \boldsymbol{v} = -\widetilde{\boldsymbol{H}}\boldsymbol{A}\boldsymbol{H}^{-1}\boldsymbol{c} - \widetilde{\boldsymbol{H}}\boldsymbol{b}. \end{cases}$$

These formulas are of theoretical importance. However, for practical purposes the system (25) is solved using factorization techniques.[5]

Example 18. Let $f(\boldsymbol{x}) = x_1^2 + 2x_1x_2 + 4x_2^2 + 5x_1$. Suppose that we want to minimize $f(\boldsymbol{x})$ subject to the constraint $x_1 + 3x_2 = 4$. In this case

$$\boldsymbol{H} = \begin{pmatrix} 2 & 2 \\ 2 & 8 \end{pmatrix}, \qquad \boldsymbol{c} = \begin{pmatrix} 5 \\ 0 \end{pmatrix}, \qquad \boldsymbol{A} = \begin{pmatrix} 1 & 3 \end{pmatrix}, \qquad \boldsymbol{b} = \begin{pmatrix} 4 \end{pmatrix}.$$

The block matrix becomes

$$\left(\begin{array}{cc|c} 2 & 2 & 1 \\ 2 & 8 & 3 \\ \hline 1 & 3 & 0 \end{array}\right)$$

and the system to be solved is

$$\begin{pmatrix} 2 & 2 & 1 \\ 2 & 8 & 3 \\ 1 & 3 & 0 \end{pmatrix} \begin{pmatrix} x_1 \\ x_2 \\ v \end{pmatrix} = \begin{pmatrix} -5 \\ 0 \\ 4 \end{pmatrix}.$$

The solution is $x_1 = -\frac{37}{14}$, $x_2 = \frac{31}{14}$, $v = -\frac{58}{14}$. The minimum point is thus

$$\boldsymbol{x} = \left(-\frac{37}{14}, \frac{31}{14}\right).$$

The minimum value is $\dfrac{329}{196}$. □

Other techniques, not using KKT theory, are possible in quadratic programming. Assume, as in LP theory, that the $\boldsymbol{x}$-variables are partitioned into two sets, corresponding to a partition of $\boldsymbol{A}$ as

$$\boldsymbol{A} = \begin{pmatrix} \boldsymbol{B} & \boldsymbol{N} \end{pmatrix},$$

with $\boldsymbol{B}$ of type $m \times m$ and invertible, and $\boldsymbol{N}$ of type $m \times (n-m)$. Correspondingly there are partitions

$$\boldsymbol{x} = \begin{pmatrix} \boldsymbol{x}_B \\ \boldsymbol{x}_N \end{pmatrix} \qquad \boldsymbol{c} = \begin{pmatrix} \boldsymbol{c}_B \\ \boldsymbol{c}_N \end{pmatrix} \qquad \boldsymbol{H} = \begin{pmatrix} \boldsymbol{H}_{BB} & \boldsymbol{H}_{BN} \\ \boldsymbol{H}_{BN}^T & \boldsymbol{H}_{NN} \end{pmatrix}.$$

[5] The matrix in (25) can be shown to be non-singular under weaker assumptions than that $\boldsymbol{H}$ is invertible. Cf. the discussion after example 22.

We have

$$\boldsymbol{A}\boldsymbol{x} = \boldsymbol{b} \iff \boldsymbol{B}\boldsymbol{x}_B + \boldsymbol{N}\boldsymbol{x}_N = \boldsymbol{b} \iff \boldsymbol{x}_B = \boldsymbol{B}^{-1}\boldsymbol{b} - \boldsymbol{B}^{-1}\boldsymbol{N}\boldsymbol{x}_N.$$

This can be wiewed as a solution of the system $\boldsymbol{A}\boldsymbol{x} = \boldsymbol{b}$ in the $n-m$ variables (parameters) $\boldsymbol{x}_N$. Substitution into $f(\boldsymbol{x})$ gives an unconstrained quadratic function of these variables

$$\varphi(\boldsymbol{x}_N) =$$

$$= \frac{1}{2} \begin{pmatrix} \boldsymbol{x}_B^T & \boldsymbol{x}_N^T \end{pmatrix} \begin{pmatrix} \boldsymbol{H}_{BB} & \boldsymbol{H}_{BN} \\ \boldsymbol{H}_{BN}^T & \boldsymbol{H}_{NN} \end{pmatrix} \begin{pmatrix} \boldsymbol{x}_B \\ \boldsymbol{x}_N \end{pmatrix} + \begin{pmatrix} \boldsymbol{c}_B^T & \boldsymbol{c}_N^T \end{pmatrix} \begin{pmatrix} \boldsymbol{x}_B \\ \boldsymbol{x}_N \end{pmatrix} + d.$$

Multiplying out and rearranging, expressing $\boldsymbol{x}_B$ in $\boldsymbol{x}_N$, we obtain

$$\varphi(\boldsymbol{x}_N) =$$

$$= \tfrac{1}{2}\left(\boldsymbol{x}_B^T \boldsymbol{H}_{BB}\boldsymbol{x}_B + 2\boldsymbol{x}_B^T \boldsymbol{H}_{BN}\boldsymbol{x}_N + \boldsymbol{x}_N^T \boldsymbol{H}_{NN}\boldsymbol{x}_N\right) + \boldsymbol{c}_B^T\boldsymbol{x}_B + \boldsymbol{c}_N^T\boldsymbol{x}_N + d =$$

$$= \dots$$

$$= \tfrac{1}{2}\boldsymbol{x}_N^T \left(\boldsymbol{H}_{NN} + (\boldsymbol{B}^{-1}\boldsymbol{N})^T\boldsymbol{H}_{BB}(\boldsymbol{B}^{-1}\boldsymbol{N}) - 2(\boldsymbol{B}^{-1}\boldsymbol{N})^T\boldsymbol{H}_{BN}\right)\boldsymbol{x}_N +$$

$$+ \left(\boldsymbol{c}_N^T - \boldsymbol{c}_B^T\boldsymbol{B}^{-1}\boldsymbol{N} + (\boldsymbol{B}^{-1}\boldsymbol{b})^T\boldsymbol{H}_{BN} - (\boldsymbol{B}^{-1}\boldsymbol{b})^T\boldsymbol{H}_{BB}(\boldsymbol{B}^{-1}\boldsymbol{N})\right)\boldsymbol{x}_N +$$

$$+ \tfrac{1}{2}(\boldsymbol{B}^{-1}\boldsymbol{b})^T\boldsymbol{H}_{BB}(\boldsymbol{B}^{-1}\boldsymbol{b}) + \boldsymbol{c}_B^T\boldsymbol{B}^{-1}\boldsymbol{b} + d.$$

This is of the form

$$\varphi(\boldsymbol{x}_N) = \tfrac{1}{2}\boldsymbol{x}_N^T\widehat{\boldsymbol{H}}\boldsymbol{x}_N + \widehat{\boldsymbol{c}}^T\boldsymbol{x}_N + \widehat{d},$$

a quadratic function in the variables $\boldsymbol{x}_N$. The minimum can be found by setting $\nabla\varphi(\boldsymbol{x}_N)$ equal to zero, which gives

$$\widehat{\boldsymbol{H}}\boldsymbol{x}_N + \widehat{\boldsymbol{c}} = \boldsymbol{0}.$$

If $\widehat{\boldsymbol{H}}$ is positive definite this problem has a unique solution.

Example 19. Consider the problem of minimizing

$$f(\boldsymbol{x}) = x_1^2 + x_2^2 + x_3^2 + x_4^2$$

subject to the constraints

$$(26)\qquad \begin{cases} 2x_1 + x_2 + 3x_3 + x_4 = 7 \\ x_1 + x_2 + 9x_3 - 2x_4 = 9. \end{cases}$$

If we partition the matrix $\boldsymbol{A}$ according to

$$\boldsymbol{A} = \begin{pmatrix} \boldsymbol{B} & \boldsymbol{N} \end{pmatrix} = \left(\begin{array}{cc|cc} 2 & 1 & 3 & 1 \\ 1 & 1 & 9 & -2 \end{array}\right),$$

$\boldsymbol{B}$ is easily seen to be invertible. Solving system (26) for $\boldsymbol{x}_B = \begin{pmatrix} x_1 \\ x_2 \end{pmatrix}$ we get

$$\begin{cases} x_1 = -2 + 6x_3 - 3x_4 \\ x_2 = 11 - 15x_3 + 5x_4. \end{cases}$$

Substitution into $f(\boldsymbol{x})$ results in a function of the two variables x_3 and x_4:

$$\varphi(x_3, x_4) = (-2 + 6x_3 - 3x_4)^2 + (11 - 15x_3 + 5x_4)^2 + x_3^2 + x_4^2 =$$
$$= 262x_3^2 + 35x_4^2 - 186x_3x_4 - 354x_3 + 122x_4 + 125.$$

Here the Hessian

$$2\begin{pmatrix} 262 & -93 \\ -93 & 35 \end{pmatrix}$$

is positive definite. The (global) minimum is given by $\boldsymbol{\nabla}\varphi(x_3, x_4) = 0$, that is,

$$\begin{cases} 524x_3 - 186x_4 - 354 = 0 \\ -186x_3 + 70x_4 + 122 = 0. \end{cases}$$

The solution of this system is $x_3 = \frac{522}{521}$, $x_4 = \frac{479}{521}$, which in turn gives $x_1 = \frac{653}{521}$, $x_2 = \frac{296}{521}$. To two decimals the minimum point is

$$(1.25, 0.57, 1.00, 0.92)$$

and the minimum value is 3.74.

Exercise: check the details of the computations, e.g. using Matlab.

□

Above we have partitioned $\boldsymbol{x}$ as a sum

$$\boldsymbol{x} = \begin{pmatrix} \boldsymbol{x}_B \\ \boldsymbol{0} \end{pmatrix} + \begin{pmatrix} \boldsymbol{0} \\ \boldsymbol{x}_N \end{pmatrix} = \begin{pmatrix} \boldsymbol{B}^{-1} \\ \boldsymbol{O} \end{pmatrix} \boldsymbol{b} + \begin{pmatrix} -\boldsymbol{B}^{-1}\boldsymbol{N} \\ \boldsymbol{I} \end{pmatrix} \boldsymbol{x}_N.$$

Note that

$$\boldsymbol{A}\begin{pmatrix} \boldsymbol{B}^{-1} \\ \boldsymbol{O} \end{pmatrix} = \boldsymbol{I}, \qquad \boldsymbol{A}\begin{pmatrix} -\boldsymbol{B}^{-1}\boldsymbol{N} \\ \boldsymbol{I} \end{pmatrix} = \boldsymbol{O}.$$

There is a more general approach. Let $\boldsymbol{Y}$ and $\boldsymbol{Z}$ be any matrices of type $n \times m$ and $n \times (n-m)$ respectively, which satisfy

$$\boldsymbol{AY} = \boldsymbol{I}, \qquad \boldsymbol{AZ} = \boldsymbol{O},$$

and are such that all columns of $\boldsymbol{Y}$ and $\boldsymbol{Z}$ together are linearly independent. Then $\boldsymbol{Yb}$ is a solution of $\boldsymbol{Ax} = \boldsymbol{b}$, and for any $\boldsymbol{y} = (y_1, \ldots, y_{n-m})$

$$\boldsymbol{Zy} = \sum_{i=1}^{n-m} y_i \boldsymbol{z}_i$$

solves the corresponding homogeneous system $\boldsymbol{Ax} = \boldsymbol{0}$. The columns $\boldsymbol{z}_1, \ldots, \boldsymbol{z}_{n-m}$ of $\boldsymbol{Z}$ in fact form a basis for the null space of $\boldsymbol{A}$, implying that all solutions to the homogeneous system are of this form. Thus, any solution to $\boldsymbol{Ax} = \boldsymbol{b}$ can be written

$$\boldsymbol{x} = \boldsymbol{Yb} + \boldsymbol{Zy}.$$

Here the $n - m$ variables $y_1, \ldots, y_{n-m}$ should be considered as parameters, corresponding to $\boldsymbol{x}_N$ above. Again, we can substitute for $\boldsymbol{x}$ in $f(\boldsymbol{x})$, obtaining an unconstrained quadratic function in the y-variables.

We refrain from doing the computations. Also, we leave to more advanced courses a discussion of various ways of choosing $\boldsymbol{Y}$ and $\boldsymbol{Z}$.

The methods indicated here might be of interest even in the case of a non-quadratic objective function. They are sometimes called *reduced gradient methods*.

Let us now briefly consider the problem of minimizing

$$f(\boldsymbol{x}) = \frac{1}{2}\boldsymbol{x}^T \boldsymbol{Hx} + \boldsymbol{c}^T \boldsymbol{x} + d$$

subject to inequality constraints

$$\boldsymbol{A}\boldsymbol{x} \leq \boldsymbol{b}, \quad \boldsymbol{x} \geq \boldsymbol{0}$$

(notation as before). The Lagrange function is

$$L(\boldsymbol{x}; \boldsymbol{u}_1, \boldsymbol{u}_2) = \tfrac{1}{2}\boldsymbol{x}^T\boldsymbol{H}\boldsymbol{x} + \boldsymbol{c}^T\boldsymbol{x} + d + \boldsymbol{u}_1^T(\boldsymbol{A}\boldsymbol{x} - \boldsymbol{b}) - \boldsymbol{u}_2^T\boldsymbol{x}, \quad \boldsymbol{u}_1 \geq \boldsymbol{0},\ \boldsymbol{u}_2 \geq \boldsymbol{0}.$$

A point $\boldsymbol{x} \in \mathbf{R}^n$ is a KKT point if $\boldsymbol{u}_1 \geq \boldsymbol{0}$ and $\boldsymbol{u}_2 \geq \boldsymbol{0}$ exist such that

$$\nabla_{\boldsymbol{x}} L(\boldsymbol{x}; \boldsymbol{u}_1, \boldsymbol{u}_2) = \boldsymbol{H}\boldsymbol{x} + \boldsymbol{c} + \boldsymbol{A}^T\boldsymbol{u}_1 - \boldsymbol{u}_2 = \boldsymbol{0}$$

and

$$\boldsymbol{u}_1^T(\boldsymbol{A}\boldsymbol{x} - \boldsymbol{b}) = 0, \qquad \boldsymbol{u}_2^T\boldsymbol{x} = 0.$$

Of course $\boldsymbol{x}$ must also be feasible. Introducing slack variables $\boldsymbol{y} = \boldsymbol{b} - \boldsymbol{A}\boldsymbol{x}$, the conditions become

$$(27) \qquad \begin{cases} \boldsymbol{H}\boldsymbol{x} \quad + \boldsymbol{A}^T\boldsymbol{u}_1 - \boldsymbol{u}_2 = -\boldsymbol{c} \\ \boldsymbol{A}\boldsymbol{x} + \boldsymbol{y} \qquad\qquad\quad = \ \boldsymbol{b} \\ \boldsymbol{x} \geq \boldsymbol{0}, \boldsymbol{y} \geq \boldsymbol{0}, \boldsymbol{u}_1 \geq \boldsymbol{0}, \boldsymbol{u}_2 \geq \boldsymbol{0} \end{cases}$$

and

$$(28) \qquad \boldsymbol{u}_1^T\boldsymbol{y} = 0, \qquad \boldsymbol{u}_2^T\boldsymbol{x} = 0.$$

The system (27) is linear in the variables $\boldsymbol{x}$, $\boldsymbol{y}$, $\boldsymbol{u}_1$, $\boldsymbol{u}_2$. Neglecting (28) it could be solved by phase I of the ordinary simplex algorithm for LP problems. However, (28) puts restrictions on the variables to enter the basis; for instance, for any i the variables $(\boldsymbol{u}_1)_i$ and y_i may not belong to the basis simultaneously.

Special algorithms have been developed to handle situations like this (*complementary pivoting algorithms*). We leave the details to other literature.

Exercise

7.17 Consider the problem to minimize $q(\boldsymbol{x}) = \frac{1}{2}\boldsymbol{x}^T\boldsymbol{H}\boldsymbol{x} + \boldsymbol{c}^T\boldsymbol{x}$ subject to

$$x_1 + x_2 + x_3 = 2,$$

where

$$\boldsymbol{H} = \begin{pmatrix} 2 & 1 & 0 \\ 1 & 1 & -1 \\ 0 & -1 & 3 \end{pmatrix}, \qquad \boldsymbol{c} = \begin{pmatrix} 1 \\ 1 \\ 0 \end{pmatrix}.$$

Solve this problem

a) by using KKT theory,

b) by elimination of one variable resulting in an unconstrained problem in two variables.

7.6 Some applications

The purpose of this section is twofold; we wish to give some indication of the applications, but also to have an opportunity to present some more solved problems on constrained optimization.

Example 20. Find the least distance from the origin to the triangle in space with corners $A = (1, 1, -1)$, $B = (1, -1, 1)$ and $C = (2, 1, -2)$.

Solution: Of course we prefer to minimize the square of the distance $f(\boldsymbol{x}) = x_1^2 + x_2^2 + x_3^2$. Thus our problem is to find that point $\bar{\boldsymbol{x}}$ in the triangle which minimizes $f(\boldsymbol{x})$.

We easily find that the plane through A, B and C (which contains the triangle) has equation $x_1 + x_2 + x_3 = 1$. Thus one constraint is

$$h(\boldsymbol{x}) = x_1 + x_2 + x_3 - 1 = 0.$$

Now choose arbitrarily another plane through the points A and B, e.g. $x_2 + x_3 = 0$. For C we have $x_2 + x_3 \leq 0$, so another constraint is

$$g_1(\boldsymbol{x}) = x_2 + x_3 \leq 0.$$

We then choose a plane through B and C: $x_1 + 4x_2 + 3x_3 = 0$, and finally one through A and C: $x_1 + x_3 = 0$. Checking signs for the remaining point we arrive at the two constraints

$$g_2(\boldsymbol{x}) = -x_1 - 4x_2 - 3x_3 \leq 0, \qquad g_3(\boldsymbol{x}) = -x_1 - x_3 \leq 0.$$

The three inequality constraints together with the constraint $h(\boldsymbol{x}) = 0$ determine the triangle. We consider the problem to minimize f subject to these.

The Lagrange function is

$$L(\boldsymbol{x};\boldsymbol{u},v) = x_1^2 + x_2^2 + x_3^2 +$$
$$+ u_1(x_2 + x_3) + u_2(-x_1 - 4x_2 - 3x_3) + u_3(-x_1 - x_3) + v(x_1 + x_2 + x_3 - 1),$$
$$u_1, u_2, u_3 \geq 0.$$

Differentiating L with respect to the x-variables we obtain the KKT conditions

$$\begin{cases} 2x_1 \qquad\quad - u_2 - u_3 + v = 0 \\ 2x_2 + u_1 - 4u_2 \qquad + v = 0 \\ 2x_3 + u_1 - 3u_2 - u_3 + v = 0. \end{cases}$$

The complementary slackness conditions are

$$u_1(x_2 + x_3) = 0,$$
$$u_2(-x_1 - 4x_2 - 3x_3) = 0,$$
$$u_3(-x_1 - x_3) = 0,$$

and of course there are the feasibility conditions (the constraints themselves).

We now have to consider several cases:

I. $u_1 = 0,\ u_2 = 0,\ u_3 = 0.$

II. $u_1 > 0,\ u_2 = 0,\ u_3 = 0.$

III. $u_1 = 0,\ u_2 > 0,\ u_3 = 0.$

IV. $u_1 = 0,\ u_2 = 0,\ u_3 > 0.$

V. $u_1 = 0,\ u_2 > 0,\ u_3 > 0.$

VI. $u_1 > 0,\ u_2 = 0,\ u_3 > 0.$

VII. $u_1 > 0,\ u_2 > 0,\ u_3 = 0.$

VIII. $u_1 > 0,\ u_2 > 0,\ u_3 > 0.$

Case I means that we have to solve the system of equations

$$\begin{cases} 2x_1 + v = 0 \\ 2x_2 + v = 0 \\ 2x_3 + v = 0 \\ x_1 + x_2 + x_3 = 1. \end{cases}$$

This has the solution $x_1 = x_2 = x_3 = \frac{1}{3}$ and $v = -\frac{2}{3}$. However, the point $(\frac{1}{3}, \frac{1}{3}, \frac{1}{3})$ is not feasible (the first inequality constraint is violated).

We now turn to case II. Here the system of equations is

$$\begin{cases} 2x_1 \qquad + v = 0 \\ 2x_2 + u_1 + v = 0 \\ 2x_3 + u_1 + v = 0 \\ x_2 + x_3 = 0 \\ x_1 + x_2 + x_3 = 1. \end{cases}$$

This yields the point $\bar{x} = (1, 0, 0)$ and $u_1 = 2 > 0$, $v = -2$. The point $\bar{x}$ is feasible.

Now, the function f is convex, the functions g_1, g_2, g_3 are convex (even affine) and h is affine. Thus by the corollary to theorem 5, page 264, the point $\bar{x}$ is the global minimum of f. We do not even have to consider the remaining 6 cases!

The minimum distance to the origin is $\sqrt{f(\bar{x})} = 1$, and the closest point is $\bar{x} = (1, 0, 0)$. □

Example 21. This example comes from economics.

Someone with an available capital K wishes to buy goods $V_1, \ldots, V_n$. The price per unit of V_i is p_i. To help in the decision on how much of each article to buy one introduces in economics a so called *utility function*, a real-valued function $U(x_1, \ldots, x_n)$ of n variables, where x_i is the quantity bought of merchandise V_i. The function U is supposed to represent the "value" received by the investor when buying quantities $x_1, \ldots, x_n$ respectively of the goods. One such function in use is the *Cobb-Douglas* function

$$U(x_1, x_2, \ldots, x_n) = x_1^{\alpha_1} x_2^{\alpha_2} \ldots x_n^{\alpha_n}.$$

The exponents α_i are positive numbers with sum equal to 1.

The intention is to choose the x_i so that U is maximized subject to the restrictions

$$\text{all } x_i \geq 0 \quad \text{and} \quad x_1 p_1 + \ldots + x_n p_n = K.$$

We first note that these conditions define a compact set in $\mathbf{R}^n$, so the maximum certainly exists. Secondly, since $U(x_1, \ldots, x_n) = 0$ whenever some x_i is equal to zero, none of the constraints $x_i \geq 0$ are active at the point of maximum. Thus they can be disregarded when forming the Lagrange function.

To adapt to the theory we turn the maximum problem into a minimum problem by inserting a minus sign; we consider

$$\min -U(x_1, \ldots, x_n) \qquad \text{when} \quad x_1 p_1 + \ldots + x_n p_n = K.$$

The Lagrange function is

$$L(\boldsymbol{x}; v) = -x_1^{\alpha_1} \ldots x_n^{\alpha_n} + v(x_1 p_1 + \ldots + x_n p_n - K).$$

Differentiating we get

$$\frac{\partial L}{\partial x_i} = -\alpha_i x_1^{\alpha_1} \ldots x_i^{\alpha_i - 1} \ldots x_n^{\alpha_n} + v p_i = -\alpha_i \frac{U(\boldsymbol{x})}{x_i} + v p_i,$$

since we can assume that all x_i are strictly positive. For $\boldsymbol{x} = (x_1, \ldots, x_n)$ to be a KKT point we thus require

$$-\alpha_i \frac{U(\boldsymbol{x})}{x_i} + v p_i = 0, \qquad i = 1, \ldots, n \tag{29}$$

and

$$x_1 p_1 + \ldots + x_n p_n = K. \tag{30}$$

From (29) we obtain

$$x_i = \alpha_i \frac{U(\boldsymbol{x})}{v p_i}, \qquad i = 1, \ldots, n,$$

and putting these expressions into (30) results in

$$(\alpha_1 + \ldots + \alpha_n) \frac{U(\boldsymbol{x})}{v} = K.$$

But $\sum_1^n \alpha_i = 1$, making $v = \dfrac{U(\boldsymbol{x})}{K}$. It follows that

$$x_i = \frac{\alpha_i K}{p_i}, \qquad i = 1, \ldots, n.$$

This being the only KKT point, it is the point of minimum of $-U$ and point of maximum of U, and is thus the answer to our problem. (The reader should verify that the constraint qualification is satisfied everywhere.)

Note that a large α_i implies a large value of x_i, and a large p_i implies a small value of x_i, which is reasonable. This indicates that the choice of utility function is a suitable one. □

Remark. For a general utility function U the same computations show that for all j

$$\frac{1}{p_j} \frac{\partial U}{\partial x_j} = v,$$

where v is the Lagrange multiplier. In this context v is in economic literature called *the marginal utility of money*. The interpretation of the left hand side is the utility per unit of money to by some more of goods V_j.

Example 22. We give an example of so called *sensitivity analysis*, with the intention of showing another case where the values of Lagrange multipliers at a certain local minimum might be of interest.

Consider the problem to minimize a function $f(\boldsymbol{x})$ subject to one equality constraint $h(\boldsymbol{x}) = 0$ only (for simplicity). Assume that $\bar{\boldsymbol{x}}$ is a KKT point with multiplier $\bar{v}$, and that the second order sufficiency conditions of theorem 6 are satisfied:

$$(31) \quad \boldsymbol{d}^T \boldsymbol{\nabla}^2_{\boldsymbol{xx}} L(\bar{\boldsymbol{x}}; \bar{v}) \boldsymbol{d} > 0 \quad \text{for all } \boldsymbol{d} \neq \boldsymbol{0} \text{ satisfying } \boldsymbol{\nabla} h(\bar{\boldsymbol{x}})^T \boldsymbol{d} = 0.$$

(Assume that $\boldsymbol{\nabla} h(\bar{\boldsymbol{x}}) \neq \boldsymbol{0}$.) We perturb the constraint a little to $h(\boldsymbol{x}) = c$, with c close to zero.

Suppose that the perturbed minimization problem has a solution $\bar{\boldsymbol{x}}(c)$ which is of class C^1 as a function of c, as is the corresponding multiplier $\bar{v}(c)$. (We show below that this is the case.) Let

$$\Phi(c) = f(\bar{\boldsymbol{x}}(c))$$

denote the value of the objective function at the perturbed minimum point. By the chain rule

$$\frac{d\Phi}{dc}(c) = \nabla f(\bar{\boldsymbol{x}}(c))^T \bar{\boldsymbol{x}}'(c).$$

Similarly, since $h(\bar{\boldsymbol{x}}(c)) = c$ we have

$$1 = \frac{d}{dc} h(\bar{\boldsymbol{x}}(c)) = \nabla h(\bar{\boldsymbol{x}}(c))^T \bar{\boldsymbol{x}}'(c).$$

Putting $c = 0$ in these formulas and combining with

$$\nabla_{\boldsymbol{x}} L(\bar{\boldsymbol{x}}; \bar{v}) = \nabla f(\bar{\boldsymbol{x}}) + \bar{v} \nabla h(\bar{\boldsymbol{x}}) = \boldsymbol{0}$$

we obtain

$$\frac{d\Phi}{dc}(0) = -\bar{v}\, \nabla h(\bar{\boldsymbol{x}})^T \bar{\boldsymbol{x}}'(0) = -\bar{v}.$$

Thus the negative of the multiplier can be interpreted as the rate of change of the minimum value.[6]

For the benefit of the interested reader we give a proof that the KKT conditions for the perturbed minimization problem when c is close to zero has a solution which is of class $\mathcal{C}^1$. We are then concerned about the possibility of solving the system of equations

$$\text{(32)} \qquad \begin{cases} \nabla f(\boldsymbol{x}) + v \nabla h(\boldsymbol{x}) = \boldsymbol{0} \\ h(\boldsymbol{x}) = c \end{cases}$$

for $\boldsymbol{x}$ and v. The Jacobian at $(\bar{\boldsymbol{x}}; \bar{v})$ of the left side of the system is the $(n+1) \times (n+1)$ matrix

$$J(\bar{\boldsymbol{x}}; \bar{v}) = \begin{pmatrix} \nabla^2_{\boldsymbol{xx}} L(\bar{\boldsymbol{x}}; \bar{v}) & \nabla h(\bar{\boldsymbol{x}}) \\ \nabla h(\bar{\boldsymbol{x}})^T & 0 \end{pmatrix}.$$

By the inverse function theorem, if this matrix is non-singular then system (32) for c in a neighbourhood of 0 has a unique solution $(\bar{\boldsymbol{x}}(c); \bar{v}(c))$ close to $(\bar{\boldsymbol{x}}; \bar{v})$, and this solution is of class $\mathcal{C}^1$ as a function of c.

[6] The reader who compares the result here with example 5.19, page 191, may wonder about the minus sign. However, in the LP example the constraint is $\boldsymbol{b} - \boldsymbol{A}\boldsymbol{x} = \boldsymbol{0}$, to adapt to the definition of the Lagrange function there. Thus $\Delta \boldsymbol{b}$ there and $\boldsymbol{c}$ in the present case have opposite signs. This explains the different signs in the results.

Thus, what we need to prove is that the Jacobian at $(\bar{x}, \bar{v})$ is non-singular. This we do by proving that the following homogeneous system of equations has the zero solution only:

$$J(\bar{x};\bar{v})\begin{pmatrix} y \\ z \end{pmatrix} = \begin{pmatrix} 0 \\ 0 \end{pmatrix} \iff \begin{cases} \nabla^2_{xx}L(\bar{x};\bar{v})y + \nabla h(\bar{x})z = 0 \\ \nabla h(\bar{x})^T y \qquad\qquad = 0. \end{cases}$$

Multiplying the first equation by y^T and using the second one we obtain

$$y^T\nabla^2_{xx}L(\bar{x};\bar{v})y + y^T\nabla h(\bar{x})z = 0 \iff y^T\nabla^2_{xx}L(\bar{x};\bar{v})y = 0.$$

By the assumption (31) this implies that $y = 0$. Then $z = 0$ as well, and the proof is complete. □

Example 23. We now solve the problem introduced in example 1.5, page 11, to minimize

$$\|x - a\|^2 + \|y - b\|^2, \qquad x, y \in \mathbb{R}^3,$$

subject to the constraint $x^T y = 0$. Here a and b denote known vectors, in the present case assumed to be

$$a = (3, 1, 1), \qquad b = (-1, 3, 1).$$

Solution: The objective function can be written

$$f(x, y) = (x-a)^T(x-a) + (y-b)^T(y-b) -$$
$$= x^T x - 2a^T x + a^T a + y^T y - 2b^T y + b^T b,$$

and the Lagrange function is

$$L(x, y; v) = x^T x - 2a^T x + a^T a + y^T y - 2b^T y + b^T b + vx^T y.$$

Putting the gradient of L equal to zero gives

$$\begin{pmatrix} \nabla_x L \\ \nabla_y L \end{pmatrix} = \begin{pmatrix} 2x - 2a + vy \\ 2y - 2b + vx \end{pmatrix} = \begin{pmatrix} 0 \\ 0 \end{pmatrix}.$$

We solve

$$\begin{cases} 2x + vy = 2a \\ vx + 2y = 2b \end{cases}$$

for $\boldsymbol{x}$ and $\boldsymbol{y}$, obtaining

$$\begin{cases} \boldsymbol{x} = \dfrac{2}{4-v^2}(2\boldsymbol{a} - v\boldsymbol{b}) \\ \boldsymbol{y} = \dfrac{2}{4-v^2}(2\boldsymbol{b} - v\boldsymbol{a}). \end{cases}$$

(Since $\boldsymbol{a}$ and $\boldsymbol{b}$ are not parallel $v \neq \pm 2$.) Inserting in the constraint $\boldsymbol{x}^T\boldsymbol{y} = 0$ we obtain an equation of degree 2 for v:

$$(2\boldsymbol{a} - v\boldsymbol{b})^T(2\boldsymbol{b} - v\boldsymbol{a}) = 0 \iff (\boldsymbol{b}^T\boldsymbol{a})v^2 - 2(\boldsymbol{a}^T\boldsymbol{a} + \boldsymbol{b}^T\boldsymbol{b})v + 4\boldsymbol{a}^T\boldsymbol{b} = 0.$$

With the values given for $\boldsymbol{a}$ and $\boldsymbol{b}$ inserted we obtain

$$v^2 - 44v + 4 = 0 \iff \begin{cases} v_1 \approx 43.91 \\ v_2 \approx 0.0911. \end{cases}$$

Thus we have two KKT points:

$$\boldsymbol{x}_1 = (-0.052,\ 0.13,\ 0.044), \quad \boldsymbol{y}_1 = (0.14,\ 0,039,\ 0.044)$$

and

$$\boldsymbol{x}_2 = (3.05,\ 0.87,\ 0.96), \quad \boldsymbol{y}_2 = (-1.14,\ 2.96,\ 0.96).$$

The function values are

$$f(\boldsymbol{x}_1, \boldsymbol{y}_1) = 22.0 \quad \text{and} \quad f(\boldsymbol{x}_2, \boldsymbol{y}_2) = 0.043.$$

We should now check for points not fulfilling the constraint qualification. The single constraint has gradient

$$\nabla \boldsymbol{x}^T\boldsymbol{y} = \begin{pmatrix} \boldsymbol{y} \\ \boldsymbol{x} \end{pmatrix},$$

which is linearly independent except at the origin $(\boldsymbol{0}, \boldsymbol{0})$. Here the function f has value

$$f(\boldsymbol{0}, \boldsymbol{0}) = \|\boldsymbol{a}\|^2 + \|\boldsymbol{b}\|^2 = 22.$$

There certainly exists a minimum to our problem. (Introducing the extra conditions $\|\boldsymbol{x}\| \leq R$, $\|\boldsymbol{y}\| \leq R$ for some big R does not affect the problem and results in a compact domain in $\mathbf{R}^6$. Since f is continuous the minimum exists by theorem 4.2, page 118.) By comparison of the values of f at the only three possibilities above, we conclude that the minimum point is $(\boldsymbol{x}_2, \boldsymbol{y}_2)$. □

Example 24. We wish to give a proof of the well known *arithmetic-geometric inequality*

$$A \geq G,$$

where

$$A = \frac{1}{N}\sum_{1}^{N} x_i, \qquad G = \Big(\prod_{1}^{N} x_i\Big)^{1/N},$$

using Karush-Kuhn-Tucker theory.[7] The x_i are positive real numbers. The inequality is invariant under a change of scale $x_i \mapsto tx_i$, so it is no restriction to assume that $\prod_1^N x_i = 1$. Then, what we have to prove is that $\sum_1^N x_i \geq N$.

For this purpose we consider the minimization problem

$$\min_{\boldsymbol{x} \geq \boldsymbol{0}} \sum_{1}^{N} x_i \qquad \text{subject to} \quad \prod_{1}^{N} x_i = 1.$$

It is not obvious that this problem has a solution. But in fact we can assume that $\boldsymbol{x}$ is restricted to the compact cube

$$0 \leq x_i \leq R, \qquad i = 1, \ldots, N,$$

since if one of the variables is larger than or equal to R then $\sum_1^N x_i \geq R$, which for R large is larger than some value inside the cube. The intersection of the surface $\prod_1^N x_i - 1$ and this cube is compact. It follows that the minimization problem has a solution inside the cube, and furthermore, that the faces of the cube do not constitute active constraints at the point of minimum.

The Lagrange function of the problem is thus

$$L(\boldsymbol{x}; v) = \sum_{1}^{N} x_i + v\Big(\prod_{1}^{N} x_i - 1\Big).$$

Differentiating gives

$$\frac{\partial L}{\partial x_i} = 1 - v(x_1 \ldots x_{i-1}x_{i+1} \ldots x_N) = 1 - v\,\frac{\prod_1^N x_i}{x_i} = 1 - \frac{v}{x_i},$$

[7] A proof using convexity was given in example 6.5, page 205.

since $\prod_i^N x_i = 1$. Thus

$$\frac{\partial L}{\partial x_i} = 0 \quad \Longleftrightarrow \quad x_i = v.$$

Putting this into the constraint we get, since v is real and positive,

$$\prod_1^N x_i = 1 \quad \Longleftrightarrow \quad v^N = 1 \quad \Longleftrightarrow \quad v = 1.$$

Hence the only KKT point is $(1, \ldots, 1)$. The (CQ) is satisfied everywhere (check!). Consequently the minimum occurs at $(1, \ldots, 1)$. The minimum value is

$$x_1 + \ldots + x_N = 1 + \ldots + 1 = N,$$

which was to be proved. □

Miscellanous exercises

7.18 Find those stationary points of the function

$$f(\boldsymbol{x}) = 2x_1^2 + x_2^2 - 2x_1x_2 + 2x_1^3 + x_1^4$$

which are *global* minimizers.

7.19 Prove using KKT theory that $(3, 0)$ is the optimal solution to

$$\begin{aligned} \text{Minimize} \quad & x^2 + 3y^2 + 4y \\ \text{subject to} \quad & x^2 - 8x - y + 15 \leq 0 \\ & y \geq 0. \end{aligned}$$

(1997–04–04, 1)

7.20 Let X be a stochastic variable on the event set $\{A_1, \ldots, A_n\}$, with $p(A_i) = p_i \geq 0$, $\sum p_i = 1$. A measure of the uncertainty is given by the *entropy*

$$H(p_1, \ldots, p_n) = -\sum_1^n p_i \ln p_i.$$

Show that the entropy is maximal when $p_i = \dfrac{1}{n}$, $i = 1, \ldots, n$.

7.21 Find the minimum of $x_1^2 + 2x_2^2 + 2x_3^2$ subject to

$$\begin{aligned} x_1 + x_2 + 3x_3 &\geq 2 \\ x_1^2 + x_2^2 - 2x_2 &\leq 3. \end{aligned}$$

(1998–12–14, 2)

7.22 Find that point on the ellipse defined by the intersection of the surfaces $x + y = 1$ and $x^2 + 2y^2 + z^2 = 1$ which is nearest to the origin.

7.23 Solve the problem

$$\begin{aligned} \text{Minimize} \quad & x_1 + 2x_2 + 3x_3 \\ \text{subject to} \quad & x_1^2 + 2x_2^2 + 3x_3^2 \leq 20 \\ & x_1 \geq 0. \end{aligned}$$

7.24 Determine the distance between the point $(1, 1, 1)$ and the cone

$$x^2 + y^2 - z^2 = 0.$$

7.25 Solve the problem

$$\begin{aligned} \text{Maximize} \quad & 10xy - 5x^2 - 7y^2 + 40x \\ \text{subject to} \quad & x + y \leq 13 \\ & x \geq 0,\ y \geq 0. \end{aligned}$$

7.26 Prove that there is a least distance between the point $(0, 3)$ and the set defined by

$$x_1^2 + x_1x_2 + x_2^3 \leq 3 \quad \text{and} \quad x_2^2 \leq x_1,$$

and find it.

7.27 Solve the optimization problem

$$\begin{aligned} \text{Minimum of} \quad & x_1^2 + x_2^2 + x_3^2 \\ \text{subject to} \quad & x_1 + x_2 + x_3 \geq 3 \\ & x_1x_2x_3 \geq 3 \\ & x_1,\ x_2,\ x_3 \geq 0. \end{aligned}$$

(*Hint:* The constraints $x_i \geq 0$ are not active at a point of minimum.)

(1998–04–16, 6)

7.28 Solve, using KKT theory, the problem to find maximum of x when $(1-x)^3 - y \geq 0$ and $y \geq 0$. (Draw a figure!)

7.29 a) Consider the constrained optimization problem

$$\text{minimize} \quad x_1x_2 + x_3^2 \qquad \text{subject to} \quad x_1^2 + x_2^2 + x_3^3 \leq 4,\ x_3 \geq 0.$$

(Note the exponent in the last term on the left in the first constraint.) Calculate all CQ and KKT points and solve the problem. Do not forget to motivate why a minimum exists.

b) Replace the inequality $x_3 \geq 0$ above with $x_3 \leq 0$. How does it affect the answer?

(2009–01–09, 5)

7.30 Suppose that $\hat{\boldsymbol{x}}$ is an optimal solution to the problem of minimizing some function $f(\boldsymbol{x})$ subject to $x_1 + \ldots + x_n \geq b$, all $x_i \geq 0$. (Here b is a fixed positive constant.) Show that

- $\hat{x}_j > 0,\ \hat{x}_k > 0 \implies \dfrac{\partial f}{\partial x_j}(\hat{\boldsymbol{x}}) = \dfrac{\partial f}{\partial x_k}(\hat{\boldsymbol{x}}),$
- $\hat{x}_j = 0,\ \hat{x}_k > 0 \implies \dfrac{\partial f}{\partial x_j}(\hat{\boldsymbol{x}}) \geq \dfrac{\partial f}{\partial x_k}(\hat{\boldsymbol{x}}).$

(1996–12–16, 5)

7.31 Consider the problem to minimize $f(\boldsymbol{x})$ subject to $g_i(\boldsymbol{x}) \leq 0,\ i = 1, \ldots, m$.

a) Show that verifying whether a point $\bar{\boldsymbol{x}}$ is a KKT point is equivalent to finding a vector $\boldsymbol{u}$ satisfying a system of the form $\boldsymbol{A}\boldsymbol{u} = \boldsymbol{c}$, $\boldsymbol{u} \geq \boldsymbol{0}$. This can be done using phase I of the simplex algorithm.

b) Illustrate by the following problem, where $\bar{\boldsymbol{x}} = (1, 2, 5)$:

$$\begin{aligned} \text{Minimize} \quad & 2x_1^2 + x_2^2 + 2x_3^2 + x_1x_3 - x_1x_2 + x_1 + 2x_3 \\ \text{subject to} \quad & x_1^2 + x_2^2 - \ x_3 < 0 \\ & x_1 + x_2 + 2x_3 \leq 16 \\ & x_1 + x_2 \qquad\quad \geq 3 \\ & x_1, x_2, x_3 \geq 0. \end{aligned}$$

7.32 Solve the problem

$$\begin{aligned} \text{Minimize} \quad & x_1 + 2x_2 + 3x_3 \\ \text{subject to} \quad & x_1^2 + 2x_2^2 + 3x_3^2 \leq 9 \\ & x_2 \geq 0 \end{aligned}$$

using Kuhn-Tucker theory. Do not forget to motivate why the global minimum exists. (2008–10–21, 5)

7.33 Find the minimum (if it exists) of $f(x, y) = xy + 2x - 2y$ subject to the constraints $x^3 + y \geq 1$, $x \leq 1$ and $y \leq 1$. (2009–04–17, 5)

Chapter 8

SADDLE POINTS AND DUALITY

The aims of this chapter are rather modest. We wish to show how the concept of duality, which appeared earlier in the context of linear programming, can be extended to more general situations. We leave to more advanced courses the applications of this to the development of optimization algorithms.

8.1 Saddle points

The definition of a saddle point

The problem which is the basis for our discussion (the *primal* problem) in this chapter is

Problem P: to minimize $f(\boldsymbol{x})$ subject to constraints

$$\boldsymbol{x} \in X, \qquad \boldsymbol{g}(\boldsymbol{x}) \leq \boldsymbol{0}, \qquad \boldsymbol{h}(\boldsymbol{x}) = \boldsymbol{0}.$$

Here X is an open[1] subset of $\mathbf{R}^n$, and $\boldsymbol{g} = (g_1 \ldots g_m)^T$ and $\boldsymbol{h} = (h_1 \ldots h_l)^T$ denote column vectors of functions of class C^1 (for simplicity). The objective function f itself is assumed differentiable on $\mathbf{R}^n$.

[1]The assumption that X is open is for convenience in stating the theorems. In some examples a closed set will appear. The theorems can then be applied to points $\boldsymbol{x}$ in $\text{int}(X)$ only.

This is of course the same constrained problem as in chapter 7, but for the moment we prefer to use vector notation.

As before, we connect to problem P the *Lagrange function*

$$L(\boldsymbol{x};\, \boldsymbol{u}, \boldsymbol{v}) = f(\boldsymbol{x}) + \sum_{i=1}^{m} u_i g_i(\boldsymbol{x}) + \sum_{j=1}^{l} v_j h_j(\boldsymbol{x}) = f(\boldsymbol{x}) + \boldsymbol{u}^T \boldsymbol{g}(\boldsymbol{x}) + \boldsymbol{v}^T \boldsymbol{h}(\boldsymbol{x}),$$

defined for

$$\boldsymbol{x} \in X \quad \text{and} \quad (\boldsymbol{u}, \boldsymbol{v}) \in U = \{\, (\boldsymbol{u}, \boldsymbol{v}) \in \mathbf{R}^m \times \mathbf{R}^l\,;\ \boldsymbol{u} \geq \boldsymbol{0} \,\}.$$

Note that if $\boldsymbol{x}$ is feasible then $L(\boldsymbol{x}; \boldsymbol{u}, \boldsymbol{v}) \leq f(\boldsymbol{x})$.

DEFINITION 1. *A point $(\bar{\boldsymbol{x}};\, \bar{\boldsymbol{u}}, \bar{\boldsymbol{v}}) \in X \times U$ is called a* **saddle point** *of L if*

$$(1) \quad L(\bar{\boldsymbol{x}};\, \boldsymbol{u}, \boldsymbol{v}) \leq L(\bar{\boldsymbol{x}};\, \bar{\boldsymbol{u}}, \bar{\boldsymbol{v}}) \leq L(\boldsymbol{x};\, \bar{\boldsymbol{u}}, \bar{\boldsymbol{v}}) \quad \textit{for all} \quad (\boldsymbol{x};\, \boldsymbol{u}, \boldsymbol{v}) \in X \times U.$$

Remark. The terminology derives from the well known function of two variables $f(x, y) = x^2 - y^2$, whose graph has the appearence in the figure.

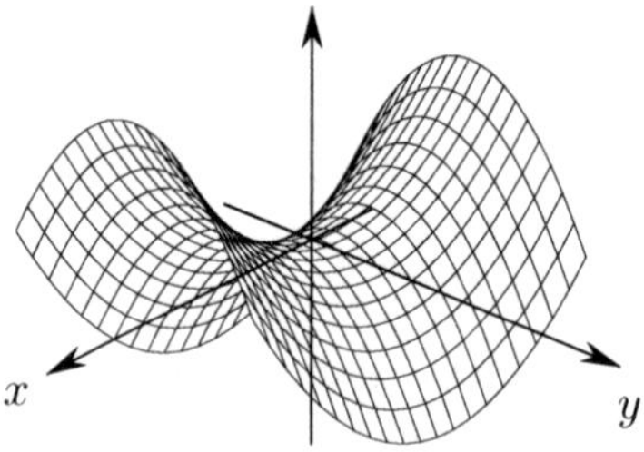

The following lemma in particular shows that if $(\bar{\boldsymbol{x}};\, \bar{\boldsymbol{u}}, \bar{\boldsymbol{v}})$ is a saddle point of the Lagrange function then $\bar{\boldsymbol{x}}$ is a feasible point for problem P.

LEMMA 1. *The point $(\bar{\boldsymbol{x}};\, \bar{\boldsymbol{u}}, \bar{\boldsymbol{v}})$ in $X \times U$ is a saddle point of L if and only if the following three conditions are true.*

1. $L(\bar{\boldsymbol{x}};\, \bar{\boldsymbol{u}}, \bar{\boldsymbol{v}}) = \min_{\boldsymbol{x} \in X} L(\boldsymbol{x};\, \bar{\boldsymbol{u}}, \bar{\boldsymbol{v}})$.
2. $\boldsymbol{g}(\bar{\boldsymbol{x}}) \leq \boldsymbol{0}$ and $\boldsymbol{h}(\bar{\boldsymbol{x}}) = \boldsymbol{0}$.
3. $\bar{\boldsymbol{u}}^T \boldsymbol{g}(\bar{\boldsymbol{x}}) = 0$.

PROOF. The first condition is merely a reformulation of the right inequality in the definition of a saddle point. Thus, what we need to prove is that the left inequality in (1) is equivalent to conditions 2 and 3.

Assume first that $(\bar{\boldsymbol{x}};\, \bar{\boldsymbol{u}}, \bar{\boldsymbol{v}})$ is a saddle point. Written out in full, the left inequality in (1) states that

$$f(\bar{\boldsymbol{x}}) + \boldsymbol{u}^T \boldsymbol{g}(\bar{\boldsymbol{x}}) + \boldsymbol{v}^T \boldsymbol{h}(\bar{\boldsymbol{x}}) \le f(\bar{\boldsymbol{x}}) + \bar{\boldsymbol{u}}^T \boldsymbol{g}(\bar{\boldsymbol{x}}) + \bar{\boldsymbol{v}}^T \boldsymbol{h}(\bar{\boldsymbol{x}}) \quad \text{for all } \boldsymbol{u} \ge \boldsymbol{0} \text{ and all } \boldsymbol{v}. \tag{2}$$

Here $f(\bar{\boldsymbol{x}})$ cancels out. Now we consider one $(\boldsymbol{u}, \boldsymbol{v})$-variable at a time, keeping the other ones fixed. Letting $\boldsymbol{u} = \bar{\boldsymbol{u}}$ and $v_j = \bar{v}_j$ for $j \ne k$ results in

$$v_k h_k(\bar{\boldsymbol{x}}) \le \bar{v}_k h_k(\bar{\boldsymbol{x}}) \qquad \text{for all } \; v_k \in \mathbf{R}.$$

This implies that $h_k(\bar{\boldsymbol{x}}) = 0$, valid of course for all k. Inserting $\boldsymbol{h}(\bar{\boldsymbol{x}}) = \boldsymbol{0}$ in (2) and letting $u_j = \bar{u}_j$ for $j \ne k$ results in

$$u_k g_k(\bar{\boldsymbol{x}}) \le \bar{u}_k g_k(\bar{\boldsymbol{x}}) \qquad \text{for all } u_k \ge 0.$$

This implies that $g_k(\bar{\boldsymbol{x}}) \le 0$, again for all k. Now the second condition is proved. Finally, to prove the third one we put $\boldsymbol{u} = \boldsymbol{0}$ in (2). Then we obtain $\boldsymbol{0} \le \bar{\boldsymbol{u}}^T \boldsymbol{g}(\bar{\boldsymbol{x}})$. The opposite inequality follows from the already proved second condition and the fact that $\bar{\boldsymbol{u}} \ge \boldsymbol{0}$.

For the proof in the reverse direction we assume that conditions 2 and 3 are satisfied. Then we directly obtain (2), i.e., the left inequality in (1). □

Saddle points and the KKT conditions

For a saddle point $(\bar{\boldsymbol{x}};\, \bar{\boldsymbol{u}}, \bar{\boldsymbol{v}}) \in X \times U$ of L lemma 1 shows that

- $\bar{\boldsymbol{x}}$ is a feasible point for problem P,
- $L(\bar{\boldsymbol{x}}; \bar{\boldsymbol{u}}, \bar{\boldsymbol{v}}) = f(\bar{\boldsymbol{x}})$.

From the last item vi conclude that if $\bar{\boldsymbol{x}}$ is feasible then

$$f(\bar{\boldsymbol{x}}) = L(\bar{\boldsymbol{x}}; \bar{\boldsymbol{u}}, \bar{\boldsymbol{v}}) \le L(\boldsymbol{x}; \bar{\boldsymbol{u}}, \bar{\boldsymbol{v}}) \le f(\boldsymbol{x}).$$

Hence $\bar{\boldsymbol{x}}$ solves problem P. Then by theorem 7.4, page 251, $\bar{\boldsymbol{x}}$ is a KKT point, assuming that condition (CQ) is fulfilled at $\bar{\boldsymbol{x}}$. However, we can reach the same conclusion directly from the lemma without any (CQ) condition, since the lemma states that

- the complementary slackness conditions are satisfied,
- $\bar{\boldsymbol{x}}$ is an optimal solution to the problem $\min_{\boldsymbol{x} \in X} L(\boldsymbol{x};\ \bar{\boldsymbol{u}}, \bar{\boldsymbol{v}})$.

This last result implies that

$$\boldsymbol{\nabla}_{\boldsymbol{x}} L(\bar{\boldsymbol{x}}; \bar{\boldsymbol{u}}, \bar{\boldsymbol{v}}) = \boldsymbol{0}.$$

The next theorem summarizes these conclusions.

THEOREM 1. *Assume that* $(\bar{\boldsymbol{x}};\ \bar{\boldsymbol{u}}, \bar{\boldsymbol{v}})$ *is a saddle point of* L. *Then*

1. $\bar{\boldsymbol{x}}$ *solves problem* P,
2. $\bar{\boldsymbol{x}}$, $\bar{\boldsymbol{u}}$, $\bar{\boldsymbol{v}}$ *satisfy the KKT conditions.*

It is a remarkable fact here that the $\bar{\boldsymbol{u}}$ and $\bar{\boldsymbol{v}}$ appearing in the definition of a saddle point are precisely the Lagrange multipliers in the Karush-Kuhn-Tucker theorem.

The converse statement requires more assumptions, e.g. convexity.

THEOREM 2. *Assume that*

1. *the set* X *is convex,*
2. *the functions* f *and* g_i, $i = 1, \ldots, m$, *are convex on* X,
3. *the functions* h_j, $j = 1, \ldots, l$, *are affine on* X,
4. $\bar{\boldsymbol{x}}$, $\bar{\boldsymbol{u}}$, $\bar{\boldsymbol{v}}$ *satisfy the KKT conditions, and* $\bar{\boldsymbol{x}}$ *is feasible for problem* P.

Then $(\bar{\boldsymbol{x}};\ \bar{\boldsymbol{u}}, \bar{\boldsymbol{v}})$ *is a saddle point of the Lagrange function* L.

PROOF. We rely on lemma 1.

The assumptions imply that $\boldsymbol{x} \mapsto L(\boldsymbol{x};\ \bar{\boldsymbol{u}}, \bar{\boldsymbol{v}})$ is a sum of convex functions and hence convex. Since $\boldsymbol{\nabla}_{\boldsymbol{x}} L(\bar{\boldsymbol{x}}; \bar{\boldsymbol{u}}, \bar{\boldsymbol{v}}) = \boldsymbol{0}$ this function has a global minimum at $\bar{\boldsymbol{x}}$. This proves the first condition in the lemma. The other two are merely statements of the feasibility of $\bar{\boldsymbol{x}}$ and the complementary slackness conditions. $\square$

8.2 The dual problem

We now introduce an optimization problem in the $(\boldsymbol{u}, \boldsymbol{v})$ variables.

Put

$$\theta(\boldsymbol{u}, \boldsymbol{v}) = \inf_{\boldsymbol{x} \in X} f(\boldsymbol{x}) + \boldsymbol{u}^T \boldsymbol{g}(\boldsymbol{x}) + \boldsymbol{v}^T \boldsymbol{h}(\boldsymbol{x}) = \inf_{\boldsymbol{x} \in X} L(\boldsymbol{x}; \boldsymbol{u}, \boldsymbol{v}),$$

which might possibly be $-\infty$.[2] The set of feasible points for problem D below is exactly those $(\boldsymbol{u}, \boldsymbol{v}) \in U$ for which this minimum is finite.

For later reference we note that $\theta(\boldsymbol{u}, \boldsymbol{v})$ is a concave function, regardless of any assumtions on f, since it is defined by minimization of concave (in fact affine) functions of $(\boldsymbol{u}, \boldsymbol{v})$.

DEFINITION 2. *The* **(Lagrangian) dual problem** *to problem P above is*

Problem D: to maximize $\theta(\boldsymbol{u}, \boldsymbol{v})$ subject to $(\boldsymbol{u}, \boldsymbol{v}) \in U$.

Obviously

$$\theta(\boldsymbol{u}, \boldsymbol{v}) \leq L(\boldsymbol{x};\, \boldsymbol{u}, \boldsymbol{v}) \leq f(\boldsymbol{x}) \tag{3}$$

when $\boldsymbol{x}$ and $(\boldsymbol{u}, \boldsymbol{v})$ are feasible for problems P and D respectively. This is the counterpart of inequality (5.13), page 181, in linear programming.

One consequence of inequality (3) is that, unless the sets of feasible points are empty, it provides lower bounds for the values of $f(\boldsymbol{x})$, and upper bounds for the values of $\theta(\boldsymbol{u}, \boldsymbol{v})$. In particular, if problem P is unbounded then problem D has no feasible points, and vice versa.

Furthermore, it follows from (3) that, assuming feasible points exist,

$$\sup_{(\boldsymbol{u}, \boldsymbol{v}) \in U} \theta(\boldsymbol{u}, \boldsymbol{v}) \leq \inf_{\boldsymbol{x} \in S} f(\boldsymbol{x}). \tag{4}$$

In the LP case there is equality here by the duality theorem. That theorem has however no counterpart in our present more general case, and in fact strict inequality may occur, as we shall see later. We call the difference between the right and left side in (4) the **duality gap**. Sometimes the term *strong duality* is used when there is equality in (4).

[2]This is why we write *inf* instead of *min*.

Of course, if there are $\bar{\boldsymbol{x}}$ and $(\bar{\boldsymbol{u}}, \bar{\boldsymbol{v}})$, feasible for problems P and D respectively, such that $f(\bar{\boldsymbol{x}}) = \theta(\bar{\boldsymbol{u}}, \bar{\boldsymbol{v}})$, then it follows that $\bar{\boldsymbol{x}}$ is optimal for P and $(\bar{\boldsymbol{u}}, \bar{\boldsymbol{v}})$ is optimal for D, and

$$f(\bar{\boldsymbol{x}}) = \theta(\bar{\boldsymbol{u}}, \bar{\boldsymbol{v}}) = L(\bar{\boldsymbol{x}}; \bar{\boldsymbol{u}}, \bar{\boldsymbol{v}}).$$

THEOREM 3. *The point $(\bar{\boldsymbol{x}};\, \bar{\boldsymbol{u}}, \bar{\boldsymbol{v}})$ in $X \times U$ is a saddle point of L if and only if $\bar{\boldsymbol{x}}$ is feasible for problem P and*

$$f(\bar{\boldsymbol{x}}) = \theta(\bar{\boldsymbol{u}}, \bar{\boldsymbol{v}}).$$

PROOF. Assume first that $(\bar{\boldsymbol{x}};\, \bar{\boldsymbol{u}}, \bar{\boldsymbol{v}})$ is a saddle point of L. Then lemma 1 shows that $\bar{\boldsymbol{x}}$ is feasible for P and $(\bar{\boldsymbol{u}}, \bar{\boldsymbol{v}})$ feasible for D. Furthermore

$$\theta(\bar{\boldsymbol{u}}, \bar{\boldsymbol{v}}) = L(\bar{\boldsymbol{x}};\, \bar{\boldsymbol{u}}, \bar{\boldsymbol{v}}) = f(\bar{\boldsymbol{x}}),$$

where the first equality is due to condition 1 and the second equality to conditions 2 and 3 in lemma 1.

For the proof in the reverse direction it is enough to prove conditions 1–3 in lemma 1. The second one is obvious, since $\bar{\boldsymbol{x}}$ is assumed to be feasible for P. By (3) we get

$$f(\bar{\boldsymbol{x}}) = \theta(\bar{\boldsymbol{u}}, \bar{\boldsymbol{v}}) \leq f(\bar{\boldsymbol{x}}) + \bar{\boldsymbol{u}}^T \boldsymbol{g}(\bar{\boldsymbol{x}}) + \bar{\boldsymbol{v}}^T \boldsymbol{h}(\bar{\boldsymbol{x}}) \leq f(\bar{\boldsymbol{x}}).$$

Consequently there is equality everywhere; in particular $\bar{\boldsymbol{u}}^T \boldsymbol{g}(\bar{\boldsymbol{x}}) = 0$. This proves the third condition in the lemma. Finally, by this equality

$$L(\bar{\boldsymbol{x}};\, \bar{\boldsymbol{u}}, \bar{\boldsymbol{v}}) = \theta(\bar{\boldsymbol{u}}, \bar{\boldsymbol{v}}) = \min_{\boldsymbol{x} \in X} L(\boldsymbol{x};\, \bar{\boldsymbol{u}}, \bar{\boldsymbol{v}}),$$

which proves the first condition. □

Example 1. Suppose that the primal problem is to find the minimum of $x_1^2 + x_2^2$ when $-x_1 - x_2 + 4 \leq 0$ and $x_1, x_2 \geq 0$. We let the closed first quadrant be the set X, and consequently we have only one constraint:

$$g(x_1, x_2) = -x_1 - x_2 + 4 \leq 0.$$

The objective function is the square of the distance from (x_1, x_2) to the origin. By inspection of the figure we realize that the minimum is 8 and occurs at $\bar{x} = (2, 2)$.

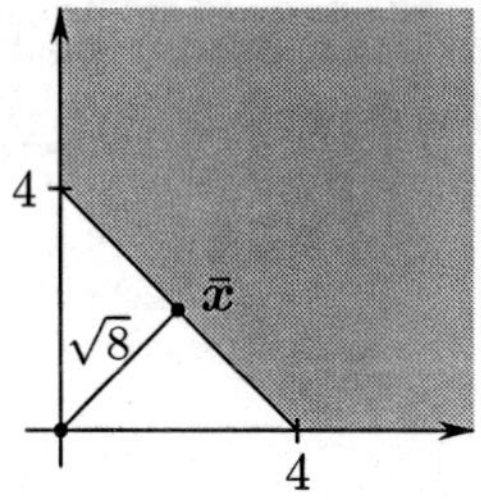

Now consider the dual problem. The Lagrange function is

$$L(x_1, x_2; u) = x_1^2 + x_2^2 + u(-x_1 - x_2 + 4), \qquad (x_1, x_2) \in X, \quad u \geq 0.$$

The function θ is dependent on one variable u only, namely

$$\theta(u) = \min_{x_1, x_2 \geq 0} x_1^2 + x_2^2 + u(-x_1 - x_2 + 4).$$

By separating the x_i-variables and completing squares we get

$$\theta(u) = \min_{x_1 \geq 0} \left(x_1 - \frac{u}{2}\right)^2 + \min_{x_2 \geq 0} \left(x_2 - \frac{u}{2}\right)^2 - \frac{1}{2}u^2 + 4u,$$

leading to

$$\theta(u) = -\frac{1}{2}u^2 + 4u, \qquad u \geq 0.$$

The dual problem is to find

$$\max_{u \geq 0} \theta(u).$$

By standard one variable analysis this maximum is equal to 8, and occurs when $u = 4$.

Since the minimum value of problem P and the maximum value of problem D are equal, $(2, 2; 4)$ is by theorem 3 a saddle point of the Lagrange function. □

Remark. The choice of which constraints should be handled via the set X affects the function θ and thus the dual problem. The reader might verify that if we let $X = \mathbf{R}^2$ and incorporate all constraints in L, the dual problem becomes

$$\max_{u, u_1, u_2 \geq 0} -\frac{1}{4}(u + u_1)^2 - \frac{1}{4}(u + u_2)^2 + 4u.$$

Example 2. Let problem P be the LP problem to minimize $\boldsymbol{c}^T\boldsymbol{x}$ when $\boldsymbol{A}\boldsymbol{x} = \boldsymbol{b}$, $\boldsymbol{x} \geq \boldsymbol{0}$. Let X be the set $\{\, \boldsymbol{x} \,;\, \boldsymbol{x} \geq \boldsymbol{0} \,\}$. Then

$$\theta(\boldsymbol{v}) = \inf_{\boldsymbol{x}\geq\boldsymbol{0}} \boldsymbol{c}^T\boldsymbol{x} + \boldsymbol{v}^T(\boldsymbol{b} - \boldsymbol{A}\boldsymbol{x}) =$$

$$= \boldsymbol{v}^T\boldsymbol{b} + \inf_{\boldsymbol{x}\geq\boldsymbol{0}} (\boldsymbol{c}^T - \boldsymbol{v}^T\boldsymbol{A})\boldsymbol{x} = \begin{cases} \boldsymbol{v}^T\boldsymbol{b} & \text{if } \ \boldsymbol{c}^T - \boldsymbol{v}^T\boldsymbol{A} \geq \boldsymbol{0} \\ -\infty & \text{if } \ \boldsymbol{c}^T - \boldsymbol{v}^T\boldsymbol{A} \not\geq \boldsymbol{0}. \end{cases}$$

The dual problem is thus to maximize $\boldsymbol{v}^T\boldsymbol{b} = \boldsymbol{b}^T\boldsymbol{v}$ subject to $\boldsymbol{A}^T\boldsymbol{v} \leq \boldsymbol{c}$. This is the same problem as the one obtained by the duality theory in linear programming. □

Example 3. We are going to give a geometrical interpretation of $\theta(\boldsymbol{u}, \boldsymbol{v})$ and the dual problem in a simple case.

Assume that problem P has one inequality constraint only and no equalities, i.e., it has the form

$$\text{minimize } \ f(\boldsymbol{x}) \qquad \text{subject to } \ \boldsymbol{x} \in X, \quad g(\boldsymbol{x}) \leq 0.$$

Put

$$\begin{cases} y = g(\boldsymbol{x}) \\ z = f(\boldsymbol{x}), \end{cases} \qquad \boldsymbol{x} \in X.$$

This defines a set G in the yz-plane. Problem P is to minimize z when $(y, z) \in G$ and $y \leq 0$. Obviously, in the figure below the solution is represented by the point Q.

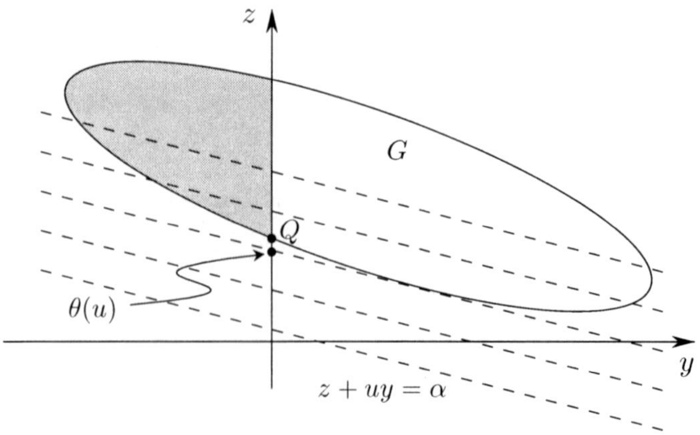

The function θ is defined by

$$\theta(u) = \min_{\boldsymbol{x} \in X} f(\boldsymbol{x}) + ug(\boldsymbol{x}) = \min_{(y,z) \in G} z + uy.$$

The geometrical interpretation of the equation $z + uy = \alpha$ is a line in the yz-plane with slope $-u \leq 0$, and with α equal to the intersection with the z-axis. For a given slope, $\theta(u)$ is the least α possible when the line is required to have a non-empty intersection with G.

Problem D is to find the maximum of $\theta(u)$ for $u \geq 0$. In the figure this maximum occurs for the line which is tangent to G at Q. The optimal values of P and D are equal.

However, the set G might have the appearence shown below. In that case there is a duality gap, and the optimal values are not equal.

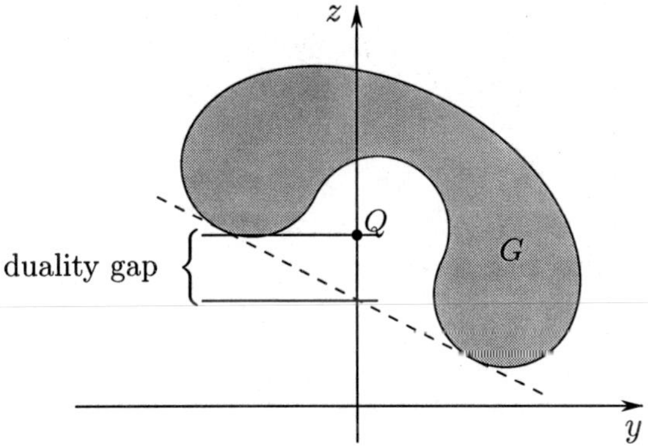

□

Remark. We can see in the geometrical interpretation above the consequence of (3) that if problem P is unbounded then problem D is unfeasible, and vice versa.

Remark. It is even more informative to consider the set

$$G' = \{\, (y, z);\ \text{there is some } x \in X \text{ such that } g(x) \leq y \text{ and } f(x) \leq z \,\}.$$

The set G above is defined by equalities instead of inequalities, so obviously G is a subset of G'. In fact G' looks like this:

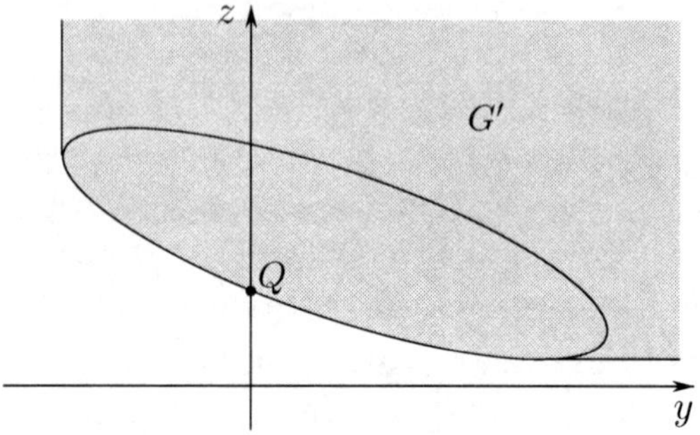

If problem P is convex one can prove from the definition that G' is a convex set. It follows that in this case there is no duality gap. (This is stated in more general terms in theorem 4 below.)

Example 4. The problem in example 1 is of the type discussed in example 3. Here the set G is defined by

$$\begin{cases} y = -x_1 - x_2 + 4 \\ z = x_1^2 + x_2^2, \end{cases} \qquad x_1 \geq 0,\ x_2 \geq 0.$$

It is seen at once that $y \leq 4$. Solving the first equation for x_2 and inserting in the second one gives

$$z = x_1^2 + (-x_1 + 4 - y)^2 = 2x_1^2 - 2(4-y)x_1 + (4-y)^2 = 2\left(x_1 - \frac{4-y}{2}\right)^2 + \frac{(4-y)^2}{2}.$$

Since $0 \leq x_1 \leq 4-y$ we conclude that

$$\frac{1}{2}(4-y)^2 \leq z \leq (4-y)^2.$$

Thus the set G has the following appearance.

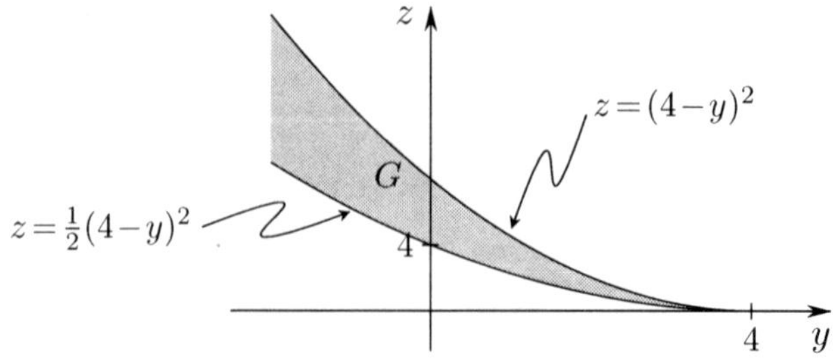

□

We now give an example in which there is a duality gap.

Example 5. Consider the problem from example 7.10, page 259, to minimize $f(x_1, x_2) = x_1$ under the constraints

$$g_1(x_1, x_2) = x_2 - x_1^3 \leq 0, \qquad g_2(x_1, x_2) = -x_2 \leq 0.$$

As in example 7.10 the minimum obviously is 0, occuring at $(0, 0)$.

The Lagrange function is

$$L(\boldsymbol{x}; \boldsymbol{u}) = x_1 + u_1(x_2 - x_1^3) + u_2(-x_2) = x_1 - u_1 x_1^3 + (u_1 - u_2)x_2.$$

Minimizing $L(\boldsymbol{x}; \boldsymbol{u})$ for $\boldsymbol{u} \geq \boldsymbol{0}$ over the set $X = \mathbf{R}^2$ results in $\theta(\boldsymbol{u}) = -\infty$ everywhere. There are in fact no feasible points for the dual problem.

□

In the next example a situation with a finite duality gap is exhibited.

Example 6. Let problem P be to minimize $f(x) = \sqrt{4 - x^2}$ when $x \in X = [-2, 2]$ and $g(x) = x^2 - 1 \leq 0$. It is easily seen that the minimum value is $f(\pm 1) = \sqrt{3}$.

To state the dual problem we must first find the minimum of

$$L(x; u) = f(x) + ug(x) = \sqrt{4 - x^2} + u(x^2 - 1), \qquad x \in X, \quad u \geq 0.$$

Since L (as a function of x) is continuous and X is compact this minimum certainly exists. It occurs at a stationary point or at one of the endpoints of X, i.e., ± 2. Now

$$\frac{\partial L}{\partial x} = -\frac{x}{\sqrt{4 - x^2}} + 2ux.$$

Putting this derivative equal to zero we obtain $x = 0$ or

$$\frac{1}{\sqrt{4 - x^2}} = 2u \iff 4 - x^2 = \frac{1}{4u^2} \iff x^2 = 4 - \frac{1}{4u^2}.$$

The last equation is solvable if and only if $u \geq \frac{1}{4}$.

Computing function values we find that

$$L(\pm 2; u) = 3u, \qquad L(0; u) = 2 - u, \qquad L\left(\pm\sqrt{4 - \frac{1}{4u^2}};\ u\right) = 3u + \frac{1}{4u}.$$

The last value is larger than the first one, hence of no interest. It follows that the minimum value of L is

$$\theta(u) = \min(3u,\, 2-u) = \begin{cases} 3u & \text{when } 0 \le u \le \frac{1}{2} \\ 2-u & \text{when } u > \frac{1}{2}. \end{cases}$$

The dual problem is to find

$$\max_{u \ge 0} \theta(u).$$

The solution to this problem is now easily found; the maximum value is $\theta(\frac{1}{2}) = \frac{3}{2}$. Consequently there is a duality gap with value $\sqrt{3} - \frac{3}{2}$.

The reader might note that the objective function f is concave. □

Remark. We leave to the reader as an exercise to verify that the set G above in this case is the curve

$$z = \sqrt{3 - y}, \qquad -1 \le y \le 3.$$

The duality gap is easily seen in the figure.

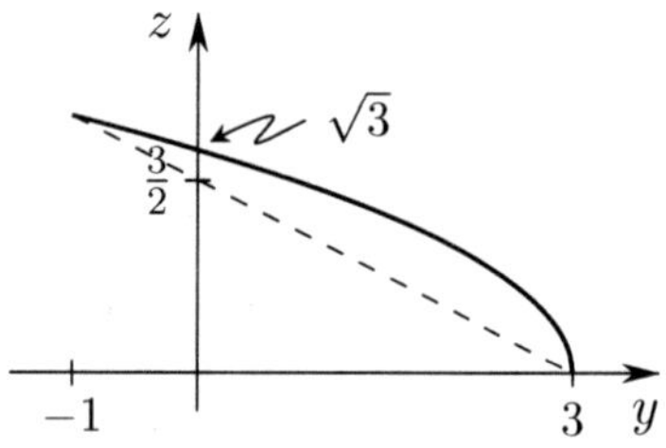

THEOREM 4. *Assume that the convexity assumptions 1–3 of theorem 2 are satisfied, and that $\bar{\boldsymbol{x}}$ is a minimum point for problem P. Assume that the constraint qualification (CQ) (see page 250) is fulfilled at $\bar{\boldsymbol{x}}$. Then there is no duality gap, i.e., the dual problem D has the same optimal value $f(\bar{\boldsymbol{x}})$.*

PROOF. By the Karush-Kuhn-Tucker theorem (theorem 7.4, page 251) $\bar{\boldsymbol{x}}$ is a KKT point. Denoting the corresponding Lagrange multipliers by $\bar{\boldsymbol{u}}$ and $\bar{\boldsymbol{v}}$, by theorem 2 $(\bar{\boldsymbol{x}}; \bar{\boldsymbol{u}}, \bar{\boldsymbol{v}})$ is a saddle point of the Lagrange function. Hence, by theorem 3, $f(\bar{\boldsymbol{x}}) = \theta(\bar{\boldsymbol{u}}, \bar{\boldsymbol{v}})$. □

Sometimes one can solve a convex minimum problem by solving its dual and then make use of this theorem.

Example 7. Let $\boldsymbol{A}$ be an $m\times n$-matrix of rank m and $\boldsymbol{b}$ an $m\times 1$-matrix, and consider the problem of minimizing $\|\boldsymbol{x}\|^2 = \boldsymbol{x}^T\boldsymbol{x}$ subject to $\boldsymbol{Ax} = \boldsymbol{b}$. (We assume this set to be non-empty.)

This is a convex programming problem, since obviously the objective function is convex and the constraints are affine. The Lagrange function is

$$L(\boldsymbol{x};\boldsymbol{v}) = \boldsymbol{x}^T\boldsymbol{x} + \boldsymbol{v}^T(\boldsymbol{Ax} - \boldsymbol{b}), \qquad \boldsymbol{x} \in \mathbf{R}^n,\ \boldsymbol{v} \in \mathbf{R}^m.$$

We first compute the dual objective function θ. The Lagrange function is a strictly convex quadratic function of $\boldsymbol{x}$, hence its minimum occurs at the stationary point. Setting the gradient equal to zero we obtain

$$\nabla_{\boldsymbol{x}} L(\boldsymbol{x};\boldsymbol{v}) = \boldsymbol{0} \quad\Longleftrightarrow\quad 2\boldsymbol{x} + \boldsymbol{A}^T\boldsymbol{v} = \boldsymbol{0} \quad\Longleftrightarrow\quad \boldsymbol{x} = -\tfrac{1}{2}\boldsymbol{A}^T\boldsymbol{v}.$$

Thus

$$\theta(\boldsymbol{v}) = \min_{\boldsymbol{x}} L(\boldsymbol{x};\boldsymbol{v}) = (-\tfrac{1}{2}\boldsymbol{A}^T\boldsymbol{v})^T(-\tfrac{1}{2}\boldsymbol{A}^T\boldsymbol{v}) + \boldsymbol{v}^T(-\tfrac{1}{2}\boldsymbol{A}\boldsymbol{A}^T\boldsymbol{v} - \boldsymbol{b}),$$

which simplifies to

$$\theta(\boldsymbol{v}) = -\tfrac{1}{4}\boldsymbol{v}^T\boldsymbol{A}\boldsymbol{A}^T\boldsymbol{v} - \boldsymbol{b}^T\boldsymbol{v}.$$

The matrix $\boldsymbol{A}\boldsymbol{A}^T$ is positive definite, since

$$\boldsymbol{x}^T\boldsymbol{A}\boldsymbol{A}^T\boldsymbol{x} = (\boldsymbol{A}^T\boldsymbol{x})^T(\boldsymbol{A}^T\boldsymbol{x}) = \|\boldsymbol{A}^T\boldsymbol{x}\|^2 > \boldsymbol{0} \qquad \text{when } \boldsymbol{x} \neq \boldsymbol{0}.$$

This makes θ a strictly concave (no surprise!) quadratic function. The dual problem is to maximize θ over $\mathbf{R}^m$.

Since we are now dealing with a concave function, the maximum occurs at the stationary point. Setting the gradient equal to zero we obtain

$$\nabla\theta(\boldsymbol{v}) = \boldsymbol{0} \quad\Longleftrightarrow\quad -\tfrac{1}{4}2\boldsymbol{A}\boldsymbol{A}^T\boldsymbol{v} - \boldsymbol{b} = \boldsymbol{0} \quad\Longleftrightarrow\quad \boldsymbol{v} = -2(\boldsymbol{A}\boldsymbol{A}^T)^{-1}\boldsymbol{b}.$$

(Since $\boldsymbol{A}\boldsymbol{A}^T$ is positive definite it is invertible.) Inserting in $\theta(\boldsymbol{v})$ and simplifying we obtain the maximum value

$$\boldsymbol{b}^T(\boldsymbol{A}\boldsymbol{A}^T)^{-1}\boldsymbol{b}.$$

By theorem 4 this is also the minimum value of the primal problem. (The condition (CQ) is fulfilled since $\boldsymbol{A}$ is of rank m.)

Put $\bar{\boldsymbol{v}} = -2(\boldsymbol{AA}^T)^{-1}\boldsymbol{b}$ and $\bar{\boldsymbol{x}} = -\frac{1}{2}\boldsymbol{A}^T\bar{\boldsymbol{v}}$. Then $\bar{\boldsymbol{x}}^T\bar{\boldsymbol{x}} = \boldsymbol{b}^T(\boldsymbol{AA}^T)^{-1}\boldsymbol{b}$. Hence $(\bar{\boldsymbol{x}};\bar{\boldsymbol{v}})$ is a saddle point of L by theorem 3. From theorem 1 we then conclude that $\bar{\boldsymbol{x}}$ is the minimum point of the primal problem. □

Example 8. Consider the quadratic programming problem

$$\text{minimize } \tfrac{1}{2}\boldsymbol{x}^T\boldsymbol{H}\boldsymbol{x} + \boldsymbol{c}^T\boldsymbol{x} \qquad \text{subject to } \boldsymbol{A}\boldsymbol{x} \leq \boldsymbol{b}.$$

We assume that $\boldsymbol{H}$ is positive semidefinite, making the objective function convex. Let $X = \mathbf{R}^n$.

Compute the objective function of the dual problem,

$$\theta(\boldsymbol{u}) = \inf_{\mathbf{R}^n} \tfrac{1}{2}\boldsymbol{x}^T\boldsymbol{H}\boldsymbol{x} + \boldsymbol{c}^T\boldsymbol{x} + \boldsymbol{u}^T(\boldsymbol{A}\boldsymbol{x} - \boldsymbol{b}). \tag{5}$$

For any $\boldsymbol{u}$ the function $\boldsymbol{x} \mapsto \frac{1}{2}\boldsymbol{x}^T\boldsymbol{H}\boldsymbol{x} + \boldsymbol{c}^T\boldsymbol{x} + \boldsymbol{u}^T(\boldsymbol{A}\boldsymbol{x} - \boldsymbol{b})$ is convex. This implies that its minima are given by the stationary points, i.e., by the solutions to the system

$$\boldsymbol{H}\boldsymbol{x} + \boldsymbol{c} + \boldsymbol{A}^T\boldsymbol{u} = \boldsymbol{0}. \tag{6}$$

If we assume $\boldsymbol{H}$ to be positive definite we have strict convexity and the minimum is unique. We can then solve for $\boldsymbol{x}$ in (6), obtaining

$$\boldsymbol{x} = -\boldsymbol{H}^{-1}(\boldsymbol{A}^T\boldsymbol{u} + \boldsymbol{c}).$$

Insertion into (5) gives, after some calculations,

$$\theta(\boldsymbol{u}) = \tfrac{1}{2}\boldsymbol{u}^T\boldsymbol{K}\boldsymbol{u} + \boldsymbol{d}^T\boldsymbol{u} - \tfrac{1}{2}\boldsymbol{c}^T\boldsymbol{H}^{-1}\boldsymbol{c},$$

where $\boldsymbol{K} = -\boldsymbol{A}\boldsymbol{H}^{-1}\boldsymbol{A}^T$ and $\boldsymbol{d} = -\boldsymbol{b} - \boldsymbol{A}\boldsymbol{H}^{-1}\boldsymbol{c}$.

The dual problem

$$\max_{\boldsymbol{u} \geq \boldsymbol{0}} \theta(\boldsymbol{u})$$

is as we now see another quadratic programming problem:

$$\text{maximize } \tfrac{1}{2}\boldsymbol{u}^T\boldsymbol{K}\boldsymbol{u} + \boldsymbol{d}^T\boldsymbol{u} - \tfrac{1}{2}\boldsymbol{c}^T\boldsymbol{H}^{-1}\boldsymbol{c} \qquad \text{subject to } \boldsymbol{u} \geq \boldsymbol{0}.$$

Here the constraints are simple; it is required to maximize over the non-negative orthant. There are algorithms for handling this kind of problem, but we will not pursue the discussion any further here.

Having found the solution to the dual problem, by theorem 4 above we also have established the minimum value of the primal problem. □

An example of sensitivity analysis

Example 9. We consider the problem $P_{\boldsymbol{c}}$ to minimize the function $f(\boldsymbol{x})$ subject to constraints

$$g_i(\boldsymbol{x}) \leq c_i, \qquad i = 1, \ldots, m.$$

The functions f and g_i are assumed to be convex on $\mathbf{R}^n$. In vector notation the constraints have the appearence $\boldsymbol{g}(\boldsymbol{x}) \leq \boldsymbol{c}$.

Assume that $\bar{\boldsymbol{x}}(\boldsymbol{c})$ solves this optimization problem. Denote the solution for $\boldsymbol{c} = \boldsymbol{0}$ by $\bar{\boldsymbol{x}}(\boldsymbol{0}) = \bar{\boldsymbol{x}}$, and the solution to the corresponding dual problem by $\bar{\boldsymbol{u}}$. Put $\Phi(\boldsymbol{c}) = f(\bar{\boldsymbol{x}}(\boldsymbol{c}))$. We want to prove the inequality

$$\Phi(\boldsymbol{c}) \geq f(\bar{\boldsymbol{x}}) - \bar{\boldsymbol{u}}^T \boldsymbol{c}, \tag{7}$$

which gives an indication of how the perturbation affects the minimum value.

From the convexity assumptions it follows that there is no duality gap. Hence

$$f(\bar{\boldsymbol{x}}) = \theta(\bar{\boldsymbol{u}}) \leq L(\boldsymbol{x}; \bar{\boldsymbol{u}}) = f(\boldsymbol{x}) + \bar{\boldsymbol{u}}^T \boldsymbol{g}(\boldsymbol{x}), \qquad \boldsymbol{x} \in \mathbf{R}^n,$$

where L denotes the Lagrange function for the $\boldsymbol{c} = \boldsymbol{0}$ problem. For any $\boldsymbol{x}$ which is feasible for problem $P_{\boldsymbol{c}}$ we then have

$$f(\bar{\boldsymbol{x}}) \leq f(\boldsymbol{x}) + \bar{\boldsymbol{u}}^T \boldsymbol{g}(\boldsymbol{x}) \leq f(\boldsymbol{x}) + \bar{\boldsymbol{u}}^T \boldsymbol{c}.$$

This can be rewritten as $f(\boldsymbol{x}) \geq f(\bar{\boldsymbol{x}}) - \bar{\boldsymbol{u}}^T \boldsymbol{c}$, and taking the minimum over all feasible $\boldsymbol{x}$ we obtain (7).

We consider problem $P_{\boldsymbol{c}}$ as a perturbation of the case when $\boldsymbol{c} = \boldsymbol{0}$. Then we can draw the following conclusions from (7).

- Suppose we relax the ith constraint somewhat by choosing a positive value for c_i. If the multiplier u_i is small this will not have much impact on the minimum value.

- Suppose we tighten the ith constraint by making c_i negative. If u_i is large we can conclude that this will have a considerable influence on the minimum value.

Since (7) is merely an inequality we can draw no conclusion e.g. in the case when c_i is positive and u_i is large. However, if we assume that Φ is a differentiable function of $\boldsymbol{c}$ we can say more.

Rewriting (7) as

$$\Phi(\boldsymbol{c}) + \bar{\boldsymbol{u}}^T \boldsymbol{c} \geq f(\bar{\boldsymbol{x}})$$

shows that the function $\boldsymbol{c} \mapsto \Phi(\boldsymbol{c}) + \bar{\boldsymbol{u}}^T \boldsymbol{c}$ has a minimum for $\boldsymbol{c} = \boldsymbol{0}$. It follows that this function has zero gradient at $\boldsymbol{c} = \boldsymbol{0}$:

$$\boldsymbol{\nabla}\Phi(\boldsymbol{0}) + \bar{\boldsymbol{u}} = \boldsymbol{0}.$$

In particular $\dfrac{d\Phi}{dc_i}(\boldsymbol{0}) = -\bar{u}_i,\ \ i = 1, \ldots, m$. Hence the Lagrange multipliers determine the rate of growth of the minimum value as function of the c_i. □

Exercises

S 8.1 Consider the problem to minimize

$$x_1^2 + 2x_2^2$$

when

$$x_1 + x_2 \geq 1 \quad \text{and} \quad x_1 \geq 0,\ x_2 \geq 0.$$

a) Compute the dual function if $X = \{\, (x_1, x_2)\,;\, x_1 \geq 0, x_2 \geq 0 \,\}$.

b) Solve the dual problem.

c) Explain why the answer to the dual problem gives the value of the primal problem.

8.2 Consider the problem to minimize

$$x_1^2 + 2x_2^2$$

when

$$x_1 + x_2 \geq 1 \quad \text{and} \quad x_1 \geq 0,\ x_2 \geq 0.$$

Compute the dual function if $X = \{\, (x_1, x_2)\,;\, x_1 \geq 0 \,\}$.

8.3 a) Let the primal problem be to minimize x_1 subject to

$$g(\boldsymbol{x}) = x_1^2 + x_2^2 - 1 \leq 0.$$

Find the dual function.

b) Find the optimal solutions to both the primal and the dual problem, and compare their objective values.

c) Do the same for the problem to minimize x_1 subject to

$$h(\boldsymbol{x}) = x_1^2 + x_2^2 - 1 = 0.$$

d) Introduce the set $X = \{ (x_1, x_2) ; x_1^2 + x_2^2 = 1 \}$, and consider the problem $\min_{\boldsymbol{x} \in X} x_1$. Now there are now no explicit constraints. Find the dual function in this case.

8.4 Consider the constrained problem from exercise 7.6, page 263, to minimize $x_1 x_2$ subject to $x_1^2 + x_2^2 \leq 4$ and $x_1^2 - x_2^2 \leq 1$.

a) Find the dual objective function. (Choose $X = \mathbf{R}^2$ and make both constraints explicit.)

b) Solve the dual problem and verify that there is no duality gap.

8.5 Consider the quadratic programming problem

$$\begin{array}{ll} \text{Minimize} & \frac{1}{2}\boldsymbol{x}^T \boldsymbol{H} \boldsymbol{x} + \boldsymbol{c}^T \boldsymbol{x} \\ \text{subject to} & \boldsymbol{A}\boldsymbol{x} \leq \boldsymbol{b}, \end{array}$$

where

$$\boldsymbol{H} = \begin{pmatrix} 1 & 2 \\ 2 & 8 \end{pmatrix}, \quad \boldsymbol{c} = \begin{pmatrix} 1 \\ -3 \end{pmatrix}, \quad \boldsymbol{A} = \begin{pmatrix} 1 & 2 \\ 3 & -2 \end{pmatrix}, \quad \boldsymbol{b} = \begin{pmatrix} 10 \\ 12 \end{pmatrix}.$$

a) In the general case show that if $\boldsymbol{H}$ is positive definite then the dual function is

$$\theta(\boldsymbol{u}) = -\tfrac{1}{2}\boldsymbol{u}^T(\boldsymbol{A}\boldsymbol{H}^{-1}\boldsymbol{A}^T)\boldsymbol{u} - (\boldsymbol{b} + \boldsymbol{A}\boldsymbol{H}^{-1}\boldsymbol{c})^T\boldsymbol{u} - \tfrac{1}{2}\boldsymbol{c}^T\boldsymbol{H}^{-1}\boldsymbol{c}.$$

Hint: Example 8.

b) Verify that $\boldsymbol{H}$ is positive definite and compute $\theta(\boldsymbol{u})$ in this particular case.

c) Solve (e.g. analytically) the dual problem

$$\begin{array}{ll} \text{Maximize} & \theta(\boldsymbol{u}) \\ \text{subject to} & \boldsymbol{u} \geq \boldsymbol{0}. \end{array}$$

d) Use the answer to solve the primal problem.

8.6 Referring to exercise 7.29a, page 290, take $X = \{ (x_1, x_2, x_3);\ x_3 \geq 0 \}$ and calculate the dual function. Solve the dual problem and find a saddle point if possible. (2009–01–09, 6b)

8.7 Consider the problem

$$\text{minimize} \quad x^2 + 2(y+3)^2 \qquad \text{subject to} \quad 2xy + 2 \leq 0,\ y \geq 1.$$

Calculate the dual function with respect to the first constraint only (that is, for $X = \{ (x, y);\ y \geq 1 \}$), solve the dual problem, and then solve the primal problem by showing no duality gap. (2007–04–10, 4)

8.8 Consider the optimization problem to minimize

$$e^{x^2} - \ln y$$

subject to

$$x^2 + y^2 \leq 1, \qquad y \geq 0.$$

Calculate the dual function if $X = \{ (x, y), ;\ y > 0 \}$. Solve the dual problem and provide a solution to the problem by showing that there is no duality gap. (2007–01–11, 4)

Chapter 9

PENALTY AND BARRIER FUNCTIONS

In this chapter we give some indications of how constrained optimization problems can be solved numerically. The search methods of chapter 3 are in general geared towards problems without constraints, and are not easily adapted to more than very simple side conditions. A wealth of methods have been devised to handle different classes of constrained problems, among them of course the simplex algorithm (see section 5.2) and others for the LP problem. The reader who wants more than a glimpse should consult other literature on the subject.

9.1 Constrained optimization in practice

Many methods for numerical optimization of unconstrained functions operate as we know by iteration of the following scheme.

- Given the current point $\boldsymbol{x}_k$, find a search direction $\boldsymbol{d}_k$.
- Choose $\boldsymbol{x}_{k+1}$ to minimize f in direction $\boldsymbol{d}_k$ from $\boldsymbol{x}_k$ (*line search*).

In cases when there are non-trivial constraints present, this procedure can of course not be applied so directly. Among other problems, in order for the line search to make sense the search direction $\boldsymbol{d}_k$ must lead to points that are feasible for the constrained problem. We are thus faced with the question of how to find search directions with this property.

We give one example of how this could be done.

Example 1. Consider a problem with affine inequality constraints: minimize $f(\boldsymbol{x})$ subject to $g_i(\boldsymbol{x}) \leq 0$, $i = 1, \ldots, m$, where

$$g_i(\boldsymbol{x}) = \boldsymbol{a}_i^T \boldsymbol{x} - c_i.$$

Let I denote the set of indices representing those constraints that are active at the current point $\boldsymbol{x}_k$. The feasible directions are as we know (example 7.3, page 236) characterized by the conditions

$$\boldsymbol{a}_i^T \boldsymbol{d} \leq 0, \qquad i \in I.$$

As search direction from $\boldsymbol{x}_k$ we choose the vector $\boldsymbol{d} = (d_1, \ldots, d_n)$ which solves the following optimization problem:

$$\begin{array}{lrll} & \text{minimize} & \nabla f(\boldsymbol{x}_k)^T \boldsymbol{d} & \\ (1) & \text{subject to} & \boldsymbol{a}_i^T \boldsymbol{d} \leq 0, & i \in I, \\ & \text{and} & -1 \leq d_j \leq 1, & j = 1, \ldots, n. \end{array}$$

The purpose of conditions $|d_j| \leq 1$ is to avoid an unbounded domain. The minimization of $\nabla f(\boldsymbol{x}_k)^T \boldsymbol{d}$ will certainly result in a non-positive value, since $\boldsymbol{0}$ is a feasible point for the problem. For an unconstrained problem the solution would of course be in the direction of the negative gradient; in general we are assured of a descent direction if there is one. If the solution proves to be $\boldsymbol{d} = \boldsymbol{0}$ there is no descent direction, and $\boldsymbol{x}_k$ is the optimal solution to the problem of minimizing $f(\boldsymbol{x})$.

Note that (1) is an LP problem; it can be solved with the simplex method. □

Generalization of the method in example 1 to non-linear constraints and to equality constraints (called the *method of Zoutendijk*) is not straightforward. You can not simply replace the $\boldsymbol{a}_i$ with $\nabla g_i(\boldsymbol{x}_k)$. In the non-linear case there might in fact not be any feasible direction at all. Not all methods can be guaranteed to converge; also you might obtain a sequence converging to a point which is not even a local minimum point.

We will not pursue these matters further here. Instead we turn our attention to the possibility of converting our optimization problem to an unconstrained one with the objective function incorporating the constraints as well as the original objective function. This new problem can be solved by algorithms developed for unconstrained problems (Newton's method, conjugate gradient, ...).

9.2 The penalty function method

9.2.1 Penalty functions

The constrained problem we wish to solve (problem P) is to find

$$\min_{\boldsymbol{x}\in S} f(\boldsymbol{x}),$$

where the set S as usual is defined by conditions

$$\boldsymbol{x} \in X, \qquad \boldsymbol{g}(\boldsymbol{x}) \leq \boldsymbol{0}, \qquad \boldsymbol{h}(\boldsymbol{x}) = \boldsymbol{0}.$$

For simplicity we assume $X = \mathbf{R}^n$ in the future. The general idea is to replace this problem with

$$\min_{\boldsymbol{x}\in\mathbf{R}^n} f(\boldsymbol{x}) + \mu\alpha(\boldsymbol{x}),$$

where μ is a positive number and α a function with the properties

- $\alpha(\boldsymbol{x}) = 0$ when $\boldsymbol{x} \in S$,
- $\alpha(\boldsymbol{x}) > 0$ when $\boldsymbol{x} \notin S$,
- α is continuous.

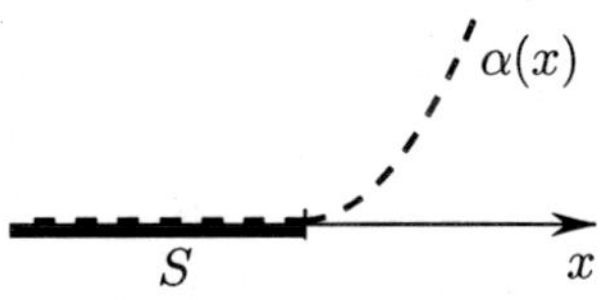

The figure gives an illustration in the one variable case.

Example 2.
a) Minimizing $f(\boldsymbol{x})$ subject to $h(\boldsymbol{x}) = 0$ could be replaced by

$$\min_{\mathbf{R}^n} f(\boldsymbol{x}) + \mu h(\boldsymbol{x})^2.$$

b) Minimizing $f(\boldsymbol{x})$ subject to $g(\boldsymbol{x}) \leq 0$ could be replaced by

$$\min_{\mathbf{R}^n} f(\boldsymbol{x}) + \mu \max(0,\ g(\boldsymbol{x}))$$

or by

$$\min_{\mathbf{R}^n} f(\boldsymbol{x}) + \mu \max(0,\ g(\boldsymbol{x}))^2.$$

The latter function has the advantage of being differentiable. □

It is obvious that a large value of μ will encourage a minimization algorithm to choose points $\boldsymbol{x}$ in S, or close to it.

DEFINITION 1. *A function α with the properties stated above is called a* **penalty function**. *We call $q_\mu = f + \mu\alpha$ an* **auxiliary function** *of problem P.*

Example 3. Often the penalty function for a problem like P is of the form

$$\alpha(\boldsymbol{x}) = \sum_{i=1}^{m} (\max(0, g_i(\boldsymbol{x}))^p + \sum_{j=1}^{l} |h_j(\boldsymbol{x})|^p$$

for some positive integer p. □

Example 4. Consider the problem to minimize $f(\boldsymbol{x}) = x_1^2 + x_2^2$ subject to the constraint $x_1 + x_2 - 1 = 0$. The solution is easily seen to be $(\frac{1}{2}, \frac{1}{2})$. Choosing the penalty function

$$\alpha(\boldsymbol{x}) = (x_1 + x_2 - 1)^2,$$

the auxiliary problem is to minimize

$$q_\mu(\boldsymbol{x}) = x_1^2 + x_2^2 + \mu(x_1 + x_2 - 1)^2$$

over the whole of $\mathbf{R}^2$. Since the auxiliary function is convex the minimum occurs where $\nabla q_\mu(\boldsymbol{x}) = \boldsymbol{0}$. We obtain the system

$$\begin{cases} 2x_1 + 2\mu(x_1+x_2-1) = 0 \\ 2x_2 + 2\mu(x_1+x_2-1) = 0 \end{cases} \iff \begin{cases} x_1 = \dfrac{\mu}{2\mu+1} \\ x_2 = \dfrac{\mu}{2\mu+1}. \end{cases}$$

As we see this solution tends to the minimum solution $(\frac{1}{2}, \frac{1}{2})$ of the original problem as $\mu \to \infty$. □

To solve problem P we now apply the following strategy. Choose a strictly increasing sequence $\{\mu_k\}_1^\infty$ of positive numbers tending to ∞ as $k \to \infty$. Solve the sequence of auxiliary problems[1]

(2) $$\min q_k(\boldsymbol{x}), \qquad \text{where} \quad q_k(\boldsymbol{x}) = f(\boldsymbol{x}) + \mu_k\alpha(\boldsymbol{x}).$$

[1]Assume that a solution exists.

Denote the (exact) solution to problem (2) by $\boldsymbol{x}_k$. We use $\boldsymbol{x}_k$ or a suitable approximation as initial point in the $(k+1)$st auxiliary problem. It is our hope that for k large $\boldsymbol{x}_k$ will be a good approximation to the minimum point of problem P. We now proceed to investigate whether this is the case.

The next lemma provides information on the behaviour of the sequence $\{\boldsymbol{x}_k\}_1^\infty$.

LEMMA 1.
a) $q_k(\boldsymbol{x}_k)$ *increases with* k.

b) $\alpha(\boldsymbol{x}_k)$ *decreases with* k.

c) $f(\boldsymbol{x}_k)$ *increases with* k.

PROOF. The following obvious formula will be used frequently:

$$f(\boldsymbol{x}_{k+1}) + \mu_k\alpha(\boldsymbol{x}_{k+1}) \geq \min_{\boldsymbol{x}} f(\boldsymbol{x}) + \mu_k\alpha(\boldsymbol{x}) = q_k(\boldsymbol{x}_k). \tag{3}$$

a) Since $\mu_{k+1} > \mu_k$ and $\alpha(\boldsymbol{x}) \geq 0$ we have

$$q_{k+1}(\boldsymbol{x}_{k+1}) = f(\boldsymbol{x}_{k+1})+\mu_{k+1}\alpha(\boldsymbol{x}_{k+1}) \geq f(\boldsymbol{x}_{k+1})+\mu_k\alpha(\boldsymbol{x}_{k+1}) \geq q_k(\boldsymbol{x}_k),$$

where the last inequality is precisely (3).

b) By (3)

$$f(\boldsymbol{x}_{k+1}) + \mu_k\alpha(\boldsymbol{x}_{k+1}) \geq f(\boldsymbol{x}_k) + \mu_k\alpha(\boldsymbol{x}_k),$$

and by the same reasoning

$$f(\boldsymbol{x}_k) + \mu_{k+1}\alpha(\boldsymbol{x}_k) \geq f(\boldsymbol{x}_{k+1}) + \mu_{k+1}\alpha(\boldsymbol{x}_{k+1}).$$

Addition of these formulas results after some cancelling in

$$\mu_k\alpha(\boldsymbol{x}_{k+1}) + \mu_{k+1}\alpha(\boldsymbol{x}_k) \geq \mu_k\alpha(\boldsymbol{x}_k) + \mu_{k+1}\alpha(\boldsymbol{x}_{k+1}),$$

which can be rewritten as

$$(\mu_{k+1} - \mu_k)\alpha(\boldsymbol{x}_k) \geq (\mu_{k+1} - \mu_k)\alpha(\boldsymbol{x}_{k+1}).$$

Since $\mu_{k+1} - \mu_k > 0$ this implies that $\alpha(\boldsymbol{x}_{k+1}) \leq \alpha(\boldsymbol{x}_k)$.

c) By (3) again and using b) we have

$$f(\boldsymbol{x}_{k+1}) + \mu_k\alpha(\boldsymbol{x}_{k+1}) \geq f(\boldsymbol{x}_k) + \mu_k\alpha(\boldsymbol{x}_k) \geq f(\boldsymbol{x}_k) + \mu_k\alpha(\boldsymbol{x}_{k+1}).$$

After cancelling we obtain $f(\boldsymbol{x}_{k+1}) \geq f(\boldsymbol{x}_k)$. The proof is finished. □

LEMMA 2. *For any $\boldsymbol{x} \in S$ and any k we have $q_k(\boldsymbol{x}_k) \leq f(\boldsymbol{x})$.*

PROOF. By the definition of $\boldsymbol{x}_k$

$$q_k(\boldsymbol{x}_k) \leq f(\boldsymbol{x}) + \mu_k \alpha(\boldsymbol{x}) = f(\boldsymbol{x}),$$

since $\alpha(\boldsymbol{x}) = 0$ when $\boldsymbol{x} \in S$. □

THEOREM 1. *Assume that the auxiliary minimization problems (2) have a solution for all $k \geq 1$, and let $\{\boldsymbol{x}_k\}_1^\infty$ be a sequence generated by successively solving them for $k = 1, 2, \ldots$. Assume that there is a subsequence converging to some limit point $\bar{\boldsymbol{x}}$. Then $\bar{\boldsymbol{x}}$ is a solution to the minimization problem P.*

PROOF. Assume for simplicity of notation that $\{\boldsymbol{x}_k\}_1^\infty$ itself is the subsequence that converges to $\bar{\boldsymbol{x}}$. Then by continuity

$$f(\boldsymbol{x}_k) \to f(\bar{\boldsymbol{x}}), \qquad \alpha(\boldsymbol{x}_k) \to \alpha(\bar{\boldsymbol{x}}).$$

The sequence $\{q_k(\boldsymbol{x}_k)\}_1^\infty$ is increasing by the first lemma and bounded above by $f(\boldsymbol{x})$ for any $\boldsymbol{x} \in S$ by the second lemma. Thus this sequence has a limit $\overline{q}$. Now it follows that

$$\mu_k \alpha(\boldsymbol{x}_k) = q_k(\boldsymbol{x}_k) - f(\boldsymbol{x}_k) \to \overline{q} - f(\bar{\boldsymbol{x}}).$$

Since $\mu_k \to \infty$ this implies that $\alpha(\boldsymbol{x}_k) \to 0$. Hence

$$\alpha(\bar{\boldsymbol{x}}) = 0,$$

and this in turn implies that $\bar{\boldsymbol{x}} \in S$, in other words that $\bar{\boldsymbol{x}}$ is a feasible point for the minimization problem P. It remains to prove that it is optimal.

We have

$$f(\boldsymbol{x}_k) \leq f(\boldsymbol{x}_k) + \mu_k \alpha(\boldsymbol{x}_k) \leq f(\boldsymbol{x}) \qquad \text{for all } \boldsymbol{x} \in S, \tag{4}$$

the last inequality by the second lemma. Letting $k \to \infty$ we obtain

$$f(\bar{\boldsymbol{x}}) \leq f(\boldsymbol{x}) \qquad \text{for all } \boldsymbol{x} \in S.$$

Thus $\bar{\boldsymbol{x}}$ is optimal. □

Remark. We leave as an exercise to the reader to prove that by (4) $q_k(\boldsymbol{x}_k) \to f(\bar{\boldsymbol{x}})$ as $k \to \infty$. As a consequence $\mu_k \alpha(\boldsymbol{x}_k) \to 0$.

Remark. Provided that the $\boldsymbol{x}_k$ remain in a compact set, by Bolzano-Weierstrass' theorem the sequence $\{\boldsymbol{x}_k\}_1^\infty$ does have a convergent subsequence. Also note that as a consequence of lemma 1c

$$f(\boldsymbol{x}_k) \to f(\bar{\boldsymbol{x}}) \qquad \text{as} \quad k \to \infty,$$

even though the whole sequence $\{\boldsymbol{x}_k\}_1^\infty$ may not converge to $\bar{\boldsymbol{x}}$.

The reader should note the assumption in theorem 1 that the auxiliary problems (2) have a solution. Also notice that $\boldsymbol{x}_k$ denotes the *exact* solution to the kth problem. What we have described is thus an idealization of the procedure which must be applied in practice. When the problems in (2) are solved numerically, the exact solution is normally only reached after an infinite number of steps. Some termination criterion must be used, and an approximate solution will therefore be obtained. Theorem 1 can be extended to take care of this more general situation.

Some other observations are in order. Assume that the optimal solution $\bar{\boldsymbol{x}}$ to problem P belongs to the *interior* of S. Since $f(\boldsymbol{x}) \leq q_1(\boldsymbol{x})$ for all $\boldsymbol{x}$, with equality when $\boldsymbol{x} \in S$, the first auxiliary problem also has $\bar{\boldsymbol{x}}$ as an optimal point. Hence we reach $\bar{\boldsymbol{x}}$ already in the first trial.

Assume instead that $\bar{\boldsymbol{x}}$ is a boundary point of S. Then we can not expect it to be a solution to the auxiliary problem $\min_{\mathbf{R}^n} q_1(\boldsymbol{x})$. In fact, $\boldsymbol{x}_1$ and all consecutive $\boldsymbol{x}_k$ can be expected to belong to the *complement* of S. (Cf. e.g. example 4 above and example 5 below.) To justify this claim consider again inequality (4), which in particular is valid for $\boldsymbol{x} = \bar{\boldsymbol{x}} \in S$. It implies that

$$f(\boldsymbol{x}_k) \leq f(\bar{\boldsymbol{x}}), \qquad k = 1, 2, \ldots .$$

If $\bar{\boldsymbol{x}}$ is a strict local minimum for problem P, the implication is that $\boldsymbol{x}_k \notin S$, or that $\boldsymbol{x}_k = \bar{\boldsymbol{x}}$.

A consequence of the last observation is that if you stop the iterations prematurely the obtained solution will be infeasible for problem P. For this reason the method is sometimes called an *exterior penalty function method.*

Example 5. We consider again that objective function which appeared frequently in the examples in chapter 3,

$$f(x_1, x_2) = 2(x_1 - 2)^4 + (2x_1 - 3x_2)^2.$$

We now wish to minimize it subject to the constraint

$$g(x_1, x_2) = x_1^2 - 2x_2 \leq 0.$$

Thus we introduce the penalty function $\alpha(x_1, x_2) = (\max(0, x_1^2 - 2x_2))^2$ and the auxiliary function

$$q_\mu(x_1, x_2) = f(x_1, x_2) + \mu\alpha(x_1, x_2).$$

Taking $\mu = 1$ and starting at $\boldsymbol{x}^{(0)} = (0,\ 2.5)$ we carry out four (outer) iterations with Fletcher-Reeves method. Successively we obtain

k	1	2	3	4
$\boldsymbol{x}^{(k)}$	$(1.7748, 1.2252)$	$(1.5446, 1.0754)$	$(1.5409, 1.0765)$	$(1.5409, 1.0765)$

We seem to have reached the minimum of q_μ.

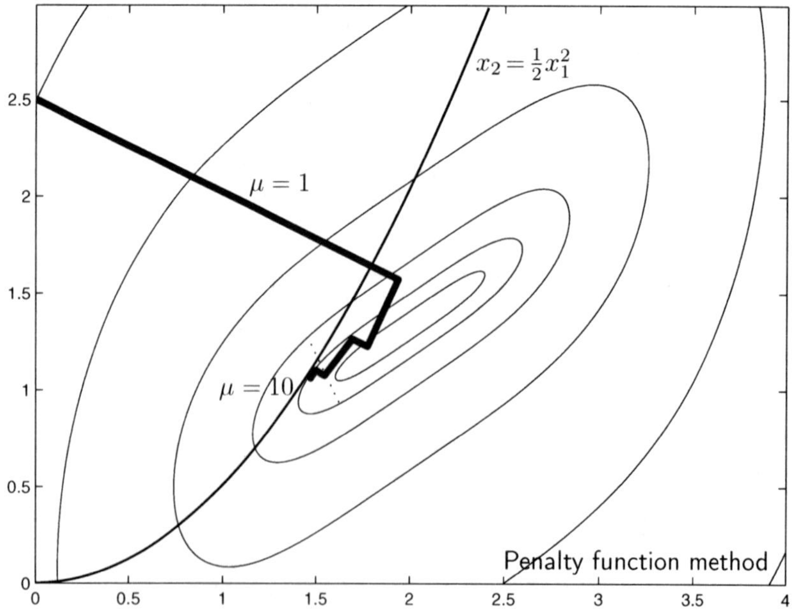

Now let $\mu = 10$, start from $(1.5409, 1.0765)$ and carry out another four iteration steps. The computational results are

k	1	2	3	4
$\boldsymbol{x}^{(k)}$	$(1.4709, 1.0624)$	$(1.4705, 1.0626)$	$(1.4703, 1.0624)$	$(1.4703, 1.0624)$

The last point $(1.4703, 1.0624)$ is the one denoted $\boldsymbol{x}_2$ in theorem 1.

The successive iterations are shown in the picture. (However the last three ones don't show on the scale used.) As expected all points lie outside the feasible domain.

Continuing with $\mu = 100$ results in $\boldsymbol{x}_3 = (1.4584, 1.0614)$. The exact minimum point of f is $\approx (1.4570, 1.0614)$. □

9.2.2 Estimation of the Lagrange multipliers

In many applications the values of the Lagrange multipliers at a local minimum point are of importance. Hence it is an interesting fact that the above scheme for computing the minimum point also provides an opportunity to obtain estimates for the multipliers.

For simplicity of notation we assume one equality and one inequality constraint only. Thus we consider the problem of minimizing $f(\boldsymbol{x})$ subject to

$$g(\boldsymbol{x}) \leq \boldsymbol{0}, \qquad h(\boldsymbol{x}) = \boldsymbol{0}.$$

All functions are assumed continuously differentiable. We assume a penalty function of the form

$$\alpha(\boldsymbol{x}) = \varphi(g(\boldsymbol{x})) + \psi(h(\boldsymbol{x})),$$

with φ and ψ continuously differentiable functions satisfying $\varphi'(y) > 0$ when $y > 0$, and any further conditions necessary to make each term in α a penalty function. One example is (see example 2)

$$\varphi(y) = \max(0, y)^2, \qquad \psi(y) = y^2.$$

Choose the sequence $\{\mu_k\}_1^\infty$ as before. The auxiliary functions to be minimized are

$$q_k(\boldsymbol{x}) = f(\boldsymbol{x}) + \mu_k\Big(\varphi(g(\boldsymbol{x})) + \psi(h(\boldsymbol{x}))\Big), \qquad k = 1, 2, \ldots.$$

Since q_k is differentiable the minimizing point $\boldsymbol{x}_k$ satisfies $\nabla q(\boldsymbol{x}_k) = \boldsymbol{0}$, i.e.

$$\nabla f(\boldsymbol{x}_k) + \mu_k \varphi'(g(\boldsymbol{x}_k))\nabla g(\boldsymbol{x}_k) + \mu_k \psi'(h(\boldsymbol{x}_k))\nabla h(\boldsymbol{x}_k) = \boldsymbol{0}. \tag{5}$$

Let

$$u_k = \mu_k \varphi'(g(\boldsymbol{x}_k)) \geq 0, \qquad v_k = \mu_k \psi'(h(\boldsymbol{x}_k)). \tag{6}$$

Assume that the sequence $\{\boldsymbol{x}_k\}_1^\infty$ from theorem 1 is convergent with limit point $\bar{\boldsymbol{x}}$. Letting $k \to \infty$ in (5) results in

$$\nabla f(\bar{\boldsymbol{x}}) + \bar{u}\, \nabla g(\bar{\boldsymbol{x}}) + \bar{v}\, \nabla h(\bar{\boldsymbol{x}}) = \boldsymbol{0},$$

where $\bar{u} \geq 0$ and $\bar{v}$ are the limits of u_k and v_k respectively.[2]

If we assume $\nabla g(\bar{\boldsymbol{x}})$ and $\nabla h(\bar{\boldsymbol{x}})$ to be linearly independent, the Lagrange multipliers in the KKT conditions for $\bar{\boldsymbol{x}}$ are unique. Then $\bar{u}$ and $\bar{v}$ above are precisely these multipliers. Hence in the iteration scheme they can be estimated from (6) for k large.

Having knowledge of the multipliers of course makes it possible to check the computed optimal solution against the KKT conditions.

9.2.3 Some computational aspects

The Hessian of the auxiliary function q_μ is

$$\nabla^2 q_\mu = \nabla^2 f + \mu \nabla^2 \alpha.$$

This matrix is of course dependent on μ, and so are its eigenvalues. In fact a careful analysis shows that some eigenvalues may become quite large when μ is large. As we know this might affect the convergence properties of a minimization algorithm unfavourably.

From this point of view a small to moderate value of μ should be used in the minimization of q_μ. On the other hand, a large value of μ is of course required if we want an effective barrier so that the minimum point of q_μ is reasonably close to the required minimum of f.

It is with the aim to resolve this conflict that we choose an increasing sequence $\{\mu_k\}_1^\infty$ tending to infinity, and solve the corresponding auxiliary problems consecutively. In this way it is hoped that the ill-conditioning can be avoided until we are already close to the minimum.

[2]It can be shown that the limits of u_k and v_k exist.

It is known that the behaviour of the steepest descent method is particularly sensitive to the condition number. This method should not be used here. In the literature Newton's method or some variant of conjugate gradient methods are recommended.

Example 6. Consider again the situation in example 4, where the auxiliary problem is to minimize

$$q_\mu(\boldsymbol{x}) = x_1^2 + x_2^2 + \mu(x_1+x_2-1)^2.$$

The Hessian is

$$\nabla^2 q_\mu(\boldsymbol{x}) = \begin{pmatrix} 2+2\mu & 2\mu \\ 2\mu & 2+2\mu \end{pmatrix}$$

with eigenvalues 2 and $2+4\mu$. The condition number $1+2\mu$ grows to infinity with μ. □

Exercises

S 9.1 Consider the following problem:

$$\begin{aligned} &\text{Minimize} && x_1^2 + x_2^2 \\ &\text{subject to} && 2x_1 + x_2 - 2 \leq 0 \\ &&& -x_2 + 1 \leq 0. \end{aligned}$$

a) Find analytically the exact solution to this problem.

b) Describe a suitable auxiliary function, with a penalty parameter μ.

c) Let $\mu = 1$. Starting from the point $(2, 6)$, solve the resulting problem by using Newton's method.

d) Set the penalty parameter μ to 10. Starting from the point you obtained in part c), solve the resulting problem.

9.2 Consider the following problem:

$$\begin{aligned} &\text{Minimize} && x_1^3 + x_2^3 \\ &\text{subject to} && x_1 + x_2 - 1 = 0. \end{aligned}$$

a) Find analytically an optimal solution to the problem.

b) Consider the following penalty problem:

$$\text{Minimize} \quad x_1^3 + x_2^3 + \mu(x_1 + x_2 - 1)^2.$$

Show that it is unbounded. Explain why this is no contradiction to theorem 9.1.

c) Add the constraints

$$0 \leq x_1 + x_2 \leq 2, \qquad -R \leq x_1 \leq R$$

(R some large number) to the problem. Let the compact set X be defined by these constraints. The penalty problem becomes

$$\begin{aligned} &\text{Minimize} && x_1^3 + x_2^3 + \mu(x_1 + x_2 - 1)^2 \\ &\text{subject to} && (x_1, x_2) \in X. \end{aligned}$$

Find analytically the optimal solution for a fixed (large) $\mu > 0$? Find the limit of the sequence of minimum points as $\mu \to \infty$? — Note that with the addition of the set X, the conclusion of theorem 9.1 does hold true.

Hint: The most simple way to solve the auxiliary problem is perhaps to study the objective function on lines $x_1 + x_2 = c$.

9.3 Consider the following problem:

$$\begin{aligned} &\text{Minimize} && e^{x_1} + x_1^2 + x_1 x_2 \\ &\text{subject to} && \tfrac{1}{2}x_1 + x_2 - 1 = 0. \end{aligned}$$

a) Describe a suitable auxiliary problem with $\mu = 4$. Starting with the point $(0, 0)$, perform three iterations with Newton's method.

b) Find an estimate for the value of the Lagrange multiplier at the minimum.

9.3 The barrier function method

For this method we consider only problems with inequality constraints, since we require the domain S to have a non-empty interior which is arbitrarily close to any point of S. Thus the problem to be considered is

$$\min_{\boldsymbol{x} \in S} f(\boldsymbol{x}), \tag{7}$$

where S is defined by $\boldsymbol{g}(\boldsymbol{x}) \leq \boldsymbol{0}$, i.e., $g_i(\boldsymbol{x}) \leq 0$, $i = 1, \ldots, m$.

We introduce the *auxiliary problem*

$$\min_{\boldsymbol{x} \in \operatorname{int}(S)} f(\boldsymbol{x}) + \varepsilon\beta(\boldsymbol{x}),$$

with ε a positive number and $\beta\colon \operatorname{int}(S) \to \mathbf{R}$ a function with the properties

- $\beta(\boldsymbol{x}) \geq 0$ when $\boldsymbol{x} \in \operatorname{int}(S)$,
- $\beta(\boldsymbol{x}) \to \infty$ as $\boldsymbol{x} \to \partial S$,
- β is continuous on $\operatorname{int}(S)$.

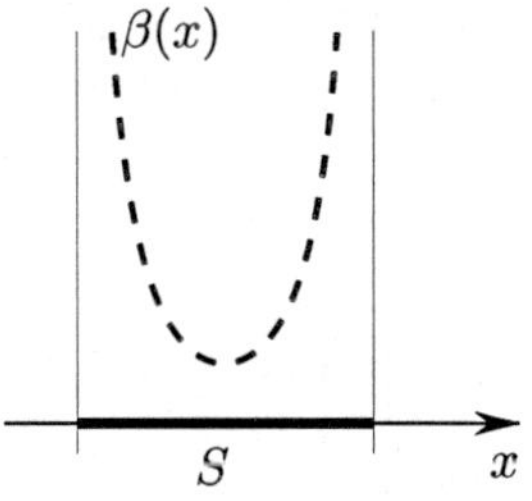

The figure gives a one variable illustration.

DEFINITION 2. *A function β with these properties is called a* **barrier function**. *The function $q_\varepsilon = f + \varepsilon\beta$ is called an* **auxiliary function** *of the problem.*

Example 7. For the constraints $g_i(\boldsymbol{x}) \leq 0$, $i = 1, \ldots, m$, the following is a possible barrier function:

$$\beta_1(\boldsymbol{x}) = \sum_{i=1}^{m} \frac{1}{g_i(\boldsymbol{x})}.$$

Another one is

$$\beta_2(\boldsymbol{x}) = -\sum_{i=1}^{m} \ln\Big(\min(1, -g_i(\boldsymbol{x}))\Big).$$

□

The barrier is supposed to prevent any search algorithm from generating points close to the boundary of S. Starting from a point in the interior of S we stay there forever. In contrast to the penalty function case all points obtained in the iterations are feasible. The method is sometimes named an *interior penalty function method*.

There is a computational issue involved here which should be noted. Experience shows that when using line search you may easily slip through a barrier such as the ones above. A remedy for this is to define $\beta(\boldsymbol{x})$ to be ∞ (or some large constant) outside the domain S. — Newton's method without line search should not be used, since the barrier will never be noticed.

Example 8. Consider the simple problem in one variable to minimize $f(x) = x$ subject to the condition $-x + 1 \leq 0$. Choose the barrier function

$$\beta(x) = \frac{-1}{-x+1}.$$

The auxiliary function is

$$q_\varepsilon(x) = x + \frac{\varepsilon}{x-1},$$

and the auxiliary problem to solve is

$$\min_{x>1} q_\varepsilon(x).$$

We have

$$q'_\varepsilon(x) = 1 - \frac{\varepsilon}{(x-1)^2} = 0 \quad \Longleftrightarrow$$

$$\Longleftrightarrow \quad x = x_\varepsilon = 1 + \sqrt{\varepsilon},$$

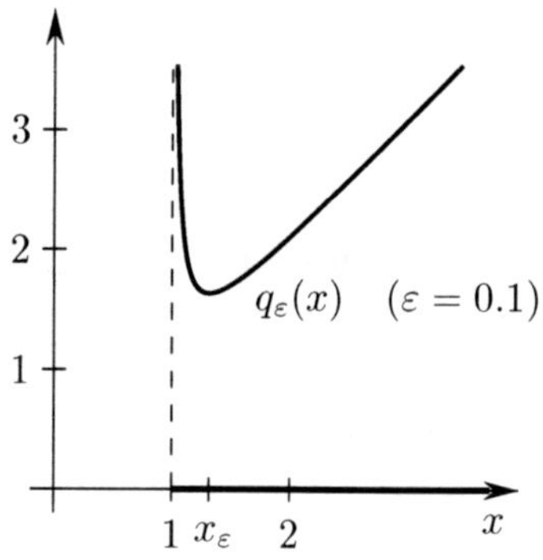

since we require a solution x_ε in the interval $]1, \infty[$.

Note that $x_\varepsilon \to 1$ as $\varepsilon \to 0$, and that the minimum value

$$q_\varepsilon(x_\varepsilon) = 1 + 2\sqrt{\varepsilon}$$

tends to 1 as ε tends to 0. □

The ideal barrier function is of course the one that is equal to 0 on $\operatorname{int}(S)$ and equal to ∞ at the boundary. Note that $\varepsilon\beta(\boldsymbol{x})$ approaches this function when ε tends to zero. For this reason choosing a value of ε close to 0 is indicated. However, for small ε the minimization of q_ε becomes an ill-conditioned problem, due to the influence of ε on the eigenvalues of the Hessian. To overcome these difficulties one introduces a sequence of auxiliary problems, exactly as in the penalty function case.

More precisely: choose a strictly decreasing sequence $\{\varepsilon_k\}_1^\infty$ of positive numbers with limit 0. Let for each k

$$q_k(\boldsymbol{x}) = f(\boldsymbol{x}) + \varepsilon_k \beta(\boldsymbol{x}). \tag{8}$$

Solve successively the problems to minimize $q_k(\boldsymbol{x})$, letting the solution $\boldsymbol{x}_k$ of one problem be the initial point of the next.

The theory of convergence in the barrier function case follows closely that for the penalty function method. We content ourselves with quoting a counterpart to theorem 1.

THEOREM 2. *Let $\{\boldsymbol{x}_k\}_1^\infty$ be the sequence generated by successively minimizing q_k in (8) for $k = 1, 2, \ldots$. Assume that there is a subsequence converging to some limit point $\bar{\boldsymbol{x}} \in S$. Then $\bar{\boldsymbol{x}}$ is a solution to the minimization problem (7).*

Exercises

9.4 Consider the problem

$$\begin{aligned} \text{Minimize} \quad & \frac{1}{3}(x_1+1)^3 + x_2 \\ \text{subject to} \quad & x_1 \geq 1 \\ & x_2 \geq 0. \end{aligned}$$

a) Explain why

$$\beta(x_1, x_2) = \frac{1}{x_1 - 1} + \frac{1}{x_2}, \qquad x_1 > 1,\ x_2 > 0$$

is an appropriate barrier function.

b) Describe the corresponding auxiliary function $q_\varepsilon(x_1, x_2)$.

c) Minimize q_ε analytically. Show that its solution converges to a solution to the problem stated above as $\varepsilon \to 0$.

9.5 Consider the following problem:

$$\begin{aligned} \text{Minimize} \quad & (x_1-5)^2 + (x_2-3)^2 \\ \text{subject to} \quad & x_1 + x_2 \leq 3 \\ & -x_1 + 2x_2 \leq 4. \end{aligned}$$

Formulate a suitable barrier problem with the initial parameter equal to 1. Use an unconstrained optimization technique, e.g. conjugate gradients starting with the point (0, 0), to solve the auxiliary problem.

Chapter 10

ALGORITHMS AND CONVERGENCE

In this chapter we discuss algorithms from a slightly abstract standpoint. The theory does not encompass optimization algorithms only, although those are of course our main concern. A result of the theory will be proofs of convergence for sequences generated by some of the algorithms of chapter 3. While this is an important property in itself of an algorithm, matters like speed of convergence, computer memory requirements and others should of course also play a role when judging its merits.

10.1 Algorithmic maps

The problem to find

$$\min_{\boldsymbol{x} \in S} f(\boldsymbol{x})$$

is in practice often solved by an *algorithm*; an iterative procedure which generates a sequence $\boldsymbol{x}_1, \boldsymbol{x}_2, \ldots$ in S (plus some stop criterion). Based on the current point $\boldsymbol{x}_k$, in the iteration step $\boldsymbol{x}_{k+1}$ is set equal to $F(\boldsymbol{x}_k)$, where F is some function from S to S. In fact, it might be claimed that the algorithm is defined by the function F and some choice of initial point $\boldsymbol{x}_1$.

Here we are going to discuss a general theory of algorithms developed mainly by Zangwill. It has proved appropriate to adopt a somewhat different view from the one described above, and consider the iteration step as a function from S to the set of *subsets of* S.

In the following definition we even allow for functions from one set to the subsets of another.

DEFINITION 1. *Let X and Y be subsets of $\mathbf{R}^n$ and $\mathbf{R}^m$ respectively. An* **algorithmic map** *from X to Y is a function $\mathbf{A}$ from the set X to the set of all subsets of Y,*

$$X \ni \boldsymbol{x} \overset{\mathbf{A}}{\longmapsto} Y' \subseteq Y.$$

We call $\mathbf{A}$ a *point-to-set* map from X to Y. If $Y = X$ we talk about a point-to-set map on X.

Instead of concentrating on a single sequence $\{\boldsymbol{x}_k\}_1^\infty$ with $\boldsymbol{x}_{k+1} = F(\boldsymbol{x}_k)$ we are now in a position to view an algorithmic map $\mathbf{A}$ as the generator of *any* sequence such that $\boldsymbol{x}_{k+1} \in \mathbf{A}(\boldsymbol{x}_k)$. Among other things this makes it possible to incorporate the uncertainties that might exist concerning the actual behavior of a specific scheme (perhaps due to numerical inaccuracies).

Example 1. On the set $\mathbf{R}^+$ of positive real numbers an algorithmic map can be defined by the rule

$$\mathbf{A}(x) = \{\, y\,;\, 0 \le y \le \frac{x}{2}\,\} = [0, \frac{x}{2}\,].$$

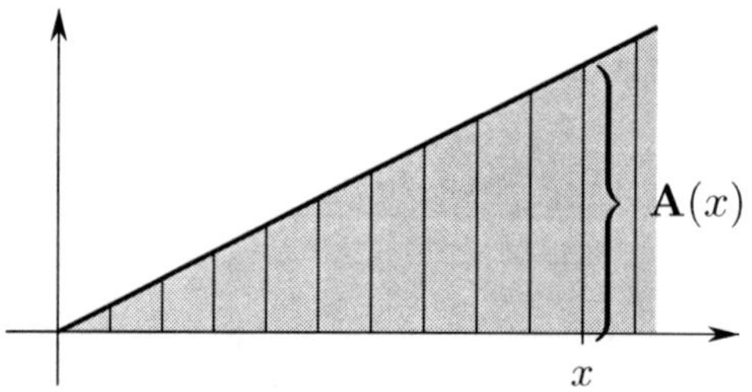

Each of the following sequences might be generated from this algorithmic map, starting at $x_1 = 50$.

$$50, 25, 12, 6, 3, 1, \ \ldots$$

$$50, 20, 10, 5, 2, 1, \ \ldots$$

$$50,\ 5,\ 0.5,\ 0.05,\ \ldots$$

□

Example 2. Suppose that we wish to generate a sequence of real numbers with limit $\bar{x} = 1$. One solution to this problem is the recursion formula

$$\begin{cases} x_{k+1} = \frac{1}{2}(x_k + 1), & k \geq 1, \\ \text{e.g. } x_1 = 4. \end{cases}$$

We could thus put $\mathbf{A}(x) = \{\, \frac{1}{2}(x+1) \,\}$ (a one point set, a *singleton*). The left figure below shows the traditional way of illustrating an iteration step for the generated sequence.

More generally we could put

$$\mathbf{A}(x) = \begin{cases} [1,\ \frac{1}{2}(x+1)] & \text{if } x \geq 1, \\ [\frac{1}{2}(x+1),\ 1] & \text{if } x < 1. \end{cases}$$

Any sequence $\{x_k\}_1^\infty$ with $x_{k+1} \in \mathbf{A}(x_k)$ will converge to $\bar{x} = 1$. See the right figure.

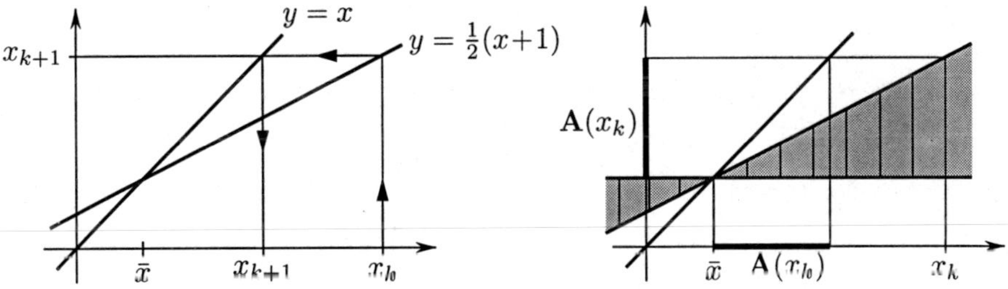

□

To every algorithmic map we associate a *solution set* Ω, representing the points in $\mathbf{R}^n$ to which we wish or expect the generated sequences to converge.[1] Examples of solution sets in different contexts are

$$\Omega = \{\, \bar{\boldsymbol{x}}\,;\ \bar{\boldsymbol{x}} \text{ is a local minimum for } f \,\},$$

$$\Omega = \{\, \bar{\boldsymbol{x}}\,;\ \boldsymbol{\nabla} f(\bar{\boldsymbol{x}}) = \boldsymbol{0} \,\},$$

$$\Omega = \{\, \bar{\boldsymbol{x}}\,;\ f(\bar{\boldsymbol{x}}) \leq b \,\}, \qquad b \text{ some given number.}$$

[1] In practice we can only expect to reach Ω in the limit as the number of iterations tends to ∞. Some other rule must be used as criterion for when to actually terminate the procedure.

DEFINITION 2. *An algorithmic map* **A** *with solution set* Ω *is said to* **converge** *over* $X' \subseteq X$ *if, whenever a sequence* $\{\boldsymbol{x}_k\}_1^\infty$ *is generated with*

$$\boldsymbol{x}_1 \in X', \qquad \boldsymbol{x}_{k+1} \in \mathbf{A}(\boldsymbol{x}_k) \quad \textit{for } k \geq 1,$$

every convergent subsequence of $\{\boldsymbol{x}_k\}_1^\infty$ *has its limit point in* Ω.

Example 3. For the algorithmic maps in example 2 let the solution set be $\Omega = \{1\}$. Then obviously these algorithms converge over the whole of **R**.

Consider now the algorithmic map

$$\mathbf{B}(x) = \begin{cases} [\,\frac{3}{2}+\frac{1}{4}x,\ 1+\frac{1}{2}x\,] & \text{if } x \geq 2, \\ \{\,\frac{1}{2}(x+1)\,\} & \text{if } x < 2, \end{cases}$$

still with $\Omega = \{1\}$. Suppose that we generate a sequence starting from some $x_1 > 2$. Then

$$x_k \to \hat{x} = 2 \quad \text{as } k \to \infty,$$

where $2 \notin \Omega$. The algorithmic map **B** converges over the set $x < 2$ only.

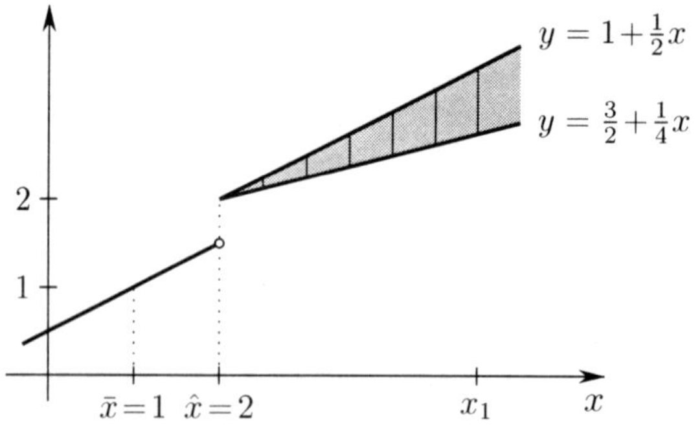

□

DEFINITION 3. *For an algorithmic map* **A** *on* X *with solution set* Ω *a function* $\alpha\colon X \to \mathbf{R}$ *is called a* **descent function** *if it is continuous and*

$$\alpha(\boldsymbol{y}) < \alpha(\boldsymbol{x}) \quad \textit{when } \boldsymbol{x} \notin \Omega \textit{ and } \boldsymbol{y} \in \mathbf{A}(\boldsymbol{x}).^2$$

[2] We are not interested in $\alpha(\boldsymbol{x})$ when $\boldsymbol{x} \in \Omega$, since we always let a generated sequence terminate when it reaches the solution set.

Obviously, if α is a descent function then for any sequence $\{\boldsymbol{x}_k\}_1^\infty$ generated by $\mathbf{A}$ the sequence $\{\alpha(\boldsymbol{x}_k)\}_1^\infty$ is strictly decreasing, as long as $\boldsymbol{x}_k \notin \Omega$.

Example 4. Examples of functions that sometimes might play the role of descent function are the objective function f itself, $\alpha(\boldsymbol{x}) = \|\nabla f(\boldsymbol{x})\|$ and $\alpha(\boldsymbol{x}) = \|\boldsymbol{x} - \bar{\boldsymbol{x}}\|$. □

Now let $X \subseteq \mathbf{R}^n$ and $Y \subseteq \mathbf{R}^m$ be closed sets, and $\mathbf{A}: X \to Y$ a point-to-set map.

DEFINITION 4. *The map* $\mathbf{A}$ *is said to be* **closed** *at a point* $\boldsymbol{x} \in X$ *if, for any sequences* $\{\boldsymbol{x}_k\}_1^\infty$ *and* $\{\boldsymbol{y}_k\}_1^\infty$,

$$\left.\begin{array}{lll} \boldsymbol{x}_k \in X, & \boldsymbol{x}_k \to \boldsymbol{x} & \text{as } k \to \infty \\ \boldsymbol{y}_k \in \mathbf{A}(\boldsymbol{x}_k), & \boldsymbol{y}_k \to \boldsymbol{y} & \text{as } k \to \infty \end{array}\right\} \quad \Longrightarrow \quad \boldsymbol{y} \in \mathbf{A}(\boldsymbol{x}).$$

We call $\mathbf{A}$ ***closed on a subset*** X' ***of*** X *if it is closed at every point* $\boldsymbol{x} \in X'$. *It is called* **closed** *if it is closed on the whole of* X.

Example 5. If $\mathbf{A}(\boldsymbol{x})$ is a one point set (a *singleton*) for all $\boldsymbol{x}$ then $\mathbf{A}$ is closed if and only if $\mathbf{A}$, considered as a point-to-point function, is continuous. Closedness is a generalization of the concept of continuity to point-to-set maps. □

Example 6. The maps in example 2 are closed. The map $\mathbf{B}$ in example 3 is not closed. Let for example

$$x_k = 2 - \frac{1}{k}, \qquad y_k = \frac{1}{2}\left(2 - \frac{1}{k} + 1\right) = \frac{3}{2} - \frac{1}{2k}.$$

Then $x_k \to 2$ and $y_k \to \frac{3}{2}$ when $k \to \infty$. However, $\frac{3}{2} \notin \mathbf{B}(2) = \{2\}$. □

In the next example we give the proof of an important theorem to be used several times later on. We are going to show that *line search is a closed map*.

Example 7. Let f be a continuous function of n variables. Given a point $\boldsymbol{x}$ in $\mathbf{R}^n$ and a search direction $\boldsymbol{d}$, we wish to minimize f in direction $\boldsymbol{d}$ from $\boldsymbol{x}$, i.e., to solve the one-dimensional minimization problem

$$\min_{\lambda \in L} f(\boldsymbol{x}+\lambda \boldsymbol{d}).$$

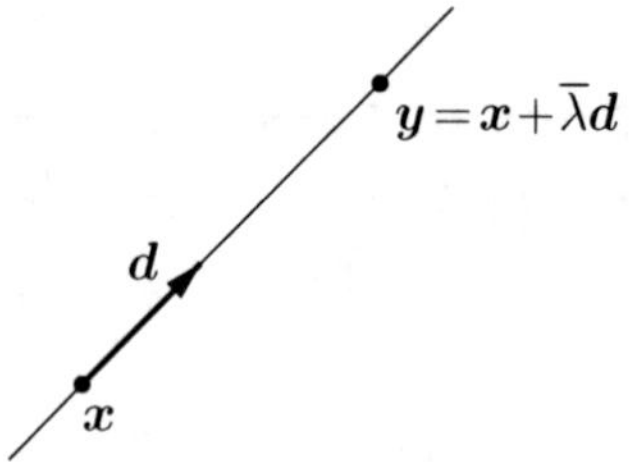

Here L is a closed interval on $\mathbf{R}$ (which may be all of $\mathbf{R}$).

This optimization problem is solved by the following algorithmic map from $\mathbf{R}^n \times \mathbf{R}^n$ to $\mathbf{R}^n$:

$$\mathbf{S}(\boldsymbol{x}, \boldsymbol{d}) =$$

$$= \{\, \boldsymbol{y} \in \mathbf{R}^n \,;\, \boldsymbol{y} = \boldsymbol{x} + \bar{\lambda}\boldsymbol{d} \ \text{ for some } \bar{\lambda} \in L \text{ and } f(\boldsymbol{y}) \le f(\boldsymbol{x}+\lambda\boldsymbol{d}) \text{ for all } \lambda \in L \,\}.$$

Note that $\mathbf{S}(\boldsymbol{x}, \boldsymbol{d})$ is exactly the set of those $\boldsymbol{y}$ on the line in direction $\boldsymbol{d}$ through $\boldsymbol{x}$ which minimize f.

We want to prove:

THEOREM 1. *The line search map* $\mathbf{S}$ *is closed at* $(\boldsymbol{x}, \boldsymbol{d})$ *if* $\boldsymbol{d} \neq \boldsymbol{0}$.

PROOF: Assume that

$$(\boldsymbol{x}_k, \boldsymbol{d}_k) \to (\boldsymbol{x}, \boldsymbol{d}) \quad \text{as} \quad k \to \infty \qquad (\boldsymbol{d} \neq \boldsymbol{0}),$$

$$\boldsymbol{y}_k \in \mathbf{S}(\boldsymbol{x}_k, \boldsymbol{d}_k) \quad \text{for all } k,$$

$$\boldsymbol{y}_k \to \boldsymbol{y} \quad \text{as} \quad k \to \infty.$$

We must prove that $\boldsymbol{y} \in \mathbf{S}(\boldsymbol{x}, \boldsymbol{d})$.

That $\boldsymbol{y}_k \in \mathbf{S}(\boldsymbol{x}_k, \boldsymbol{d}_k)$ implies

$$\boldsymbol{y}_k = \boldsymbol{x}_k + \lambda_k \boldsymbol{d}_k, \tag{1}$$

where $\lambda_k \in L$. Subtracting $\boldsymbol{x}_k$ and taking norms we get

$$\|\boldsymbol{y}_k - \boldsymbol{x}_k\| = |\lambda_k| \, \|\boldsymbol{d}_k\|.$$

It is no restriction to assume that all $\lambda_k \ge 0$. (We can pick a subsequence with those indices k for which $\lambda_k \ge 0$, or, if there are only finitely many of these, a subsequence indexed with those k for which $\lambda_k < 0$. In the latter case we have to change signs everywhere.) Furthermore, $\|\boldsymbol{d}_k\| \neq 0$ when k is large, since $\boldsymbol{d} \neq \boldsymbol{0}$. Thus

$$\lambda_k = \frac{\|\boldsymbol{y}_k - \boldsymbol{x}_k\|}{\|\boldsymbol{d}_k\|} \to \frac{\|\boldsymbol{y} - \boldsymbol{x}\|}{\|\boldsymbol{d}\|} \qquad \text{as } k \to \infty.$$

Denoting this limit by $\overline{\lambda}$ it follows from (1) that

$$\boldsymbol{y} = \boldsymbol{x} + \overline{\lambda}\boldsymbol{d},$$

and $\overline{\lambda} \in L$ since L is closed. Furthermore, by the definition of $\mathbf{S}$

$$f(\boldsymbol{y}_k) \leq f(\boldsymbol{x}_k + \lambda \boldsymbol{d}_k) \qquad \text{for all } \lambda \in L,$$

which by the continuity of f implies that

$$f(\boldsymbol{y}) \leq f(\boldsymbol{x} + \lambda \boldsymbol{d}) \qquad \text{for all } \lambda \in L.$$

This proves that $\boldsymbol{y} \in \mathbf{S}(\boldsymbol{x}, \boldsymbol{d})$. □

We close this section with an example of an algorithmic map which appears in some optimization algorithms and which is *not* closed.

Example 8. For simplicity we conduct the discussion in two dimensions only. Define an algorithmic map $\mathbf{M}\colon \mathbf{R}^2 \times \mathbf{R}^2 \to \mathbf{R}^2$ by

$\boldsymbol{y} \subset \mathbf{M}(\boldsymbol{x}, \boldsymbol{d})$ if and only if $\boldsymbol{y}$ is a solution to the problem

$$\min f(\boldsymbol{x} + \lambda \boldsymbol{d}) \quad \textit{subject to} \quad \boldsymbol{x} + \lambda \boldsymbol{d} \geq \boldsymbol{0},\ \lambda \geq 0.$$

Here f is a continuous function of two variables.

To show that $\mathbf{M}$ is not closed we must exhibit sequences $\{(\boldsymbol{x}_k, \boldsymbol{d}_k)\}_1^\infty$ converging to some $(\boldsymbol{x}, \boldsymbol{d})$ and $\{\boldsymbol{y}_k\}_1^\infty$ converging to some $\boldsymbol{y}$ such that $\boldsymbol{y}_k \in \mathbf{M}(\boldsymbol{x}_k, \boldsymbol{d}_k)$ but $\boldsymbol{y} \notin \mathbf{M}(\boldsymbol{x}, \boldsymbol{d})$. We do that as follows.

Let $f(x_1, x_2) = (x_1+2)^2 + (x_2-2)^2$. Starting at $\boldsymbol{x}_1 = (1, 0)$, let $\boldsymbol{x}_{k+1}$ be that point on the circle $(x_1-1)^2 + (x_2-1)^2 = 1$ which is midway between $(0, 1)$ and $\boldsymbol{x}_k$. Define

$$\boldsymbol{d}_k = \frac{\boldsymbol{x}_{k+1} - \boldsymbol{x}_k}{\|\boldsymbol{x}_{k+1} - \boldsymbol{x}_k\|}, \qquad k \geq 1.$$

Then obviously $\boldsymbol{x}_k$ converges to $(0, 1) = \boldsymbol{x}$, and $\boldsymbol{d}_k$ converges to $(0, 1) = \boldsymbol{d}$.

The points $\boldsymbol{y}_k$ are defined by line search from $\boldsymbol{x}_k$ in the positive direction $\boldsymbol{d}_k$, however with the additional restriction that $\boldsymbol{y}_k$ belongs to the first quadrant, by the definition of **M**. Considering the level curves of f it is obvious that the $\boldsymbol{y}_k$ are situated on the x_2-axis and that $\boldsymbol{y}_k \to (0,1) = \boldsymbol{y}$ as $k \to \infty$. However

$$f(\boldsymbol{x} + \lambda \boldsymbol{d}) = f(0, 1+\lambda)$$

has its minimum at $\lambda = 1$ (again, consider the level curves of f), that is, at $(0,2)$. Thus

$$(0,1) = \boldsymbol{y} \notin \mathbf{M}(\boldsymbol{x}, \boldsymbol{d}) = \{\,(0,2)\,\}.$$

The conclusion is that **M** is not a closed map.

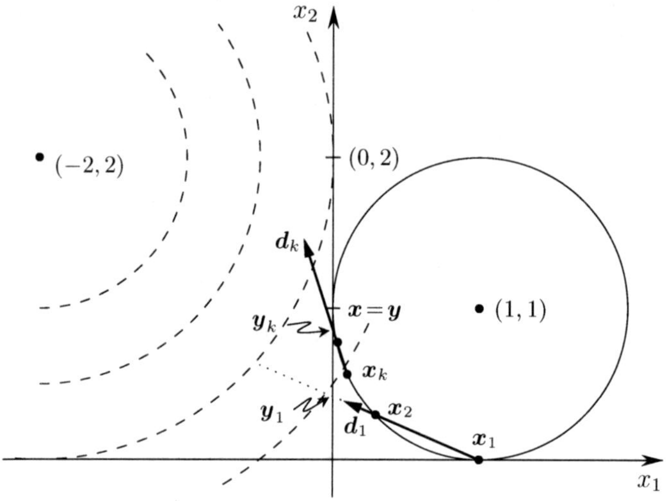

□

Remark. The reader should explain why there is no conflict between this example and the statement of theorem 1 that "line search is a closed map".

Exercises

S 10.1 Determine whether the following algorithmic maps from **R** to **R** are closed or not?

a) $\mathbf{A}(x) = \{\, y \,;\, x^2 + y^2 \le 1 \,\}$,

b) $\mathbf{B}(x) = \begin{cases} \{\, y \,;\, x^2 + y^2 \le 1 \,\} & \text{if } x \neq 0 \\ [-1, 0] & \text{if } x = 0. \end{cases}$

10.2 Are the following point-to-set maps from $\mathbf{R}^2$ to $\mathbf{R}^2$ closed?

a) $\mathbf{A}(\boldsymbol{x}) = \{\, \boldsymbol{y} \in \mathbf{R}^2 \,;\, \boldsymbol{x}^T\boldsymbol{y} \le 1 \,\}$,
b) $\mathbf{A}(\boldsymbol{x}) = \{\, \boldsymbol{y} \in \mathbf{R}^2 \,;\, \|\boldsymbol{y} - \boldsymbol{x}\| \le 1 \,\}$.

S 10.3 An algorithmic map $\mathbf{A}$ is defined by

$$(y_1, y_2) \in \mathbf{A}(x_1, x_2) \quad \Longleftrightarrow \quad \begin{cases} y_1 = 2x_1 - 1 \\ y_2 \in [x_2 - 1, x_2 + 2]. \end{cases}$$

Is the map closed?

10.4 Determine if the following maps are closed.

a) $(y_1, y_2) \in \mathbf{A}(x_1, x_2)$ means that $y_1 = x_1 - 1$ and

$$y_2 \in [-x_2{+}1, x_2{+}1] \;\text{ if }\; x_2 \ge 0$$
$$y_2 \in [x_2{+}1, -x_2{+}1] \;\text{ if }\; x_2 < 0.$$

b) $(y_1, y_2) \in \mathbf{A}(x_1, x_2)$ means that $y_1 \in [x_1 - \|\boldsymbol{x}\|,\, x_1 + \|\boldsymbol{x}\|\,]$ and $y_2 = x_2$.

10.5 Determine if the map defined for $x \ge 0$ by

$$\mathbf{A}(x) = \begin{cases} [0, x] & \text{if } x \ne 1 \\ [0, \frac{1}{2}] & \text{if } x = 1 \end{cases}$$

is closed.

S 10.6 Let X and Y be nonempty closed sets in $\mathbf{R}^p$ and $\mathbf{R}^q$, respectively. Show that the point-to-set map $\mathbf{A}\colon X \to Y$ is closed if and only if the set

$$Z = \{(\boldsymbol{x}, \boldsymbol{y}) \,;\, \boldsymbol{x} \in X,\, \boldsymbol{y} \in \mathbf{A}(\boldsymbol{x})\}$$

is closed.

10.7 Define an algorithmic map $\mathbf{A}\colon \mathbf{R}^n{\times}\mathbf{R}^n \to \mathbf{R}^n$ by the following rule. If $\boldsymbol{x}, \boldsymbol{z}$ are vectors in $\mathbf{R}^n$, then $\boldsymbol{y} \in \mathbf{A}(\boldsymbol{x}, \boldsymbol{z})$ means that

$$\boldsymbol{y} = \bar{\lambda}\boldsymbol{x} + (1{-}\bar{\lambda})\boldsymbol{z} \qquad \text{for some } \bar{\lambda} \in [0, 1]$$

and that

$$\|\boldsymbol{y}\| \le \|\lambda\boldsymbol{x} + (1{-}\lambda)\boldsymbol{z}\| \qquad \text{for all } \;\lambda \in [0, 1].$$

Show that this map $\mathbf{A}$ is closed.

10.2 The convergence theorem

Let $\mathbf{A}$ be an algorithmic map on the set X in $\mathbf{R}^n$. We are now going to discuss the convergence of sequences $\{\boldsymbol{x}_k\}_1^\infty$ generated from some starting point $\boldsymbol{x}_1 \in X$ and satisfying

$$\boldsymbol{x}_{k+1} \in \mathbf{A}(\boldsymbol{x}_k), \qquad k \geq 1.$$

The sequence is terminated as soon as some $\boldsymbol{x}_k$ belongs to the solution set Ω; if that happens the sequence is finite and there is no point in talking about convergence. This case is excluded from now on.

The important issue at the moment is to find conditions ensuring that a generated sequence $\{\boldsymbol{x}_k\}_1^\infty$, or some subsequence of it, converges to a limit point $\bar{\boldsymbol{x}}$ in Ω.

THEOREM 2. *Let $\{\boldsymbol{x}_k\}_1^\infty$ be a sequence generated by some algorithmic map $\mathbf{A}$ on X. Assume that*

- *all $\boldsymbol{x}_k$ belong to a compact subset of X,*
- *there exists a descent function α for $\mathbf{A}$,*
- *the map $\mathbf{A}$ is closed on $X \setminus \Omega$.*

Then

1. *$\alpha(\boldsymbol{x}_k) \to \alpha(\boldsymbol{x})$ for some $\boldsymbol{x}$ in Ω,*
2. *any convergent subsequence of $\{\boldsymbol{x}_k\}_1^\infty$ has its limit point in Ω.*

Remark. Note that we do not claim that the sequence $\{\boldsymbol{x}_k\}_1^\infty$ itself is convergent, only that $\{\alpha(\boldsymbol{x}_k)\}_1^\infty$ converges.

PROOF. We may assume that all $\boldsymbol{x}_k$ are outside Ω. Using the assumption that all $\boldsymbol{x}_k$ belong to a compact subset of X, by the Bolzano-Weierstrass theorem (page 117) $\{\boldsymbol{x}_k\}_1^\infty$ has a convergent subsequence $\{\boldsymbol{x}_k\}_{k\in\mathcal{K}}$. Denote by $\boldsymbol{x}$ the limit point of such a sequence. Of course $\boldsymbol{x} \in X$. We want to prove that

1° $\alpha(\boldsymbol{x}_k) \to \alpha(\boldsymbol{x})$ when $k \to \infty$ and $k \in \mathbf{Z}^+$,

2° $\boldsymbol{x} \in \Omega$.

Here 1° is obvious for the subsequence $\{\boldsymbol{x}_k\}_{k\in\mathcal{K}}$, since α is continuous. But then it is obvious also for the original sequence $\{\boldsymbol{x}_k\}_1^\infty$, since $\{\alpha(\boldsymbol{x}_k)\}_1^\infty$ is a strictly decreasing sequence.

It remains to prove that $\boldsymbol{x} \in \Omega$. Suppose to the contrary that $\boldsymbol{x} \notin \Omega$. Consider the shifted subsequence $\{\boldsymbol{x}_{k+1}\}_{k\in\mathcal{K}}$. According to Bolzano-Weierstrass' theorem it has a convergent subsequence $\{\boldsymbol{x}_{k+1}\}_{k\in\overline{\mathcal{K}}}$ (where $\overline{\mathcal{K}} \subseteq \mathcal{K}$). Denote the limit point of this subsequence by $\bar{\boldsymbol{x}}$. Then $\alpha(\bar{\boldsymbol{x}}) = \alpha(\boldsymbol{x})$ by 1°. For k belonging to $\overline{\mathcal{K}}$ we now have

$$\begin{aligned}
&\boldsymbol{x}_k \to \boldsymbol{x} \quad \text{as } k \to \infty,\\
&\boldsymbol{x}_{k+1} \in \mathbf{A}(\boldsymbol{x}_k),\\
&\boldsymbol{x}_{k+1} \to \bar{\boldsymbol{x}} \quad \text{as } k \to \infty.
\end{aligned}$$

Since $\mathbf{A}$ is assumed to be closed this implies that $\bar{\boldsymbol{x}} \in \mathbf{A}(\boldsymbol{x})$. This in turn implies that $\alpha(\bar{\boldsymbol{x}}) < \alpha(\boldsymbol{x})$, by the definition of a descent function. Now we have arrived at a contradiction, and the proof is finished. □

COROLLARY 1. *Under the assumptions of the theorem, if the solution set $\Omega = \{\hat{\boldsymbol{x}}\}$ is a singleton then the whole original sequence $\{\boldsymbol{x}_k\}_1^\infty$ converges to $\hat{\boldsymbol{x}}$.*

PROOF. Otherwise there would be infinitely many elements of $\{\boldsymbol{x}_k\}_1^\infty$ outside some ball around $\hat{\boldsymbol{x}}$. Then by the Bolzano-Weierstrass theorem these elements would contain a convergent subsequence. By theorem 2 its limit point would belong to Ω. However, then the only possibility for this limit is $\hat{\boldsymbol{x}}$, contradicting the assumption that the elements of the subsequence keep away from $\hat{\boldsymbol{x}}$. □

Example 9. Consider a sequence $\{x_k\}_1^\infty$ of positive real numbers generated by the formula

$$x_{k+1} = \frac{1}{2}\left(x_k + \frac{2}{x_k}\right)$$

and some initial value x_1. The algorithmic map involved here is a singleton,

$$\mathbf{A}(x) = \left\{\frac{1}{2}\left(x + \frac{2}{x}\right)\right\}, \qquad x > 0.$$

This map is closed, since it is defined by a continuous function. By the arithmetic-geometric inequality

$$\frac{1}{2}\Big(x+\frac{2}{x}\Big) \geq \sqrt{x\cdot\frac{2}{x}} = \sqrt{2}.$$

Hence for $k \geq 2$ we have $x_k \in [\sqrt{2}, \infty) = X$.

Let $\Omega = \{\sqrt{2}\}$ be the solution set. We now prove that $\alpha(x) = x$ is a descent function on X. Let $x \in X$ and $y \in \mathbf{A}(x)$. For $x > \sqrt{2}$ (i.e. $x \in X \setminus \Omega$) we have $\frac{2}{x} < x$. Thus

$$\alpha(y) = y = \frac{1}{2}\Big(x+\frac{2}{x}\Big) < \frac{1}{2}(x+x) = x = \alpha(x).$$

As a consequence of this, for any sequence generated by $\mathbf{A}$ the elements x_k belong to the compact interval $[\sqrt{2}, x_2]$ when $k \geq 2$. Invoking the corollary above we conclude that

$$x_k \to \sqrt{2} \quad \text{as} \quad k \to \infty,$$

whatever the initial value $x_1 > 0$. □

We are now going to use theorem 2 to prove that Newton's method for optimization converges if the starting point is sufficiently close to the minimum point.

We remind the reader that the idea behind Newton's method is the following. Assume that $f \in \mathcal{C}^2(\mathbf{R}^n)$. Near the current point $\boldsymbol{x}_k$ we approximate f by its Taylor polynomial of degree 2:

$$f(\boldsymbol{x}) \approx f(\boldsymbol{x}_k) + \boldsymbol{\nabla} f(\boldsymbol{x}_k)^T(\boldsymbol{x}-\boldsymbol{x}_k) + \frac{1}{2}(\boldsymbol{x}-\boldsymbol{x}_k)^T \boldsymbol{H}(\boldsymbol{x}_k)(\boldsymbol{x}-\boldsymbol{x}_k).$$

Denote the Taylor polynomial by $q(\boldsymbol{x})$. It has minimum where $\boldsymbol{\nabla} q(\boldsymbol{x})$ is zero. Assuming $\boldsymbol{H}(\boldsymbol{x}_k)$ to be invertible we get

$$\boldsymbol{\nabla} f(\boldsymbol{x}_k) + \boldsymbol{H}(\boldsymbol{x}_k)(\boldsymbol{x}-\boldsymbol{x}_k) = \boldsymbol{0} \quad \iff \quad \boldsymbol{x} = \boldsymbol{x}_k - \boldsymbol{H}(\boldsymbol{x}_k)^{-1}\boldsymbol{\nabla} f(\boldsymbol{x}_k).$$

The recursion formula for Newton's method is thus

$$\boldsymbol{x}_{k+1} = \boldsymbol{x}_k - \boldsymbol{H}(\boldsymbol{x}_k)^{-1}\boldsymbol{\nabla} f(\boldsymbol{x}_k). \tag{2}$$

For simplicity of notation we state and prove the following theorem in the one variable case only. The reader will realize that the proof is merely an adaption of the proof of theorem 2.1, page 36, to a new context.

THEOREM 3. *Assume that $f \in C^3(\mathbf{R})$. Put*

$$\mathbf{A}(x) = \left\{x - \frac{f'(x)}{f''(x)}\right\}.$$

Assume that $\bar{x}$ satisfies $f'(\bar{x}) = 0$, $f''(\bar{x}) \neq 0$. Choose the number ε so small that there are positive constants A and B with $AB\varepsilon < 1$ making

$$\frac{1}{|f''(x)|} \leq A \qquad \text{when } |x - \bar{x}| \leq \varepsilon,$$

$$|f'''(x)| \leq 2B \quad \text{when } |x - \bar{x}| \leq \varepsilon.$$

Then the algorithm defined by **A** *converges over $|x - \bar{x}| \leq \varepsilon$ to $\bar{x}$. In other words, the sequence $\{x_k\}_1^\infty$ generated by* **A** *converges to $\bar{x}$ if $|x_1 - \bar{x}| \leq \varepsilon$.*

PROOF. We use the corollary of theorem 2 above. The solution set is $\Omega = \{x\}$, and X is the intervall $|x - \bar{x}| \leq \varepsilon$, which is compact. The map **A** is closed, since it is a continuous function outside Ω. We now prove that

$$\alpha(x) = |x - \bar{x}|$$

is a descent function. In particular, this will show that $\mathbf{A}(x) \subseteq X$ if $x \in X$.

Let $x \in X$, $x \notin \Omega$, and let $y \in \mathbf{A}(x)$. Then, since $f'(\bar{x}) = 0$,

$$y - \bar{x} = x - \frac{f'(x)}{f''(x)} - \bar{x} + \frac{f'(\bar{x})}{f''(x)} = \frac{1}{f''(x)}\Big(f'(\bar{x}) - f'(x) - f''(x)(\bar{x} - x)\Big).$$

Taylor's formula for f' *at* x gives

$$f'(\bar{x}) = f'(x) + f''(x)(\bar{x} - x) + \frac{1}{2} f'''(\xi)\,(\bar{x} - x)^2,$$

where ξ lies between x and $\bar{x}$. Inserting this expression for $f'(\bar{x})$ above we get

$$|y - \bar{x}| = \frac{1}{|f''(x)|}\left|\frac{1}{2} f'''(\xi)\,(\bar{x} - x)^2\right|.$$

We now use the estimates in the assumptions. It follows from those that

$$|y - \bar{x}| \leq AB\,|\bar{x} - x|^2 \leq AB\varepsilon|x - \bar{x}| < |x - \bar{x}|.$$

This inequality shows that $\alpha(y) < \alpha(x)$, making α a descent function.

Since the solution set is a singleton, the statement now follows from the corollary above. □

Exercise

10.8 Consider the map $\mathbf{A}$, where $\mathbf{A}(x) = \{\sqrt{x}\,\}$.

a) Show that $\alpha(x) = |x - 1|$ is a descent function for $\mathbf{A}$.

b) Prove that the algorithm defined by the map $\mathbf{A}$ converges to $\overline{x} = 1$, whatever the starting point $x > 0$.

10.3 Composite maps

In many algorithms the iteration step can be viewed as the composition of two operations. For example, in the steepest descent method (and many others) we have the two operations

D: Choice of search direction $\boldsymbol{d}$ at $\boldsymbol{x}$.

S: Minimization in direction $\boldsymbol{d}$ from $\boldsymbol{x}$.

We are now going to define and study the properties of composite algorithmic maps like $\mathbf{A} = \mathbf{SD}$. We are of course dealing with a generalization of the concept of composite functions.

DEFINITION 5. *Let $X \subseteq \mathbf{R}^n$, $Y \subseteq \mathbf{R}^p$ and $Z \subseteq \mathbf{R}^q$, and let*

$$\mathbf{B}: X \to Y, \qquad \mathbf{C}: Y \to Z$$

be point-to-set maps. The **composite map** $\mathbf{A} = \mathbf{CB}: X \to Z$ *is defined by*

$$\mathbf{A}(\boldsymbol{x}) = \bigcup_{\boldsymbol{y} \in \mathbf{B}(\boldsymbol{x})} \mathbf{C}(\boldsymbol{y}).$$

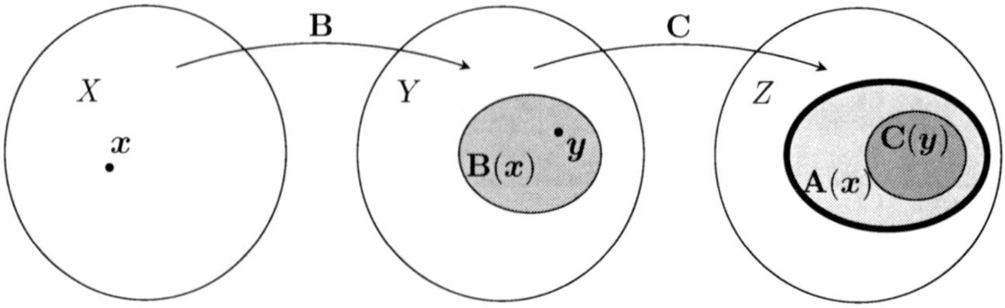

THEOREM 4. *Assume that*

1. **B** *is closed at* $\boldsymbol{x}$,
2. **C** *is closed over* $\mathbf{B}(\boldsymbol{x})$,
3. *the set* Y *is compact.*

Then the composite map $\mathbf{A} = \mathbf{CB}$ *is closed at* $\boldsymbol{x}$.

PROOF. Let $\{\boldsymbol{x}_k\}_1^\infty$ be a sequence in X with limit point $\boldsymbol{x}$. Assume that $\boldsymbol{z}_k \in \mathbf{A}(\boldsymbol{x}_k)$ and that $\boldsymbol{z}_k \to \boldsymbol{z}$ as $k \to \infty$. We must prove that $\boldsymbol{z} \in \mathbf{A}(\boldsymbol{x})$.

By the definition of a composite map

$$\boldsymbol{z}_k \in \bigcup_{\boldsymbol{y} \in \mathbf{B}(\boldsymbol{x}_k)} \mathbf{C}(\boldsymbol{y}).$$

Thus $\boldsymbol{z}_k \in \mathbf{C}(\boldsymbol{y}_k)$ for some $\boldsymbol{y}_k \in \mathbf{B}(\boldsymbol{x}_k)$. Since Y is compact the sequence $\{\boldsymbol{y}_k\}_1^\infty$ has a convergent subsequence $\{\boldsymbol{y}_k\}_{k \in \mathcal{K}}$, by Bolzano-Weierstrass' theorem, page 117. Denote its limit by $\boldsymbol{y}$. Then, by the definition of a closed map, since **B** is closed at $\boldsymbol{x}$ we have $\boldsymbol{y} \in \mathbf{B}(\boldsymbol{x})$, and since **C** is closed at $\boldsymbol{y}$ we have $\boldsymbol{z} \in \mathbf{C}(\boldsymbol{y})$. Consequently $\boldsymbol{z} \in \mathbf{A}(\boldsymbol{x})$. □

We now state some alternative assumptions which result in the composite map being closed.

THEOREM 5. *Assume that* **B** *is a point-to-point function and that*

1. **B** *is continuous at* $\boldsymbol{x}$.
2. **C** *is closed at* $\mathbf{B}(\boldsymbol{x})$.

Then the composite map $\mathbf{A} = \mathbf{CB}$ *is closed at* $\boldsymbol{x}$.

PROOF. We can copy the proof of theorem 4. However we do not need Bolzano-Weierstrass' theorem, since in this case it follows directly from the continuity of **B** that $\boldsymbol{y}_k = \mathbf{B}(\boldsymbol{x}_k)$ has a limit $\boldsymbol{y} = \mathbf{B}(\boldsymbol{x})$ when $k \to \infty$. □

Example 10. In the *steepest descent* method for the search of a minimum of a continuously differentiable function f the algorithm can be described as follows.

Start at some point $\boldsymbol{x}_1$ and repeat the following steps until some termination criterion is satisfied:

1. *Choose the search direction $\boldsymbol{d}_k = -\boldsymbol{\nabla} f(\boldsymbol{x}_k)$.*
2. *Find the minimum of f in direction $\boldsymbol{d}_k$ from $\boldsymbol{x}_k$. More precisely: put $\boldsymbol{x}_{k+1} = \boldsymbol{x}_k + \lambda_k \boldsymbol{d}_k$, where λ_k solves the problem $\min_{\lambda \geq 0} f(\boldsymbol{x}_k + \lambda \boldsymbol{d}_k)$.*

The first step uses the algorithmic map

$$\mathbf{D}(\boldsymbol{x}) = \{(\boldsymbol{x}, -\boldsymbol{\nabla} f(\boldsymbol{x}))\}$$

from $\mathbf{R}^n$ to $\mathbf{R}^n \times \mathbf{R}^n$. Note that $\mathbf{D}$ not only identifies the search direction, but also "remembers" the point $\boldsymbol{x}$. The second step is the line search algorithm $\mathbf{S}$, discussed in example 7. This is a point-to-set map from $\mathbf{R}^n \times \mathbf{R}^n$ to $\mathbf{R}^n$. Hence, the steepest descent algorithmic map is the composite map

$$\mathbf{A} = \mathbf{SD}.$$

The map $\mathbf{D}$ is in fact a continuous point-to-point function. In theorem 1 we proved that $\mathbf{S}$ is closed when $\boldsymbol{d} \neq \boldsymbol{0}$. Consequently, by theorem 5 above the map $\mathbf{A}$ is closed outside the set $\Omega = \{\, \boldsymbol{x}\,;\ \boldsymbol{\nabla} f(\boldsymbol{x}) = \boldsymbol{0}\,\}$. □

We are now in a position to prove a theorem on the global convergence of the steepest descent method.

THEOREM 6. *Assume that the sequence generated by the steepest descent method belongs to a bounded set. Then any convergent subsequence converges to a stationary point of the objective function.*

PROOF. The algorithmic map $\mathbf{A}$ of the steepest descent method is defined in example 10 above, together with the solution set

$$\Omega = \{\, \boldsymbol{x}\,;\ \boldsymbol{\nabla} f(\boldsymbol{x}) = \boldsymbol{0}\,\}.$$

There we showed that it is closed over the set $\mathbf{R}^n \setminus \Omega$. By theorem 2 all we need to do more is to find a descent function for $\mathbf{A}$.

In fact f itself serves this purpose. To prove this we consider the directional derivative in direction $\boldsymbol{d} = -\boldsymbol{\nabla} f(\boldsymbol{x})$ when $\boldsymbol{x} \notin \Omega$. We have

$$f'(\boldsymbol{x}; \boldsymbol{d}) = \boldsymbol{\nabla} f(\boldsymbol{x})^T \boldsymbol{d} = -\|\boldsymbol{\nabla} f(\boldsymbol{x})\|^2 < 0.$$

This implies that $f(\boldsymbol{y}) < f(\boldsymbol{x})$ if $\boldsymbol{y} \in \mathbf{A}(\boldsymbol{x})$ and $\boldsymbol{x} \notin \Omega$. Consequently, f is a descent function, and the proof is finished. □

Example 11. Newton's method for the minimization of a twice continuously differentiable function is defined via the recursion formula (2). In theorem 3 it is proved that it converges if the starting point is chosen close enough to the minimum solution. However it is not globally convergent.

We now discuss the modification introduced in subsection 3.4.2, page 65, which makes sure that a descent direction is always chosen, and which enjoys global convergence. The idea is to modify the Hessian $\boldsymbol{H}(\boldsymbol{x})$ by adding a multiple of the identity matrix, ensuring that the resulting matrix is positive definite. More precisely, fix a number $\delta > 0$ and let $\varepsilon \geq 0$ be the smallest number making all eigenvalues of the matrix $\boldsymbol{H}(\boldsymbol{x}) + \varepsilon \boldsymbol{I}$ greater than or equal to δ. Choose the search direction

$$\boldsymbol{d} = -(\boldsymbol{H}(\boldsymbol{x}) + \varepsilon \boldsymbol{I})^{-1} \nabla f(\boldsymbol{x}), \tag{3}$$

and minimize f along the ray from $\boldsymbol{x}$ in direction $\boldsymbol{d}$.

The algorithmic map $\mathbf{A}$ defined by this procedure is the composite of the search direction finding map

$$\mathbf{D}(\boldsymbol{x}) = \{(\boldsymbol{x}, \boldsymbol{d})\},$$

where $\boldsymbol{d}$ is defined by (3), and the line search map $\mathbf{S}$ from example 7. The map $\mathbf{D}$ is a continuous point-to-point function, since the eigenvalues of a matrix depend continuously on its elements. That $\mathbf{S}$ is closed at any $(\boldsymbol{x}, \boldsymbol{d})$ with $\boldsymbol{d} \neq \boldsymbol{0}$ we know already. Consequently, by theorem 5, $\mathbf{A} = \mathbf{SD}$ is closed outside the set $\Omega = \{\, \boldsymbol{x}\,;\ \nabla f(\boldsymbol{x}) = \boldsymbol{0} \,\}$.

The convergence of the method now follows precisely as in the steepest descent case (theorem 6), if we can prove that a descent function for $\mathbf{A}$ can be found. Also in this case the objective function f itself will do. To prove this let $\boldsymbol{x} \in \mathbf{R}^n \setminus \Omega$ and let $\boldsymbol{d} \neq \boldsymbol{0}$ be defined by (3). The directional derivative in direction $\boldsymbol{d}$ is

$$f'(\boldsymbol{x}; \boldsymbol{d}) = \nabla f(\boldsymbol{x})^T \boldsymbol{d} = -\nabla f(\boldsymbol{x})^T (\boldsymbol{H}(\boldsymbol{x}) + \varepsilon \boldsymbol{I})^{-1} \nabla f(\boldsymbol{x}).$$

Since $\boldsymbol{H}(\boldsymbol{x}) + \varepsilon \boldsymbol{I}$ is positive definite $f'(\boldsymbol{x}; \boldsymbol{d}) < 0$. Hence

$$f(\boldsymbol{y}) < f(\boldsymbol{x}) \qquad \text{when} \quad \boldsymbol{y} \in \mathbf{A}(\boldsymbol{x}), \quad \boldsymbol{x} \notin \Omega.$$

By theorem 2 the algorithm converges over $\mathbf{R}^n \setminus \Omega$. □

Spacer steps

Some algorithms for finding a minimum may be so complex that they are difficult to analyze. If one in such an algorithm infinitely often inserts a step from an algorithm (e.g. steepest descent) that is known to converge, it can be proved that the overall algorithm converges under simple conditions. This is the content of the theorem below. The step which is inserted is called a *spacer step*.

Let X be a subset of $\mathbf{R}^n$. Denote by $\mathbf{B}\colon X \to X$ the algorithmic map generating the spacer step. It is supposed to satisfy the conditions of theorem 2. In particular there is a descent function α for $\mathbf{B}$. We now introduce the algorithm $\mathbf{C}\colon X \to X$ defined by the point-to-set map

$$\mathbf{C}(\boldsymbol{x}) = \{\, \boldsymbol{y} \in X\,;\, \alpha(\boldsymbol{y}) \le \alpha(\boldsymbol{x})\,\}. \tag{4}$$

This algorithm represents all the complex steps that are separated by the spacer steps. The only restriction on these is that the value of the descent function α must not increase. The composite algorithm $\mathbf{CB}$ then converges according to the following theorem.

THEOREM 7. *Let $\{\boldsymbol{x}_k\}_1^\infty$ be a sequence generated by the algorithmic map* $\mathbf{CB}$. *Assume that*

1. *the algorithm $\mathbf{B}\colon X \to X$ is closed outside the solution set Ω, and has a descent function α,*
2. *the algorithm $\mathbf{C}\colon X \to X$ is defined by (4),*
3. *the set $Y = \{\, \boldsymbol{x} \in X\,;\, \alpha(\boldsymbol{x}) \le \alpha(\boldsymbol{x}_1)\,\}$ is compact.*

Then the limit point of any convergent subsequence of $\{\boldsymbol{x}_k\}_1^\infty$ belongs to Ω.

PROOF. The algorithmic map $\mathbf{C}$ is closed, since

$$\begin{cases} \boldsymbol{x}_k \to \boldsymbol{x} & \text{as } k \to \infty \\ \boldsymbol{y}_k \to \boldsymbol{y} & \text{as } k \to \infty \\ \alpha(\boldsymbol{y}_k) \le \alpha(\boldsymbol{x}_k) \end{cases}$$

obviously implies that

$$\alpha(\boldsymbol{y}) \le \alpha(\boldsymbol{x}).$$

For any $\boldsymbol{x} \in X$ let $\widetilde{\mathbf{B}}(\boldsymbol{x}) = Y \cap \mathbf{B}(\boldsymbol{x})$. Then the composite map $\mathbf{A} = \mathbf{C}\widetilde{\mathbf{B}}$ is closed on $X \setminus \Omega$ by theorem 4. Hence theorem 2 can be applied and the result follows. □

Example 12. The normal quasi-Newton and conjugate gradient methods converge, since they regularly insert a step of the steepest descent algorithm. □

Miscellanous exercises

10.9 Define the point-to-set map $\mathbf{A}: \mathbf{R}^n \to \mathbf{R}^n$ as follows. An $m \times n$ matrix $\boldsymbol{B}$, an m-vector $\boldsymbol{b}$ and an n-vector $\boldsymbol{x}$ are fixed. We let $\boldsymbol{y} \in \mathbf{A}(\boldsymbol{x})$ mean that $\boldsymbol{y}$ is an optimal solution to the problem to minimize $\boldsymbol{x}^T \boldsymbol{z}$ (where $\boldsymbol{z}$ is the variable) subject to

$$\boldsymbol{B}\boldsymbol{z} = \boldsymbol{b}, \quad \boldsymbol{z} \geq \boldsymbol{0}.$$

Show that this map $\mathbf{A}$ is closed.

10.10 Let ε be a positive number, and f a continuous function on $\mathbf{R}^n$. Consider the following algorithmic map (*inexact line search*)

$$\mathbf{S}(\boldsymbol{x}, \boldsymbol{d}) = \{\, \boldsymbol{y}\,;\ \boldsymbol{y} = \boldsymbol{x} + \alpha\boldsymbol{d} \text{ for some } \alpha \geq 0,\ f(\boldsymbol{y}) \leq \min_{\beta \geq 0} f(\boldsymbol{x}+\beta\boldsymbol{d}) + \varepsilon \,\}.$$

Show that this map is closed at any $(\boldsymbol{x}, \boldsymbol{d})$ with $\boldsymbol{d} \neq \boldsymbol{0}$.

(1999–04–09, 6)

Appendix A

Some concepts from matrix theory[1]

This appendix brings up some properties of positive definite matrices which are needed frequently in the text. We also discuss so called *LU-factorization* of matrices.

The reader is assumed to be familiar with the basic concepts of linear algebra, including eigenvectors, eigenvalues and diagonalization of matrices and quadratic forms. The fundamental theorem of linear algebra is recalled here for the reader's convenience.

THEOREM 1. *Let* $\boldsymbol{A}$ *be a quadratic* $n \times n$ *matrix. The following conditions on* $\boldsymbol{A}$ *are equivalent.*

- *The columns of* $\boldsymbol{A}$ *form a basis for* $\mathbf{R}^n$.
- *The rows of* $\boldsymbol{A}$ *form a basis for* $\mathbf{R}^n$.
- *The rank of* $\boldsymbol{A}$ *is equal to* n.
- $\boldsymbol{A}$ *is invertible.*
- *The homogeneous system of equations* $\boldsymbol{Ax} = \boldsymbol{0}$ *has the zero solution only.*
- *The system of equations* $\boldsymbol{Ax} = \boldsymbol{b}$ *has a unique solution for any* $\boldsymbol{b}$.
- $\det \boldsymbol{A} \neq 0$.

[1]This appendix is based on an earlier version written by Stefan Diehl and Gunnar Sparr.

We also quote the spectral theorem for symmetric matrices.

THEOREM 2. *Let* $\boldsymbol{H}$ *be a symmetric matrix. Then*

a) there exists an orthogonal basis of eigenvectors of $\boldsymbol{H}$,

b) there is an orthogonal matrix $\boldsymbol{Q}$ *and a diagonal matrix* $\boldsymbol{\Lambda}$ *such that*

$$\boldsymbol{Q}^{-1}\boldsymbol{H}\boldsymbol{Q} = \boldsymbol{Q}^T\boldsymbol{H}\boldsymbol{Q} = \boldsymbol{\Lambda}.$$

In $\boldsymbol{Q}$ *the columns are eigenvectors of* $\boldsymbol{H}$, *in* $\boldsymbol{\Lambda}$ *the diagonal elements are the eigenvalues.*

A.1 Positive definite matrices

DEFINITION 1. *Let* $\boldsymbol{H}$ *be a symmetric* $n \times n$ *matrix.* $\boldsymbol{H}$ *is called*[2]

> ***positive definite*** *if* $\boldsymbol{x}^T\boldsymbol{H}\boldsymbol{x} > 0$ *for all* $\boldsymbol{x} \neq \boldsymbol{0}$,
>
> ***positive semidefinite*** *if* $\boldsymbol{x}^T\boldsymbol{H}\boldsymbol{x} \geq 0$ *for all* $\boldsymbol{x}$,
>
> ***indefinite*** *if* $\boldsymbol{x}^T\boldsymbol{H}\boldsymbol{x} > 0$ *for some* $\boldsymbol{x} \in \mathbf{R}^n$ *and* $\boldsymbol{x}^T\boldsymbol{H}\boldsymbol{x} < 0$ *for some* $\boldsymbol{x} \in \mathbf{R}^n$,
>
> ***negative definite*** *if* $-\boldsymbol{H}$ *is positive definite,*
>
> ***negative semidefinite*** *if* $-\boldsymbol{H}$ *is positive semidefinite.*

These concepts are defined for symmetric matrices only. Whenever the terminology is used it is understood without saying that the matrix in question is symmetric.

Let $\boldsymbol{H} = (h_{ij})$ and let $\boldsymbol{e}_i$ denote the ith canonical basis vector (the elements are zero except at the ith position, where there is a 1). The multiplication

$$\boldsymbol{e}_i^T\boldsymbol{H}\boldsymbol{e}_i = h_{ii}$$

shows that if $\boldsymbol{H}$ is positive definite then all diagonal elements h_{ii} are strictly positive. Also, if elements of both signs appear in the diagonal of $\boldsymbol{H}$ the matrix is indefinite.

[2]The reader might have encountered this terminology in connection with quadratic forms. As can be seen from the definition the connection is simple: the matrix $\boldsymbol{H}$ is positive definite if and only if the quadratic form $\boldsymbol{x}^T\boldsymbol{H}\boldsymbol{x}$ is positive definite, etc.

THEOREM 3. *A symmetric matrix $\boldsymbol{H}$ is positive definite if and only if all of its eigenvalues $\lambda_1, \ldots, \lambda_n$ are strictly positive.*

PROOF. In the quadratic form $\boldsymbol{x}^T\boldsymbol{H}\boldsymbol{x}$ we change variables to an eigenvector basis. Using the notation of theorem 2 the change is executed by putting $\boldsymbol{x} = \boldsymbol{Q}\hat{\boldsymbol{x}}$, where $\hat{\boldsymbol{x}}$ denotes coordinates in the new basis. We obtain

$$\boldsymbol{x}^T\boldsymbol{H}\boldsymbol{x} = \hat{\boldsymbol{x}}^T\boldsymbol{Q}^T\boldsymbol{H}\boldsymbol{Q}\hat{\boldsymbol{x}} = \hat{\boldsymbol{x}}^T\boldsymbol{\Lambda}\hat{\boldsymbol{x}} = \lambda_1\hat{x}_1^2 + \ldots + \lambda_n\hat{x}_n^2.$$

Here the right hand side is strictly positive for all $\hat{\boldsymbol{x}} \neq \boldsymbol{0}$ if and only if all $\lambda_i > 0$. Since $\boldsymbol{x} = \boldsymbol{0}$ if and only if $\hat{\boldsymbol{x}} = \boldsymbol{0}$ the theorem is proved. □

Remark. By the same proof a matrix $\boldsymbol{H}$ is positive *semi*definite if and only if all eigenvalues are ≥ 0.

COROLLARY 1. *If $\boldsymbol{H}$ is positive definite then* $\det \boldsymbol{H} > 0$.

PROOF. This follows immediately from the theorem, since

$$\det \boldsymbol{H} = \prod_{i=1}^{n} \lambda_i.$$

□

We can improve considerably on the corollary, but for that we need some new notation. We denote by $\boldsymbol{H}_k$ the $k \times k$ submatrix in the upper left corner of $\boldsymbol{H}$:

$$\begin{pmatrix} h_{11} & \cdots & h_{1k} & \cdots & h_{1n} \\ \vdots & & \vdots & & \vdots \\ h_{k1} & \cdots & h_{kk} & \cdots & h_{kn} \\ \vdots & & \vdots & & \vdots \\ h_{n1} & \cdots & h_{nk} & \cdots & h_{nn} \end{pmatrix}.$$

LEMMA 1. *If* $\boldsymbol{H}$ *is positive definite then* $\det \boldsymbol{H}_k > 0$ *for all* k.

PROOF. Let $\boldsymbol{x} = (\tilde{\boldsymbol{x}}, \boldsymbol{0})$, where $\tilde{\boldsymbol{x}}$ denotes the first k components of $\boldsymbol{x}$ and the last $n-k$ ones are zero. Then

$$\boldsymbol{x}^T \boldsymbol{H} \boldsymbol{x} = \tilde{\boldsymbol{x}}^T \boldsymbol{H}_k \tilde{\boldsymbol{x}}.$$

From the positive definiteness of $\boldsymbol{H}$ it follows that $\tilde{\boldsymbol{x}}^T \boldsymbol{H}_k \tilde{\boldsymbol{x}} > 0$ for all $\tilde{\boldsymbol{x}} \neq \boldsymbol{0}$. Thus $\boldsymbol{H}_k$ is positive definite, making $\det \boldsymbol{H}_k > 0$ by the corollary above. □

Remark. In fact the conditions in the lemma are equivalent. The proof that $\det \boldsymbol{H}_k > 0$ for all k implies that $\boldsymbol{H}$ is positive definite will be given later (Sylvester's criterion, see theorem 7 below).

Exercises

S A.1 Prove that if a symmetric matrix $\boldsymbol{H}$ is positive definite then it is invertible and its inverse $\boldsymbol{H}^{-1}$ is positive definite as well.

Hint: All you need is the definition of a positive definite matrix.

A.2 Let f be a function of class $\mathcal{C}^2$ with Hessian $\boldsymbol{H}(\boldsymbol{x})$. Show that

a) $\boldsymbol{H}(\boldsymbol{x})$ positive definite $\Longrightarrow$ $\boldsymbol{d} = -\boldsymbol{H}(\boldsymbol{x})\nabla f(\boldsymbol{x})$ is a descent direction for f at $\boldsymbol{x}$. (Used in quasi-Newton methods.)

b) $\boldsymbol{H}(\boldsymbol{x})$ positive definite $\Longrightarrow$ $\boldsymbol{d} = -\boldsymbol{H}(\boldsymbol{x})^{-1}\nabla f(\boldsymbol{x})$ is a descent direction for f at $\boldsymbol{x}$. (Used in Newton's method.)

A.3 Let $\boldsymbol{H}$ be an $n \times n$ symmetric, positive semidefinite matrix, and suppose that $\boldsymbol{x}^T \boldsymbol{H} \boldsymbol{x} = 0$ for some $\boldsymbol{x} \in \mathbf{R}^n$. Show that $\boldsymbol{H}\boldsymbol{x} = \boldsymbol{0}$.

Hint: Consider the diagonalization of the quadratic form $\boldsymbol{x}^T \boldsymbol{H} \boldsymbol{x}$ via the transformation $\boldsymbol{x} = \boldsymbol{Q}\boldsymbol{y}$, where the columns of $\boldsymbol{Q}$ are the normalized eigenvectors of $\boldsymbol{H}$.

A.2 LU-factorization

In this section we discuss matters where $\boldsymbol{A}$ is an arbitrary quadratic matrix, not necessarily symmetric. The notation $\boldsymbol{A}_k$ for the upper left $k \times k$ submatrix of $\boldsymbol{A}$ introduced on page 349 is retained in the general case.

THEOREM 4. *Assume that* $\det \boldsymbol{A}_k \neq 0$ *for all* k.

1. *There is a unique factorization* $\boldsymbol{A} = \boldsymbol{LU}$, *with* $\boldsymbol{L}$ *a lower triangular matrix with all diagonal elements equal to 1, and* $\boldsymbol{U}$ *an upper triangular matrix;*

$$\boldsymbol{L} = \begin{pmatrix} 1 & 0 & \cdots & 0 \\ l_{21} & 1 & \cdots & 0 \\ \vdots & & \ddots & \vdots \\ l_{n1} & l_{n2} & \cdots & 1 \end{pmatrix}, \qquad \boldsymbol{U} = \begin{pmatrix} d_1 & u_{12} & \cdots & u_{1n} \\ 0 & d_2 & \cdots & u_{2n} \\ \vdots & & \ddots & \vdots \\ 0 & 0 & \cdots & d_n \end{pmatrix}.$$

2. *When using Gaussian elimination to solve the linear system* $\boldsymbol{Ax} = \boldsymbol{b}$ *these matrices appear as follows:*

   - *The columns* $\boldsymbol{l}_1, \ldots, \boldsymbol{l}_n$ *of* $\boldsymbol{L}$ *contain the multipliers needed in the elimination; specifically, column* $\boldsymbol{l}_j$ *contains those needed to eliminate* x_j *from rows* $j+1, \ldots, n$.
   - *The result of the eliminations is* $\boldsymbol{U}$.
   - *The diagonal elements in* $\boldsymbol{U}$ *(the* **pivot elements***) are*

$$d_k = \frac{\det \boldsymbol{A}_k}{\det \boldsymbol{A}_{k-1}}, \qquad k = 1, \ldots, n \tag{1}$$

   (where $\det \boldsymbol{A}_0$ *should be interpreted as 1).*

3. *In the special case of* $\boldsymbol{A}$ *being symmetric* $\boldsymbol{U} = \boldsymbol{DL}^T$, *where* $\boldsymbol{D}$ *is the diagonal matrix whose diagonal entries are* $d_1, \ldots, d_n$. *Hence*

$$\boldsymbol{A} = \boldsymbol{LDL}^T$$

   when $\boldsymbol{A}$ *is symmetric.*

We postpone the proof for a while and give some examples.

Example 1. Using the LU-factorization a system of linear equations can be transformed into two systems involving only triangular matrices:

$$\boldsymbol{A}\boldsymbol{x} = \boldsymbol{b} \quad \Longleftrightarrow \quad \boldsymbol{L}\boldsymbol{U}\boldsymbol{x} = \boldsymbol{b} \quad \Longleftrightarrow \quad \begin{cases} \boldsymbol{L}\boldsymbol{y} = \boldsymbol{b} \\ \boldsymbol{U}\boldsymbol{x} = \boldsymbol{y}. \end{cases}$$

These systems can be solved easily. This reduction is of course particularly advantageous when the system is to be solved for several different $\boldsymbol{b}$.

□

Example 2. Suppose we have a 3×3 matrix

$$\boldsymbol{A} = \begin{pmatrix} a_{11} & a_{12} & a_{13} \\ a_{21} & a_{22} & a_{23} \\ a_{31} & a_{32} & a_{33} \end{pmatrix}.$$

Elimination of x_1 from the second and third row in a system $\boldsymbol{A}\boldsymbol{x} = \boldsymbol{b}$ is done by row operations; the first row is multiplied by $\dfrac{a_{21}}{a_{11}}$ and subtracted from the second one, and similarly for the third row with the multiplier $\dfrac{a_{31}}{a_{11}}$. Collecting these multipliers in a column matrix with 1 as the first entry results in

$$\boldsymbol{l}_1 = \begin{pmatrix} 1 \\ a_{21}/a_{11} \\ a_{31}/a_{11} \end{pmatrix}.$$

This will be the first column of $\boldsymbol{L}$ in the factorization. The result of the row operations on $\boldsymbol{A}$ is

$$\boldsymbol{B} = \begin{pmatrix} a_{11} & a_{12} & a_{13} \\ 0 & a_{22} - a_{12}\dfrac{a_{21}}{a_{11}} & a_{23} - a_{13}\dfrac{a_{21}}{a_{11}} \\ 0 & a_{32} - a_{12}\dfrac{a_{31}}{a_{11}} & a_{33} - a_{13}\dfrac{a_{31}}{a_{11}} \end{pmatrix} \stackrel{\text{def}}{=} \begin{pmatrix} a_{11} & a_{12} & a_{13} \\ 0 & \tilde{a}_{11} & \tilde{a}_{12} \\ 0 & \tilde{a}_{21} & \tilde{a}_{22} \end{pmatrix}.$$

We discern in the first row what is to become the first row $\boldsymbol{u}_1^T$ of $\boldsymbol{U}$.

The procedure is now repeated. Carrying out row operations on the third row in $\boldsymbol{B}$ by use of the second one, the multiplier is $\dfrac{\tilde{a}_{21}}{\tilde{a}_{11}}$, and we introduce the column matrix

$$l_2 = \begin{pmatrix} 0 \\ 1 \\ \tilde{a}_{21}/\tilde{a}_{11} \end{pmatrix}.$$

The matrix resulting from the row operations on $\boldsymbol{B}$ is

$$\boldsymbol{C} = \begin{pmatrix} a_{11} & a_{12} & a_{13} \\ 0 & \tilde{a}_{11} & \tilde{a}_{12} \\ 0 & 0 & \tilde{a}_{22} - \tilde{a}_{12}\dfrac{\tilde{a}_{21}}{\tilde{a}_{11}} \end{pmatrix}.$$

The second row here is $\boldsymbol{u}_2^T$. No more computations are in fact required now; we let

$$l_3 = \begin{pmatrix} 0 \\ 0 \\ 1 \end{pmatrix}, \qquad \boldsymbol{L} = \begin{pmatrix} l_1 & l_2 & l_3 \end{pmatrix} \quad \text{and} \quad \boldsymbol{U} = \boldsymbol{C}.$$

The reader may check that in fact this makes $\boldsymbol{LU} = \boldsymbol{A}$.

Of course, for the computations to work the denominators (the *pivot elements*) a_{11} and $\tilde{a}_{11}$ above must not be equal to zero. The proof of this will be contained in the proof of the theorem. □

We now turn to the proof.

Proof. Let $\boldsymbol{A} = (a_{ij})$. We proceed recursively, proving the first two statements of the theorem simultaneously. Since $a_{11} = \det \boldsymbol{A}_1 \neq 0$ by assumption, we may choose column matrices l_1 and $\boldsymbol{u}_1$ of length n exactly as in example 2:

$$l_1 = \begin{pmatrix} 1 \\ a_{21}/a_{11} \\ \vdots \\ a_{n1}/a_{11} \end{pmatrix}, \qquad \boldsymbol{u}_1 = \begin{pmatrix} a_{11} \\ a_{12} \\ \vdots \\ a_{1n} \end{pmatrix}.$$

Multiplying $\boldsymbol{l}_1$ and $\boldsymbol{u}_1^T$ the result is an $n \times n$ matrix whose first row and first column agree with those of $\boldsymbol{A}$,

$$\boldsymbol{l}_1\boldsymbol{u}_1^T = \begin{pmatrix} a_{11} & a_{12} & \cdots & a_{1n} \\ a_{21} & a_{12}\dfrac{a_{21}}{a_{11}} & \cdots & a_{1n}\dfrac{a_{21}}{a_{11}} \\ \vdots & \vdots & & \vdots \\ a_{n1} & a_{12}\dfrac{a_{n1}}{a_{11}} & \cdots & a_{1n}\dfrac{a_{n1}}{a_{11}} \end{pmatrix}.$$

Thus $\boldsymbol{A} - \boldsymbol{l}_1\boldsymbol{u}_1^T$ is of the form

$$\boldsymbol{A} - \boldsymbol{l}_1\boldsymbol{u}_1^T = \begin{pmatrix} 0 & 0 \quad \cdots & 0 \\ 0 & & \\ \vdots & \widetilde{\boldsymbol{A}} & \\ 0 & & \end{pmatrix}.$$

The submatrix $\widetilde{\boldsymbol{A}}$ here also appears in the result of the row operations on $\boldsymbol{A}$:

$$\boldsymbol{B} = \begin{pmatrix} a_{11} & a_{12} \quad \cdots & a_{1n} \\ 0 & & \\ \vdots & \widetilde{\boldsymbol{A}} & \\ 0 & & \end{pmatrix}.$$

We can now repeat the above procedure, working on $\widetilde{\boldsymbol{A}}$ instead of $\boldsymbol{A}$. (Cf. example 2.) Then it is important that the assumption that all $\det \widetilde{\boldsymbol{A}}_k \neq 0$ is satisfied. We will verify that in a moment.

After n iteration steps we have constructed n columns $\boldsymbol{l}_i$ and n columns $\boldsymbol{u}_j$ such that

$$\boldsymbol{A} - \boldsymbol{l}_1\boldsymbol{u}_1^T - \ldots - \boldsymbol{l}_n\boldsymbol{u}_n^T = \boldsymbol{O} \quad \Longleftrightarrow \quad \boldsymbol{A} = \begin{pmatrix} \boldsymbol{l}_1 & \ldots & \boldsymbol{l}_n \end{pmatrix} \begin{pmatrix} \boldsymbol{u}_1^T \\ \vdots \\ \boldsymbol{u}_n^T \end{pmatrix} = \boldsymbol{LU}.$$

The matrices $\boldsymbol{L}$ and $\boldsymbol{U}$ defined in this manner obviously are lower and upper triangular respectively.

It must be checked that the subdeterminants of the successively constructed matrices remain non-zero. We now verify that this is the case in the iteration step from $\boldsymbol{A}$ to $\widetilde{\boldsymbol{A}}$ described above. Since determinants are not affected by row operations,

$$\text{(2)} \qquad \det \boldsymbol{A}_{k+1} = \det \boldsymbol{B}_{k+1} = a_{11} \det \widetilde{\boldsymbol{A}}_k, \qquad k = 1, \ldots, n-1.$$

From the assumptions in the theorem it follows that $\det \widetilde{\boldsymbol{A}}_k \neq 0$ for all k.

For the diagonal elements in $\boldsymbol{U}$ we in particular obtain $d_1 = a_{11}$ and

$$d_2 = \widetilde{a}_{11} = \det \widetilde{\boldsymbol{A}}_1 = \frac{\det \boldsymbol{A}_2}{a_{11}} = \frac{\det \boldsymbol{A}_2}{\det \boldsymbol{A}_1}.$$

Since by (2)

$$\frac{\det \boldsymbol{A}_{k+1}}{\det \boldsymbol{A}_k} = \frac{\det \widetilde{\boldsymbol{A}}_k}{\det \widetilde{\boldsymbol{A}}_{k-1}},$$

the similar result (1) will be obtained iteratively for all d_k.

It is now time to prove the uniqueness of the factorization. Suppose that some $\boldsymbol{A}$ can be factorized in two ways, equal to

$$\boldsymbol{L}_1 \boldsymbol{U}_1 = \boldsymbol{L}_2 \boldsymbol{U}_2.$$

Then

$$\boldsymbol{L}_2^{-1} \boldsymbol{L}_1 = \boldsymbol{U}_2 \boldsymbol{U}_1^{-1},$$

where the left side is a lower triangular matrix and the right side is an upper triangular one. Hence both sides are diagonal matrices. The diagonal entries on the left are all equal to one. Thus

$$\boldsymbol{L}_2^{-1} \boldsymbol{L}_1 = \boldsymbol{U}_2 \boldsymbol{U}_1^{-1} = \boldsymbol{I},$$

from which the uniqueness follows.

The first two items in the statement of the theorem have now been proved. It remains to consider the special case when $\boldsymbol{A}$ is symmetric. The matrix $\boldsymbol{U}$ can be factorized as

$$\boldsymbol{U} = \boldsymbol{D}\boldsymbol{M},$$

where $\boldsymbol{M}$ is upper triangular with diagonal entries equal to 1, and $\boldsymbol{D}$ is a diagonal matrix. When $\boldsymbol{A}$ is symmetric we get

$$\boldsymbol{L}\boldsymbol{D}\boldsymbol{M} = \boldsymbol{A} = \boldsymbol{A}^T = \boldsymbol{M}^T \boldsymbol{D} \boldsymbol{L}^T.$$

By the recently proved uniqueness of the factorization it follows that $\boldsymbol{M} = \boldsymbol{L}^T$. □

Remark. Note that the fact that $\det \boldsymbol{A} \neq 0$ does not imply that $\det \boldsymbol{A}_k \neq 0$ for $k < n$. The matrix $\boldsymbol{A} = \begin{pmatrix} 0 & 1 \\ 1 & 0 \end{pmatrix}$ is a counterexample.

Some comments on numerical computation

When LU-factorization, perhaps as a substitute for matrix inversion, is used as a subroutine in some optimization algorithm it is of interest to estimate how the number of floating point operations (*flops*) depends on the size n. We use the same convention as Matlab: additions, subtractions, multiplications, divisions count one flop each.

Example 3. Consider the procedure of solving a system $\boldsymbol{U}\boldsymbol{x} = \boldsymbol{b}$ with $\boldsymbol{U}$ triangular, done by back substitution. In the kth equation

$$d_k x_k + u_{k,k+1} x_{k+1} + \ldots + u_{kn} x_n = b_k$$

you insert the previously calculated $x_{k+1}, \ldots, x_n$ and solve for x_k. This requires $n-k$ multiplications of the x_i with their respective coefficient, then equally many subtractions in the right hand side, and finally one division by the pivot element d_k. In total this step requires $2(n-k)+1$ flops. For the whole procedure the number of flops is

$$\sum_{k=1}^{n} \Big(2(n-k)+1\Big) = 2\Big((n-1)+\ldots+1+0\Big) + n = 2\,\frac{(n-1)n}{2} + n = n^2.$$

As a consequence, the number of flops required to solve the two triangular systems in example 1 is $2n^2$, once the LU-factorization is computed. □

We now state a result on the number of flops required to factorize a matrix $\boldsymbol{A}$.

THEOREM 5. *The factorization of the $n \times n$ matrix $\boldsymbol{A}$ in theorem 4 requires $\dfrac{2n^3}{3} + O(n^2)$ flops.*

Here $O(n^2)$ denotes any function which when divided by n^2 remains bounded as $n \to \infty$.

PROOF. We refer back to the proof of theorem 4. In the step from $\boldsymbol{A}$ to $\widetilde{\boldsymbol{A}}$ the computation of $\boldsymbol{l}_1$ requires $n-1$ divisions, the multiplication of $\boldsymbol{l}_1$ and $\boldsymbol{u}_1^T$ requires $(n-1)^2$ multiplications, and the formation of $\widetilde{\boldsymbol{A}}$ consists of $(n-1)^2$ subtractions, in total $(n-1)+2(n-1)^2$ flops. This is for the first iteration step only. As we continue the size of the problem diminishes. In all, the number of flops required is

$$(n-1)+(n-2)+\ldots+2+1+2\left[(n-1)^2+(n-2)^2+\ldots+2^2+1^2\right]=$$

$$=\sum_{k=1}^{n-1} k+2\sum_{k=1}^{n-1} k^2=\frac{(n-1)n}{2}+2\,\frac{(n-1)n(2n-1)}{6}=\frac{2n^3}{3}+O(n^2).$$

Thus the theorem is proved. □

Remark. In the case of a symmetric matrix it is not difficult to show that $\frac{n^3}{3}+O(n^2)$ flops suffice.

The algorithm presented in the proof of theorem 4 is not numerically reliable, since the pivot elements can be small. To handle this problem the LU decomposition is combined with *partial* or *full pivoting.* This means that a permutation of rows (and columns) is also carried out. The interested reader is referred to a course in matrix theory or numerical linear algebra.

A.3 More on positive definite matrices

Armed with the LU-factorization theorem from section A.2 we are now in a position to prove some important theorems about positive definite matrices.

THEOREM 6. *The symmetric $n\times n$ matrix $\boldsymbol{H}$ is positive definite if and only if it can be factorized as*

$$\boldsymbol{H}=\boldsymbol{L}\boldsymbol{D}\boldsymbol{L}^T,$$

with

$$\mathbf{L} = \begin{pmatrix} 1 & 0 & \cdots & 0 \\ l_{21} & 1 & \cdots & 0 \\ \vdots & & \ddots & \vdots \\ l_{n1} & l_{n2} & \cdots & 1 \end{pmatrix}, \qquad \mathbf{D} = \begin{pmatrix} d_1 & 0 & \cdots & 0 \\ 0 & d_2 & \cdots & 0 \\ \vdots & & \ddots & \vdots \\ 0 & 0 & \cdots & d_n \end{pmatrix}$$

$$\text{and } d_i > 0, \quad i = 1, \ldots, n.$$

The numbers d_i are the pivot elements in the Gaussian elimination.

PROOF. Assuming that $\mathbf{H}$ is positive definite the factorization follows from lemma 1 and theorem 4.

For the proof in the reverse direction we use the definition of a positive definite matrix. Letting $\tilde{\boldsymbol{x}} = \mathbf{L}^T\boldsymbol{x}$ we obtain from the factorization of $\mathbf{H}$ that

$$\begin{aligned} \boldsymbol{x}^T\mathbf{H}\boldsymbol{x} &= \boldsymbol{x}^T\mathbf{L}\mathbf{D}\mathbf{L}^T\boldsymbol{x} = (\mathbf{L}^T\boldsymbol{x})^T\mathbf{D}(\mathbf{L}^T\boldsymbol{x}) = \\ &= \tilde{\boldsymbol{x}}^T\mathbf{D}\tilde{\boldsymbol{x}} = d_1\tilde{x}_1^2 + \ldots + d_n\tilde{x}_n^2. \end{aligned} \tag{3}$$

Since all d_i are strictly positive $\boldsymbol{x}^T\mathbf{H}\boldsymbol{x} > 0$ for all $\boldsymbol{x} \neq \mathbf{0}$. □

Writing out the coordinate change in (3) more explicitly,

$$\boldsymbol{x}^T\mathbf{H}\boldsymbol{x} = (\mathbf{L}^T\boldsymbol{x})^T\mathbf{D}(\mathbf{L}^T\boldsymbol{x}) =$$

$$= d_1(x_1+l_{21}x_2+\ldots+l_{n1}x_n)^2 + d_2(x_2+l_{32}x_3+\ldots+l_{n2}x_n)^2 + \ldots + d_nx_n^2,$$

we conclude that the matrices $\mathbf{L}$ and $\mathbf{D}$ can be obtained by completing squares in the quadratic form $\boldsymbol{x}^T\mathbf{H}\boldsymbol{x}$. This is most conveniently done in matrix form, via row operations (Gaussian elimination) in $\mathbf{H}$, since

$$\mathbf{H}\boldsymbol{x} = \boldsymbol{b} \quad \Longleftrightarrow \quad \mathbf{L}\mathbf{D}\mathbf{L}^T\boldsymbol{x} = \boldsymbol{b} \quad \Longleftrightarrow \quad \mathbf{L}^T\boldsymbol{x} = \mathbf{D}^{-1}\mathbf{L}^{-1}\boldsymbol{b}.$$

Example 4. The following computations show how row operations on a matrix $\mathbf{H}$ produce both the elements of $\mathbf{D}$ and of $\mathbf{L}$. The pivot elements

d_i have been boxed when they appear.

$$\boldsymbol{H} = \begin{pmatrix} \boxed{1} & 2 & -3 \\ 2 & 6 & -10 \\ -3 & -10 & 23 \end{pmatrix} \xrightarrow{d_1=1} \begin{pmatrix} 1 & 2 & -3 \\ 0 & \boxed{2} & -4 \\ 0 & -4 & 14 \end{pmatrix} \xrightarrow{d_2=2} \begin{pmatrix} 1 & 2 & -3 \\ 0 & 1 & -2 \\ 0 & -4 & 14 \end{pmatrix} \longrightarrow$$

$$\longrightarrow \begin{pmatrix} 1 & 2 & -3 \\ 0 & 1 & -2 \\ 0 & 0 & \boxed{6} \end{pmatrix} \xrightarrow{d_3=6} \begin{pmatrix} 1 & 2 & -3 \\ 0 & 1 & -2 \\ 0 & 0 & 1 \end{pmatrix} = \boldsymbol{L}^T.$$

Since all d_i are strictly positive this matrix $\boldsymbol{H}$ is positive definite. □

Remark. By formula (1) for the diagonal elements in theorem 4 we obtain that $\det \boldsymbol{H}_k = \prod_{i=1}^{k} d_i$ for all k. Hence

$$\det \boldsymbol{H}_k \neq 0 \quad \text{for all } k \qquad \Longleftrightarrow \qquad d_k \neq 0 \quad \text{for all } k,$$

and the $\boldsymbol{H} = \boldsymbol{LDL}^T$ factorization exists if the pivot elements encountered in the computation are non-zero.[3]

This may be a good opportunity to remind the reader of *Sylvester's law of inertia*. It states that the number of positive/negative/zero eigenvalues of $\boldsymbol{H}$ is equal to the number of positive/negative/zero pivot elements d_k. For example, if one pivot element is negative and the others are positive, there is one negative eigenvalue and the others are positive, and the matrix is indefinite.

We now prove *Sylvester's criterion* for a symmetric matrix to be positive definite.

THEOREM 7. (SYLVESTER) *The symmetric matrix $\boldsymbol{H}$ is positive definite if and only if*

$$\det \boldsymbol{H}_k > 0 \quad \textit{for all } k.$$

PROOF. Assuming that $\boldsymbol{H}$ is positive definite, all subdeterminants are strictly positive by lemma 1.

[3] Actually the last one may be zero.

Assume for the proof in the reverse direction that $\det \boldsymbol{H}_k > 0$ for all k. By theorem 4 $\boldsymbol{H} = \boldsymbol{L}\boldsymbol{D}\boldsymbol{L}^T$ with the diagonal elements in $\boldsymbol{D}$ equal to

$$d_k = \frac{\det \boldsymbol{H}_k}{\det \boldsymbol{H}_{k-1}} > 0, \qquad k = 1, \ldots, n.$$

By theorem 6 this implies that $\boldsymbol{H}$ is positive definite. □

Example 5. For the matrix $\boldsymbol{H}$ in example 4 $\det \boldsymbol{H}_1 = 1 > 0$, $\det \boldsymbol{H}_2 = 2 > 0$ and $\det \boldsymbol{H}_3 = 12 > 0$, implying that $\boldsymbol{H}$ positive definite. □

A matrix $\boldsymbol{H}$ is positive *semi*definite if and only if $\boldsymbol{H} + \varepsilon \boldsymbol{I}$ is positive definite for all $\varepsilon > 0$. By Sylvester's criterion this is equivalent to

$$\det(\boldsymbol{H} + \varepsilon \boldsymbol{I})_k = \det(\boldsymbol{H}_k + \varepsilon \boldsymbol{I}_k) > 0 \quad \text{for all positive } \varepsilon,$$

and for all k. Obviously this implies that $\det \boldsymbol{H}_k \geq 0$ for all k.

It may come as a surprise that this result can not be reversed. The matrix

$$\boldsymbol{H} = \begin{pmatrix} 1 & 0 & 2 \\ 0 & 0 & 0 \\ 2 & 0 & 1 \end{pmatrix}$$

is a counterexample. Here $\det \boldsymbol{H}_1 = 1$, $\det \boldsymbol{H}_2 = \det \boldsymbol{H}_3 = 0$, but the matrix is indefinite, as the reader may verify. However, the following theorem is sometimes useful.

THEOREM 8. *Assume that* $\det \boldsymbol{H}_k > 0$ *for* $k = 1, \ldots, n-1$, *and that* $\det \boldsymbol{H}_n = 0$ *(i.e.,* $\det \boldsymbol{H} = 0$*). Then the symmetric matrix* $\boldsymbol{H}$ *is positive semidefinite.*

PROOF. It is easy to verify that the proof of theorem 4 works also when $\det \boldsymbol{H}_n = 0$. For the diagonal elements of $\boldsymbol{D}$ we now obtain

$$d_1 > 0, \; \ldots, \; d_{n-1} > 0, \; d_n = 0.$$

Hence the change of coordinates $\tilde{\boldsymbol{x}} = \boldsymbol{L}^T \boldsymbol{x}$ in (3) results in

$$\boldsymbol{x}^T \boldsymbol{H} \boldsymbol{x} = d_1 \tilde{x}_1^2 + \ldots + d_{n-1} \tilde{x}_{n-1}^2,$$

proving that $\boldsymbol{H}$ is positive semidefinite. □

We finally give a modification of theorem 6, called *Cholesky factorization.*

THEOREM 9. *The matrix $\mathbf{H}$ is positive definite if and only if there is a factorization*

$$\mathbf{H} = \widehat{\mathbf{L}}\widehat{\mathbf{L}}^T,$$

where $\widehat{\mathbf{L}}$ is lower triangular with positive elements on the diagonal.

PROOF. Assume first that $\mathbf{H}$ is positive definite. By theorem 6 $\mathbf{H} = \mathbf{L}\mathbf{D}\mathbf{L}^T$ with strictly positive diagonal elements in $\mathbf{D}$. Hence we can define a diagonal matrix $\sqrt{\mathbf{D}}$ by taking the positive square root of every diagonal element of $\mathbf{D}$. Then

$$\mathbf{H} = \mathbf{L}\mathbf{D}\mathbf{L}^T = \mathbf{L}\sqrt{\mathbf{D}}\sqrt{\mathbf{D}}\mathbf{L}^T = (\mathbf{L}\sqrt{\mathbf{D}})(\mathbf{L}\sqrt{\mathbf{D}})^T = \widehat{\mathbf{L}}\widehat{\mathbf{L}}^T,$$

where $\widehat{\mathbf{L}} = \mathbf{L}\sqrt{\mathbf{D}}$.

Now assume that $\mathbf{H}$ has a Cholesky factorization. Then

$$\boldsymbol{x}^T\mathbf{H}\boldsymbol{x} = \boldsymbol{x}^T\widehat{\mathbf{L}}\widehat{\mathbf{L}}^T\boldsymbol{x} = \left(\widehat{\mathbf{L}}^T\boldsymbol{x}\right)^T\left(\widehat{\mathbf{L}}^T\boldsymbol{x}\right) = \left\|\widehat{\mathbf{L}}^T\boldsymbol{x}\right\|^2.$$

It follows that $\boldsymbol{x}^T\mathbf{H}\boldsymbol{x} \geq 0$ for all $\boldsymbol{x}$, with equality if and only if $\widehat{\mathbf{L}}^T\boldsymbol{x} = \mathbf{0}$, that is, if and only if $\boldsymbol{x} = \mathbf{0}$, since $\widehat{\mathbf{L}}$ has strictly positive diagonal elements. Thus by definition $\mathbf{H}$ is positive definite.

Hence the theorem is proved. □

Example 6. For the matrix $\mathbf{H}$ in example 4 the computations there imply the following Cholesky factorization:

$$\mathbf{H} = \mathbf{L}\mathbf{D}\mathbf{L}^T = \begin{pmatrix} 1 & 0 & 0 \\ 2 & 1 & 0 \\ -3 & -2 & 1 \end{pmatrix}\begin{pmatrix} 1 & 0 & 0 \\ 0 & 2 & 0 \\ 0 & 0 & 6 \end{pmatrix}\begin{pmatrix} 1 & 2 & -3 \\ 0 & 1 & -2 \\ 0 & 0 & 1 \end{pmatrix} =$$

$$= \begin{pmatrix} 1 & 0 & 0 \\ 2 & 1 & 0 \\ -3 & -2 & 1 \end{pmatrix}\begin{pmatrix} 1 & 0 & 0 \\ 0 & \sqrt{2} & 0 \\ 0 & 0 & \sqrt{6} \end{pmatrix}\begin{pmatrix} 1 & 0 & 0 \\ 0 & \sqrt{2} & 0 \\ 0 & 0 & \sqrt{6} \end{pmatrix}\begin{pmatrix} 1 & 2 & -3 \\ 0 & 1 & -2 \\ 0 & 0 & 1 \end{pmatrix} =$$

$$= \widehat{\mathbf{L}}\widehat{\mathbf{L}}^T,$$

where

$$\widehat{\mathbf{L}} = \begin{pmatrix} 1 & 0 & 0 \\ 2 & \sqrt{2} & 0 \\ -3 & -2\sqrt{2} & \sqrt{6} \end{pmatrix}.$$

As an alternative, the Cholesky factorization can be obtained by writing

$$\begin{pmatrix} l_{11} & 0 & 0 \\ l_{21} & l_{22} & 0 \\ l_{31} & l_{32} & l_{33} \end{pmatrix} \begin{pmatrix} l_{11} & l_{21} & l_{31} \\ 0 & l_{22} & l_{32} \\ 0 & 0 & l_{33} \end{pmatrix} = \begin{pmatrix} 1 & 2 & -3 \\ 2 & 6 & -10 \\ -3 & -10 & 23 \end{pmatrix}.$$

Carrying out the matrix multiplication we obtain a system of equations for the matrix elements:

$$\begin{cases} l_{11}^2 & = & 1 \\ l_{11}l_{21} & = & 2 \\ l_{11}l_{31} & = & -3 \\ l_{21}^2 + l_{22}^2 & = & 6 \\ l_{21}l_{31} + l_{22}l_{32} & = & -10 \\ l_{31}^2 + l_{32}^2 + l_{33}^2 & = & 23. \end{cases}$$

This system can be solved recursively. Remembering always to pick the positive square root the result is

$$l_{11} = 1, \quad l_{21} = 2, \quad l_{31} = -3, \quad l_{22} = \sqrt{2}, \quad l_{32} = -2\sqrt{2}, \quad l_{33} = \sqrt{6}.$$

□

Exercises

S A.4 Verify that the matrix

$$\boldsymbol{H} = \begin{pmatrix} 3 & 2 & 0 \\ 2 & 6 & 1 \\ 0 & 1 & 2 \end{pmatrix}$$

is positive definite by means of

a) Gaussian elimination (completion of squares),

b) Sylvester's criterion,

c) the existence of a Cholesky factorization,

d) computation of eigenvalues (numerically).

A.5 Sylvester's criterion says that

$$\boldsymbol{H} \text{ positive definite} \iff \det \boldsymbol{H}_k > 0, \quad k = 1, \dots, n,$$

where $\boldsymbol{H}_k$ is the $k \times k$ submatrix in the upper left corner of $\boldsymbol{H}$.

a) Show that

$$\boldsymbol{H} \text{ negative definite} \iff \begin{cases} \det \boldsymbol{H}_k < 0, & k \text{ odd} \\ \det \boldsymbol{H}_k > 0, & k \text{ even} \end{cases} \qquad 1 \le k \le n.$$

b) Show that Sylvester's criterion can also be formulated:

$$\boldsymbol{H} \text{ positive definite} \iff \det \widetilde{\boldsymbol{H}}_k > 0, \quad k = 1, \dots, n,$$

where $\widetilde{\boldsymbol{H}}_k$ is the $k \times k$ submatrix in the *lower right* corner of $\boldsymbol{H}$.

Hint: Consider the quadratic form $\boldsymbol{x}^T \boldsymbol{H} \boldsymbol{x}$ and make a transformation of variables $\boldsymbol{x} = \widetilde{\boldsymbol{I}} \boldsymbol{y} = (\ y_n \ \ \dots \ \ y_1 \ )^T$.

Note: By using other permutation matrices instead of $\widetilde{\boldsymbol{I}}$ we can prove that any ordering of the variables will do.

S A.6 a) Obtain expressions for all first and second order derivatives of the function of two variables

$$f(x) = 100(x_2 - x_1^2)^2 + (1 - x_1)^2.$$

Verify that the minimizer $\boldsymbol{x}^* = (1, 1)$ satisfies $\nabla f(\boldsymbol{x}^*) = \boldsymbol{0}$ and that $\boldsymbol{H}(\boldsymbol{x}^*)$ is positive definite.

b) Show that $\det \boldsymbol{H}(\boldsymbol{x}) = 0$ if and only if $\boldsymbol{x}$ satisfies the condition

$$x_2 - x_1^2 = 0.005.$$

As a consequence prove that $\boldsymbol{H}(\boldsymbol{x})$ is positive definite for all $\boldsymbol{x}$ such that $f(\boldsymbol{x}) < 0.0025$.

Hint: Use exercise A.5 b.

Suggested further reading

The present text has, to a more or less extent, benefitted from most of the following books. Of course they cover much more material, often with a different approach to matters. In particular, some of them contain a lot of optimization algorithms for diverse applications.

Andersson, R., Gårding, L., Tengstrand, A., Tillämpad linjär algebra. (Lunds universitets matematiska institution 1975)

Bazaraa, M.S., Sherali, H.D., Shetty, C.M., Nonlinear programming, Theory and Algorithms, 2nd ed. (Wiley 1993)

Boyd, S., Vandenberghe, L., Convex Optimization. (Mimeographed, 1999)

Cooper, L., Steinberg, D.I., Introduction to Methods of Optimization. (Saunders 1970)

Fletcher, R., Practical Methods of Optimization, 2nd ed. (Wiley 1993)

Gill, P.E., Murray, W., Wright, M.H., Practical Optimization. (Academic Press 1981)

Greig, D.M., Optimisation. (Longman 1980)

Hestenes, M.R., Optimization Theory: the finite dimensional case. (Wiley 1975)

Luenberger, D.G., Introduction to Linear and Nonlinear Programming. (Addison-Wesley 1973)

Lundgren, J., Rönnqvist, M., Värbrand, P., Optimeringslära.[4]
(Studentlitteratur, 2008)

Wright, S.J., Primal-Dual Interior-Point Methods.
(SIAM 1997)

[4]Version in English 2010.

Answers to the exercises

Chapter 1

1.1

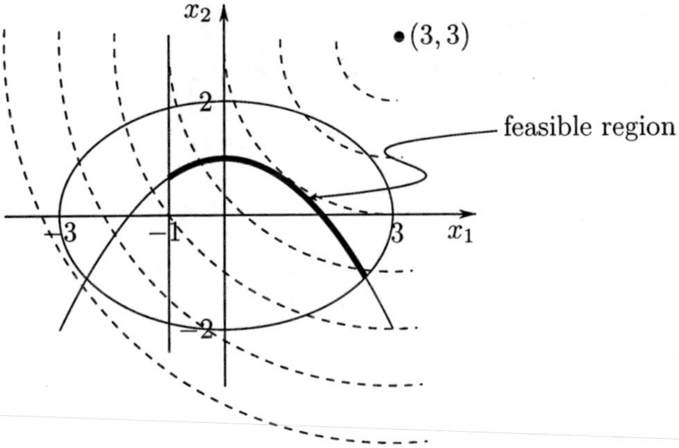

Minimum at (1.14, 0.56).

1.2 a) $\boldsymbol{c}$.

b) $\boldsymbol{A}$, $\boldsymbol{I}$.

c) $$\begin{pmatrix} \dfrac{\partial^2 f}{\partial x_1^2} & \cdots & \dfrac{\partial^2 f}{\partial x_1 \partial x_n} \\ \vdots & & \vdots \\ \dfrac{\partial^2 f}{\partial x_n \partial x_1} & \cdots & \dfrac{\partial^2 f}{\partial x_n^2} \end{pmatrix}.$$

d) Gradient: $\boldsymbol{A}\boldsymbol{x} + \boldsymbol{A}^T\boldsymbol{x}$, Hessian: $\boldsymbol{A} + \boldsymbol{A}^T$.

e) Gradient: $\boldsymbol{H}\boldsymbol{x} + \boldsymbol{c}$, Hessian: $\boldsymbol{H}$.

f) Gradient: $2\boldsymbol{x}$, Hessian: $2\boldsymbol{I}$.

1.3 a) $\varphi'(\alpha) = \nabla f(\boldsymbol{x}(\alpha))^T \boldsymbol{x}'(\alpha) = \boldsymbol{x}'(\alpha)^T \nabla f(\boldsymbol{x}(\alpha))$,

b) $\varphi'(0) = \boldsymbol{d}^T \nabla f(\hat{\boldsymbol{x}})$, $\quad \varphi''(0) = \boldsymbol{d}^T \nabla^2 f(\hat{\boldsymbol{x}}) \boldsymbol{d}$.

1.5 b) $d' = d - \frac{1}{2} \bar{\boldsymbol{x}}^T \boldsymbol{H} \bar{\boldsymbol{x}}$.

1.7 $\boldsymbol{g}(\boldsymbol{a} + \boldsymbol{\delta}) = \boldsymbol{g}(\boldsymbol{a}) + \nabla \boldsymbol{g}(\boldsymbol{a})^T \boldsymbol{\delta} + O(\|\boldsymbol{\delta}\|^2) = \boldsymbol{g}(\boldsymbol{a}) + \boldsymbol{H}(\boldsymbol{a})\boldsymbol{\delta} + O(\|\boldsymbol{\delta}\|^2)$.

1.8 a) Positive definite. b) indefinite.

1.9 Local maximum at $(-1, -1)$, local minimum at $(1, 0)$. The other stationary points, $(0, 0)$ and $(0, -1)$, are saddle points.

1.10 a) Stationary point: $(\frac{6}{5}, \frac{6}{5}, \frac{17}{5})$, local minimum.

b) Stationary points: $(0, 0)$ saddle point, $(\pm 1, 0)$ local minima.

Chapter 2

2.1 $F_{\min} = 0.633$. Uncertainty interval: a) [0.944, 1.236], b) [0.750, 1.100].

2.2 a) [0.43, 0.53].

b) 6 and 8 respectively.

2.3 a) $[0.236, 0.416]$, b) $[0.324, 0.391]$.

2.4 a) $F(\lambda) = \big((\hat{x}_1 + \lambda d_1)^3 + \hat{x}_2 + \lambda d_2\big)^2 + 2\big(\hat{x}_2 + \lambda d_2 - \hat{x}_1 - \lambda d_1 - 4\big)^4$.

b) Interval of uncertainty: $[-0.154,\ 0.154]$.

c) Interval of uncertainty: $[-2,\ -1.639]$.

2.5

q	0.1	0.01	0.001	0.0001
Uniform	19	199	1999	19999
Dichotomous	8	14	20	28
Golden section	6	11	16	21
Fibonacci	6	11	16	20

2.6 a) 0.3517, b) $[0.312, 0.375]$.

2.7 0.6529.

Chapter 3

3.1 $\lambda_k = -\dfrac{\boldsymbol{d}_k^T(\boldsymbol{H}\boldsymbol{x}_k + \boldsymbol{c})}{\boldsymbol{d}_k^T \boldsymbol{H} \boldsymbol{d}_k}.$

3.3 a) $(2,-1)$, b) $(-1,-4)$ and $(4,-1)$, c) $k \geq 6$.

3.4 b) $\boldsymbol{x}_2 = (1.051,\ -0.201)$, $\boldsymbol{x}_3 = (0.325,\ 0.100)$.

3.5 37.

3.6 a) The level curves are ellipses with center at $\bar{\boldsymbol{x}} = (2,1)$.

c) If $\boldsymbol{x}_1$ lies on any of the lines $x_1 = 2$ or $x_2 = 1$ we obtain convergence in one step only, otherwise not in finitely many steps.

3.7 a) Minimum at $\bar{\boldsymbol{x}} = (1,1)$, $f(\bar{\boldsymbol{x}}) = 0$.

b) $\boldsymbol{x}_2 = (\frac{9}{5}, \frac{16}{5})$, $f(\boldsymbol{x}_1) = 2.5$, $f(\boldsymbol{x}_2) = 0.32$.

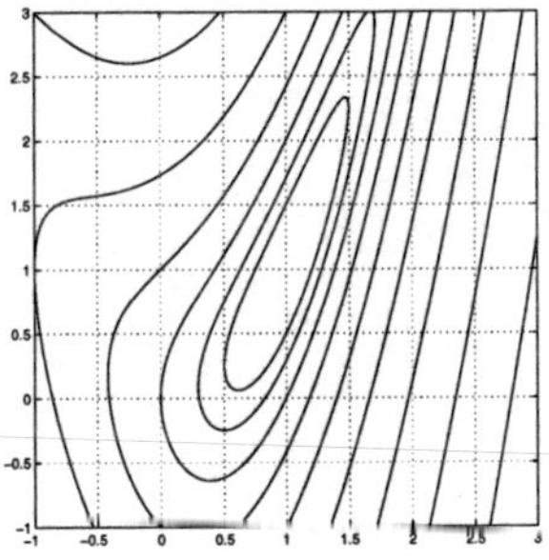

Exercise 3.7

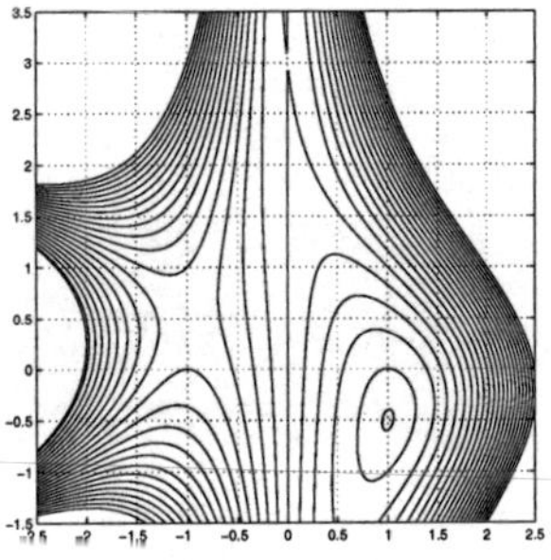

Exercise 3.9

3.9 b) $\boldsymbol{H}(1,1) = \begin{pmatrix} 8 & 5 \\ 5 & 2 \end{pmatrix}$, indefinite,

$\boldsymbol{d}_1 = (-1,1)$, $\boldsymbol{x}_2 = (0,2)$, $f(\boldsymbol{x}_1) = 0$, $f(\boldsymbol{x}_2) = 0$.

c) $\boldsymbol{H}(1,1) + 2\boldsymbol{I} = \begin{pmatrix} 10 & 5 \\ 5 & 4 \end{pmatrix}$, positive definite,

$\boldsymbol{d}_1 = (0.2,\ -1)$, $\boldsymbol{x}_2 = (1.2,\ 0)$, $f(\boldsymbol{x}_1) = 0$, $f(\boldsymbol{x}_2) = -1.872$, $(f_{\min} = -2.25)$.

3.11 $\varepsilon > \sqrt{2} - 1$. If $\varepsilon = 1$ then $\boldsymbol{d}_1 = (1,-1)$.

3.12 a) No, b) $\boldsymbol{L} = \begin{pmatrix} 1 & 0 & 0 \\ 2 & 1 & 0 \\ 0 & 3 & 1 \end{pmatrix}$, c) $\boldsymbol{x}^{(2)} = (-8,4,-2)$.

3.15 b) E.g. $\boldsymbol{d}_3 = (4, -3, 0)$, c) $\boldsymbol{x}^{(2)} = \frac{1}{5}(2, 6, 5)$, d) $\boldsymbol{x}^{(4)} = (-7, 5, 0)$.

3.16 b) $\boldsymbol{d}_1 = (1, 0, 0)$, $\boldsymbol{d}_2 = (-1, 1, 0)$, $\boldsymbol{d}_3 = (1, -1, 1)$.

c) $\boldsymbol{x}_2 = (1, 0, 0)$, $\boldsymbol{x}_3 = (0, 1, 0)$, $\boldsymbol{x}_4 = (2, -1, 2) = \bar{\boldsymbol{x}}$.

3.17 a) Minimum at $(2, 1)$.

3.18 a) Minimum at $(2, 1)$.

3.19 $\boldsymbol{x}_3 = (-1, -1, -1, -1)$.

3.20 $\boldsymbol{x} = (3 \quad 1.5 \quad -2.5)^T$.

3.21 $C = 4.0394$, $k = 0.3326$.

3.22 a) Order 1, b) order 1, c) linear, d) superlinear,

e) superlinear, f) superlinear, g) order e.

3.27 a) $(0.894, -0.447)$.

b) $(0.559, -4.030)$.

c) $\boldsymbol{x}^{(2)} = (0.559, -3.030)$.

3.28 a) D, b) G, c) E, d) H.

3.29 a) $\varepsilon > \sqrt{5} - 2$.

b) $\boldsymbol{x}^{(1)} = (-0.25, -3.25)$, $f(\boldsymbol{x}^{(0)}) = 1.33$, $f(\boldsymbol{x}^{(1)}) = -2.88$.

c) $\boldsymbol{x}^{(2)} = (-0.15, -2.59)$, $f(\boldsymbol{x}^{(2)}) = -3.58$.

3.31 a) Global minimum -81 at $(0, \pm 3)$.

b) $\boldsymbol{x}_3 = (0, -\frac{1}{7207083})$. Converges to $(0, 0)$ (a saddle point).

c) Choose e.g. $\varepsilon = 36$ in the modified Newton's algorithm.

3.32 a) $(0, 1)$, global minimum, b) $\boldsymbol{x}_{2k+2} = \begin{pmatrix} 0 - 2 \cdot 5^{-(k+1)} \\ 1 - 3 \cdot 5^{-(k+1)} \end{pmatrix}$.

Chapter 4

4.1 a) $\mathrm{cl}(S) = \{\,\boldsymbol{x} \in \mathbf{R}^2\,;\, x_1^2 \le x_2 \le 1\,\}$,

$\mathrm{int}(S) = \{\,\boldsymbol{x} \in \mathbf{R}^2\,;\, x_1^2 < x_2 < 1\,\}$,

$\partial S = \{\,\boldsymbol{x}\,;\, x_2 = x_1^2,\ -1 \le x_1 \le 1\,\} \cup \{\,\boldsymbol{x}\,;\, x_2 = 1,\ -1 \le x_1 \le 1\,\}$.

b) $\mathrm{cl}(S) = S$,

$\mathrm{int}(S) = \{\,\boldsymbol{x}\,;\, 0 < x_1 < 1,\ 0 < x_2 < 1 - x_1\,\}$,

$\partial S =$

$= \{\,(x_1, 0)\,;\, 0 \le x_1 \le 1\} \cup \{\,(x_1, 1-x_1)\,;\, 0 \le x_1 \le 1\} \cup \{\,(0, x_2)\,;\, 0 \le x_2 \le 1\}$.

c) $\mathrm{cl}(S) = \{\,\boldsymbol{x} \in \mathbf{R}^2\,;\, x_2 = 2x_1,\ 0 \le x_1 \le 2\,\}$,

$\mathrm{int}(S) = \emptyset$,

$\partial S = \mathrm{cl}(S)$.

d) The closure is S itself, the boundary is S and the interior is empty.

e) The closure is S itself, the boundary is S and the interior is empty.

4.5 b) and d) are convex.

4.10

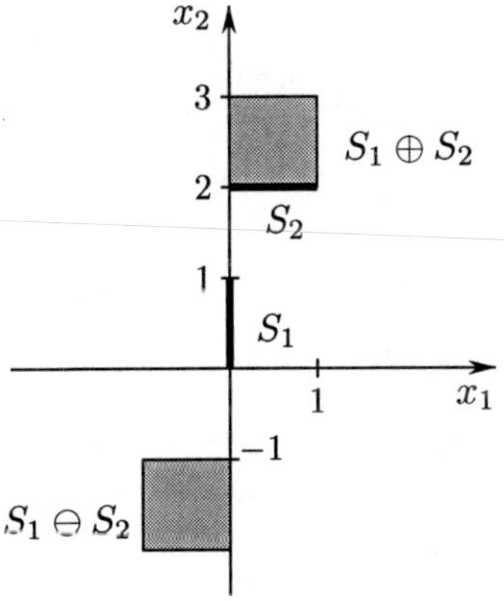

4.11

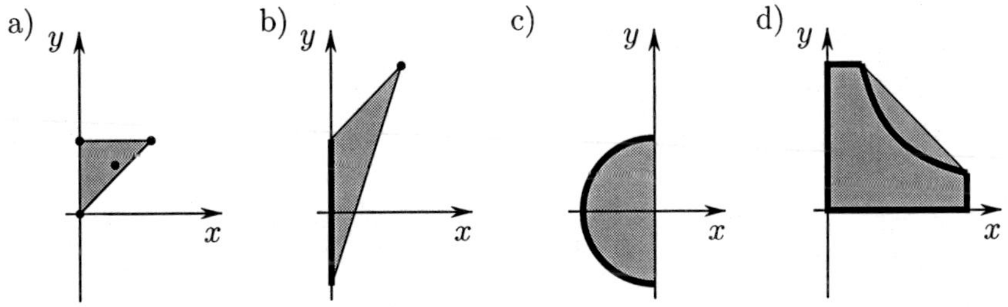

4.12 $\frac{6}{5} \leq a \leq \frac{26}{11}$.

4.13 No equality.

4.14 $0 \leq a < 1$.

4.15 a) Yes, b) no, c) no.

4.16 $-1 \leq a \leq \frac{17}{3}$.

4.17 $a = 4$.

4.21 a) $C^* = \{ (x_1, x_2) \,;\, x_2 \leq -x_1, x_1 \leq 0 \}$,

b) $C^* = \emptyset$,

c) $C^* = \{ \boldsymbol{y} \,;\, \boldsymbol{a}_i^T \boldsymbol{y} \leq 0 \text{ for all } i \}$, where $\boldsymbol{a}_i$ are the columns of $\boldsymbol{A}$.

4.23 $1 \leq a \leq 3$.

4.24 $9 \leq a \leq 12$.

4.27 $-8 \leq a \leq 24$.

4.28 No.

Chapter 5

5.1 Extreme points: $(2,0)$, $(1,1)$, $(0,1)$, $(0,0)$.

$(1, \frac{1}{2}) = \lambda_1(2,0) + \lambda_2(1,1) + \lambda_3(0,1) + \lambda_4(0,0)$,

$$\text{where } \begin{cases} \lambda_1 = \frac{1}{2} - \frac{1}{2}t \\ \lambda_2 = \quad\; t \\ \lambda_3 = \frac{1}{2} - \; t \\ \lambda_4 = \quad\; \frac{1}{2}t, \end{cases} \quad \text{for any } t \text{ such that } 0 \leq t \leq \tfrac{1}{2}.$$

5.2 a) $(1, a)$, where $1 \leq a \leq \sqrt{3}$,

b) $\alpha(0,1)$, $\alpha > 0$,

c) (d_1, d_2), where $d_1 \geq 0$, $d_2 \geq 0$,

d) no directions.

Extreme directions:

a) $(1,1)$ and $(1, \sqrt{3})$,

b) $\alpha(0,1)$, $\alpha > 0$,

c) $(d_1, 0)$ and $(0, d_2)$, where $d_1, d_2 \geq 0$.

5.4 Maximum $= 16$ at $(1,3)$.

5.5 $\boldsymbol{x} = (1\ \ 9\ \ 0)^T$ is a basic feasible solution with $f(\boldsymbol{x}) = -22$.
Minimum is $f(\boldsymbol{x}^*) = -24$ for $\boldsymbol{x}^* = (0\ \ 10\ \ 2)^T$.

5.6 Minimum is $-\frac{60}{7}$ for $x_2 = \frac{4}{7}$, $x_3 = \frac{12}{7}$, $x_1 = x_4 = x_5 = 0$.

5.7 No minimum. Arbitrary large negative values are possible.

5.8 a) The basic variables are x_2, x_6, x_7. The current point is $\boldsymbol{x} = (0, 4, 0, 0, 0, \frac{7}{2}, 5)$. The value is $-\frac{17}{2}$.

b) The basic variables are x_2, x_3, x_6. The new point is $\boldsymbol{x} = (0, \frac{3}{2}, \frac{5}{2}, 0, 0, 1, 0)$. The value is -16.

c) No. (In fact, minimum is $-\frac{101}{6}$ at $(\frac{1}{6}, \frac{7}{6}, \frac{8}{3}, 0, 0, 0, 0)$.)

5.9 Minimum is 45 for $x_1 = 0$, $x_2 = 0$, $x_3 = 5$.

5.10 Maximum is $\frac{13}{2}$ for $x_1 = x_2 = 1$, $x_3 = \frac{1}{2}$, $x_4 = 0$.

5.12 The set is empty.

5.13 a) The dual problem is $D:$

$$\begin{array}{ll} \text{Maximize} & y_1 - 3y_2 \\ \text{subject to} & 2y_1 + y_2 \le 3 \\ & -y_1 - 2y_2 \le -2 \\ & y_1 - y_2 \le 1 \\ & -y_1 + 2y_2 \le 1 \\ & y_2 \ge 0 \ \ (y_1 \text{ unrestricted}). \end{array}$$

b) Maximum is $\frac{1}{3}$ for $y_1 = \frac{4}{3}$, $y_2 = \frac{1}{3}$.

c) Minimum is $\frac{1}{3}$ for $\boldsymbol{x} = (t, \frac{2}{3} + t, \frac{5}{3} - t, 0)$, where $0 \le t \le \frac{5}{3}$.

5.14 Yes.

5.15 a) The dual problem is $D:$

$$\begin{array}{ll} \text{Minimize} & 3x_1 + 4x_2 \\ \text{subject to} & x_1 + 2x_2 + x_3 \ge 4 \\ & x_1 + 2x_2 - x_3 \ge -2 \\ & x_1 + x_2 \ge -1 \\ & \boldsymbol{x} \ge \boldsymbol{0}. \end{array}$$

b) An optimal solution to the dual problem is $\hat{\boldsymbol{x}} = (0, \frac{1}{2}, 3)$.

c) Yes.

5.16 $-2 \le c \le 1$.

5.17 The optimum production levels are $x_1 = 10$, $x_3 = 5$ and $x_2 = x_4 = 0$, which give the profit 105 €.

5.18 $b = \frac{17}{4}$.

5.19 a) Minimum $= -7$ at $\boldsymbol{x} = (0, \frac{5}{2}, \frac{1}{2}, \frac{5}{2}, 0)$.

b) $k \geq -11$.

5.20 2.033 € per litre.

5.21 a) $\boldsymbol{x} = (2, 1, 0, 1, 0)$, $\boldsymbol{s} = (0, 0)$, $f = -6$.

b) $\boldsymbol{x} = (1, 0, 0, 2, 0)$, $\boldsymbol{s} = (0, 1)$, $f = -7$. Unique.

c) Maximize $-7y_1 + 10y_2 + 2y_3$ when

$$\begin{aligned} -\ y_1 + \ y_2 \qquad\quad &\leq -1 \\ -2y_1 + 3y_2 + y_3 &\leq -1 \\ -3y_1 + 4y_2 \qquad\quad &\leq -1 \\ -3y_1 + 5y_2 + y_3 &\leq -3 \\ -5y_1 + 6y_2 \qquad\quad &\leq -3 \\ y_1 \geq 0, \quad y_2 \geq 0, \quad &y_3 \text{ free.} \end{aligned}$$

d) $\boldsymbol{y} = (1, 0, 0)$.

5.22 E.g. $(0.1, 5.8, 0)$ or $(0, 6, 0)$.

Chapter 6

6.2 b, c.

6.3 $\{\, x;\, x \leq -\sqrt{2}\,\} \cup \{\, x;\, x \geq \sqrt{2}\,\}$.

6.9 a, b, c are convex.

6.10 $a \geq 5$.

6.11 a) Convex, b) neither convex nor concave.

6.15 b) $\{\text{ subgradients at } \bar{\boldsymbol{u}}\,\} = H(\{\, \boldsymbol{A}\boldsymbol{x}_i - \boldsymbol{b},\ i = 1, \dots, k\,\})$, where $\boldsymbol{x}_1, \dots, \boldsymbol{x}_k$ are those extreme points of X for which $g(\bar{\boldsymbol{u}}) = \min \boldsymbol{c}^T\boldsymbol{x} + \bar{\boldsymbol{u}}(\boldsymbol{A}\boldsymbol{x} - \boldsymbol{b})$.

6.16 $g(\boldsymbol{u}) = \min\bigl(-3u_1 - 4u_2,\ -2 - u_1 - 6u_2,\ -\frac{13}{2} + \frac{1}{2}u_1 - 3u_2,\ -\frac{1}{2} - \frac{3}{2}u_1 - u_2\bigr)$.

6.19 $(\frac{1}{2}, 0)$.

6.20 b) $\begin{cases} a = 5/2 \\ b = e^{-16} \end{cases}$ and $\begin{cases} a = 3/2 \\ b = e^{-12}. \end{cases}$

6.21 a) $\boldsymbol{A}^T\boldsymbol{A}\boldsymbol{x} = \boldsymbol{A}^T\boldsymbol{b}$.

b) If $\boldsymbol{A}^T\boldsymbol{A}$ is invertible: $\boldsymbol{x} = (\boldsymbol{A}^T\boldsymbol{A})^{-1}\boldsymbol{A}^T\boldsymbol{b}$.

6.24 a) Not convex, b) convex.

6.27 Convex.

6.28 Not convex.

6.30 $a \leq 0,\ b \leq 0$, or $a \leq 0,\ a + b \geq 1$, or $b \leq 0,\ a + b \geq 1$.

6.32 b) Maximum is $\sqrt{207}$, at $(1, 2)$.

6.34 b) Convex, c) convex.

Chapter 7

7.1 See the solution, page 396.

7.2 Minimum at $(4, 3)$.

7.3 $-\frac{5}{3} \leq a \leq -1$.

7.4 b) $(0, 3)$,

c) KKT points: $(\pm\sqrt{\frac{3}{2}}, \frac{3}{2})$, $(\pm\sqrt{6}, -\sqrt{3})$.
Minimum value $-6\sqrt{3}$ at $(\pm\sqrt{6}, -\sqrt{3})$.

7.5 $c \leq -5$.

7.6 Minimum at $\pm(\sqrt{2}, -\sqrt{2})$.

7.7 $a + b + c + 2\sqrt{ab} + 2\sqrt{ac} + 2\sqrt{bc}$.

7.8 The maximum is $17^{3/4}$, the minimum is $-17^{3/4}$.

7.9 Minimum at $(-1, 0, -2)$.

7.10 $\bar{x}_j = \dfrac{b\sqrt{c_j a_j}}{a_j \sum \sqrt{c_k a_k}}$, $j = 1, \ldots, n$.

7.11 Minimum at $(1, 0)$.

7.12 Maximum at $(\frac{2}{3}, \frac{14}{9})$.

7.13 Minimum at $(\frac{2}{7}, \frac{2}{7}, -\frac{3}{7})$.

7.15 b) The KKT points are $(0,0)$ and $(-1,1)$.

c) No conclusion about $(0,0)$, local minimum at $(-1,1)$. No conclusion about a global minimum.

7.16 a) Only one KKT point: $(-5,-4,1)$. — No conclusion from 2nd order conditions.

7.17 Minimum equal to $\frac{3}{2}$ at $(-1,2,1)$.

7.18 $(0,0)$ and $(-1,-1)$. Note that since we seek a *global* minimum point, it is not sufficient to investigate the stationary points.

7.21 Minimum is $\frac{2}{3}$, at $\boldsymbol{x} = (\frac{1}{3}, \frac{1}{6}, \frac{1}{2})$.

7.22 $(\frac{1}{3}, \frac{2}{3}, 0)$.

7.23 Minimum at $(0,-2,-2)$.

7.24 $\sqrt{\frac{3-2\sqrt{2}}{2}} = 1 - \frac{1}{\sqrt{2}}$ for $\left(\frac{2+\sqrt{2}}{4}, \frac{2+\sqrt{2}}{4}, \frac{2+2\sqrt{2}}{4}\right)$.

7.25 Maximum at $(8,5)$.

7.26 $\sqrt{5}$ (at $(1,1)$).

7.27 Minimum is $3^{5/3}$ at $(\sqrt[3]{3}, \sqrt[3]{3}, \sqrt[3]{3})$.

7.28 1.

7.29 a) No CQ points. KKT points are $(0,0,0)$ and $\pm(\sqrt{2}, -\sqrt{2}, 0)$. Minimum at $\pm(\sqrt{2}, -\sqrt{2}, 0)$.

b) No minimum exists.

7.31 b) $\bar{\boldsymbol{x}}$ is *not* a KKT point.

7.32 Minimum at $(-\frac{3}{2}, 0, -\frac{3}{2})$.

7.33 Minimum -2 at $(0,1)$.

Chapter 8

8.1 a) $\theta(u) = -\frac{3}{8}u^2 + u, \quad u \geq 0.$

b) Maximum is $\frac{2}{3}$ at $u = \frac{4}{3}$.

c) Minimum is $\frac{2}{3}$ at $x_1 = \frac{2}{3}$, $x_2 = \frac{1}{3}$.

8.2 $\theta(u_1, u_2) = -\frac{3}{8}u_1^2 - \frac{1}{8}u_2^2 - \frac{1}{4}u_1u_2 + u_1$ when $u_1 \geq 0,\ u_2 \geq 0.$

8.3 a) $$\theta(u) = \begin{cases} -\dfrac{1}{4u} - u & \text{for } u > 0, \\ -\infty & \text{for } u = 0 \end{cases}$$

b) In both cases the optimal value is -1.

c) $$\theta(v) = \begin{cases} -\dfrac{1}{4v} - v & \text{for } v > 0, \\ -\infty & \text{for } v \leq 0 \end{cases}$$ optimal value is -1.

d) $\theta = -1$.

8.4 a) $$\theta(u_1, u_2) = \begin{cases} -4u_1 - u_2 & \text{if } u_1^2 - u_2^2 \geq \frac{1}{4} \\ -\infty & \text{if } u_1^2 - u_2^2 < \frac{1}{4}. \end{cases}$$

b) Maximum is -2 for $u_1 = \frac{1}{2}$, $u_2 = 0$.

8.5 b) $\theta(\boldsymbol{u}) = -\frac{1}{2}(u_1^2 + 6u_1u_2 + 25u_2^2) - (11u_1 + 25u_2) - \frac{29}{8}.$

c) Maximum is $-\frac{29}{8}$ at $(0, 0)$.

d) Minimum is $-\frac{29}{8}$ at $(-\frac{14}{4}, \frac{5}{4})$.

8.6 $$\theta(u) = \begin{cases} -4u & \text{if } u \geq \frac{1}{2} \\ -\infty & \text{otherwise.} \end{cases}$$ Saddle points $(\pm\sqrt{2}, \mp\sqrt{2};\ 0, \frac{1}{2})$.

8.7 $$\theta(u) = \begin{cases} 32 + 2u - u^2 & \text{if } 0 \leq u \leq \sqrt{2} \\ -\infty & \text{if } u > \sqrt{2}. \end{cases}$$

Maximum 33 at $u = 1$. Minimum at $x = -1$, $y = 1$.

8.8 $\theta(u) = \frac{3}{2} + \frac{1}{2}\ln 2 + \frac{1}{2}\ln u - u$. Minimum value 1 at $(0, 1)$.

Chapter 9

9.1 a) Minimum at $(0, 1)$.

b) $F(x, \mu) = x_1^2 + x_2^2 + \mu\Big(\max(0,\ 2x_1+x_2-2)^2 + \max(0,\ -x_2+1)^2\Big)$.

c) See the solution, page 401.

d) See the solution.

9.2 a) Minimum for $(\frac{1}{2}, \frac{1}{2})$.

c) For large μ the optimal solution is $\left(\dfrac{1}{\sqrt{1+\frac{3}{2\mu}}+1}, \dfrac{1}{\sqrt{1+\frac{3}{2\mu}}+1}\right)$.

The limit is $(\frac{1}{2}, \frac{1}{2})$.

9.3 a) Choose $q_\mu(x_1, x_2) = e^{x_1} + x_1^2 + x_1x_2 + \mu(\frac{1}{2}x_1 + x_2 - 1)^2$. After three iterations the obtained point is $(-1.419,\ 1.887)$.

b) $v \approx 1.419$.

9.4 b) $q_\varepsilon(x_1, x_2) = \dfrac{1}{3}(x_1+1)^3 + x_2 + \varepsilon\Big(\dfrac{1}{x_1-1} + \dfrac{1}{x_2}\Big)$.

c) The solution to the auxiliary problem is $(\sqrt{1+\sqrt{\varepsilon}}, \sqrt{\varepsilon})$. Exact minimum point is $(1, 0)$.

9.5 Choose $q_\varepsilon(x_1, x_2) = (x_1-5)^2 + (x_2-3)^2 + \varepsilon\left(\dfrac{1}{3-x_1-x_2} + \dfrac{1}{4+x_1-2x_2}\right)$.

Two iterations with the conjugate gradient method and with $\varepsilon = 1$ results in $(2.31, 0.26)$. — Exact minimum is at $(2.5, 0.5)$.

Chapter 10

10.1 a) Closed, b) not closed.

10.2 a) Yes, b) yes.

10.3 Yes.

10.4 Both are closed.

10.5 Not closed

Solutions to selected exercises

Chapter 1

1.4 By the chain rule $F'(\lambda) = \nabla f(\boldsymbol{x}+\lambda\boldsymbol{d})^T\,\boldsymbol{d}$. Hence

$$F'(\lambda^*) = 0 \quad\Longleftrightarrow\quad \nabla f(\boldsymbol{x}^*)^T\,\boldsymbol{d} = 0 \quad\Longleftrightarrow\quad \nabla f(\boldsymbol{x}^*) \perp \boldsymbol{d}.$$

1.9 The stationary points are given by the system of equations

$$\begin{cases} f_1'(\boldsymbol{x}) = 6x_1^2 - 6x_1 - 6x_2(x_1 - x_2 - 1) - 6x_1x_2 = 0 \\ f_2'(\boldsymbol{x}) = -6x_1(x_1 - x_2 - 1) + 6x_1x_2 = 0. \end{cases}$$

The last equation can be factorized as $-6x_1(x_1 - 2x_2 - 1) = 0$, hence splits in the two cases $x_1 = 0$ and $x_1 - 2x_2 - 1 = 0$. In the first case the first equation yields $x_2 = 0$ or $x_2 = -1$. In the second case the system simplifies to

$$\begin{cases} x_1(x_1 - 1 - 2x_2) + x_2^2 + x_2 = 0 \\ x_1 - 2x_2 - 1 = 0 \end{cases} \quad\Longleftrightarrow\quad \begin{cases} x_2^2 + x_2 = 0 \\ x_1 = 2x_2 + 1 \end{cases} \quad\Longleftrightarrow$$

$$\Longleftrightarrow \quad \begin{cases} x_2 = 0 \text{ or } -1 \\ x_1 = 2x_2 + 1. \end{cases}$$

Thus we have obtained the following stationary points:

$$(0,0), \quad (0,-1), \quad (1,0), \quad (-1,-1).$$

To determine their character we compute the Hessians

$$\boldsymbol{H}(\boldsymbol{x}) = \begin{pmatrix} f_{11}''(\boldsymbol{x}) & f_{21}''(\boldsymbol{x}) \\ f_{12}''(\boldsymbol{x}) & f_{22}''(\boldsymbol{x}) \end{pmatrix}$$

at these points:

$$\boldsymbol{H}(0,0) = \begin{pmatrix} -6 & 6 \\ 6 & 0 \end{pmatrix} \qquad \boldsymbol{H}(0,-1) = \begin{pmatrix} 6 & -6 \\ -6 & 0 \end{pmatrix}$$

$$\boldsymbol{H}(1,0) = \begin{pmatrix} 6 & -6 \\ -6 & 12 \end{pmatrix} \qquad \boldsymbol{H}(-1,-1) = \begin{pmatrix} -6 & 6 \\ 6 & -12 \end{pmatrix}.$$

We find that the matrices $\boldsymbol{H}(0,0)$ and $\boldsymbol{H}(0,-1)$ are indefinite, corresponding to saddle points and hence not local extreme points. At the point $(1,0)$ the Hessian is positive definite, and this point is a local minimum. Finally, $(-1,-1)$ is a local maximum, since the Hessian is negative definite.

Answer: Stationary points are $(0,0)$, $(0,-1)$, $(-1,-1)$, $(1,0)$. The first two ones are saddle points, the third one is a local maximum and the fourth one a local minimum.

Chapter 2

2.1 a)

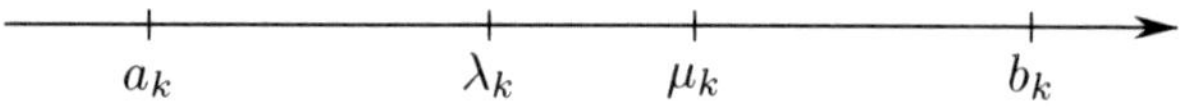

The formulas for choosing λ_k and μ_k are

$$\lambda_k = a_k + (1-\alpha)(b_k - a_k)$$

$$\mu_k = a_k + \alpha(b_k - a_k)$$

where $\alpha = \dfrac{\sqrt{5}-1}{2} \approx 0.618$. At each iteration step we evaluate the objective function at these points, and use the result to decide on a new interval of uncertainty. If λ_k is chosen as the left endpoint of this interval, we know without computing that μ_k becomes λ_{k+1}. Hence we need only compute μ_{k+1} and $f(\mu_{k+1})$. (Correspondingly if μ_k is chosen.) — The results of the computations are shown in the following table.

k	a_k	b_k	λ_k	μ_k	$F(\lambda_k)$	$F(\mu_k)$	Comments
1	0	2	**0.764**	**1.236**	**0.644**	**0.641**	Since $F(\lambda_1) > F(\mu_1)$ the new interval of uncertainty is $[\lambda_1, b_1]$.
2	0.764	2	1.236	**1.528**	0.641	**0.668**	New interval of uncertainty is $[a_2, \mu_2]$. New evaluations are printed in boldface.
3	0.764	1.528	**1.056**	1.236	**0.633**	0.641	New interval of uncertainty is $[a_3, \mu_3]$.
4	0.764	1.236	**0.944**	1.056	**0.633**	0.633	Of course, when $F(\lambda_k) \approx F(\mu_k)$ are so close any choice will do.
5	0.944	1.236					

Our final interval of uncertainty is [0.944, 1.236]. We can take the midpoint as an approximation of the minimum point, $\bar{\lambda} = 1.09$, with an error estimate of ± 0.15. We find that $F(\bar{\lambda}) = 0.633$, which becomes our estimate of the minimum value.

b) In Fibonacci search the formulas for λ_k and μ_k are

$$\lambda_k = a_k + (1-\alpha_{n,k})(b_k - a_k)$$
$$\mu_k = a_k + \alpha_{n,k}(b_k - a_k)$$

where

$$\alpha_{n,k} = \frac{F_{n-k}}{F_{n-k+1}} \quad \text{and} \quad 1 - \alpha_{n,k} = \frac{F_{n-k-1}}{F_{n-k+1}}$$

are quotients of Fibonacci numbers. We are supposed to carry out 4 iterations, which means that we have to make 5 function evaluations, hence $n = 5$ above.

The Fibonacci numbers (starting with F_0) are $1, 1, 2, 3, 5, 8$, and the quotients needed in the computations are shown in the following table.

	$1-\alpha$	α
$k=1$	3/8	5/8
$k=2$	2/5	3/5
$k=3$	1/3	2/3
$k=4$	1/2	1/2

The results of the computations are as follows.

k	a_k	b_k	λ_k	μ_k	$F(\lambda_k)$	$F(\mu_k)$	Comments
1	0	2	**0.750**	**1.250**	**0.646**	**0.642**	Since $F(\lambda_1) > F(\mu_1)$ the new interval of uncertainty is $[\lambda_1, b_1]$.
2	0.750	2	1.250	**1.500**	0.642	**0.665**	New interval of uncertainty is $[a_2, \mu_2]$. New evaluations are printed in boldface.
3	0.750	1.500	**1.000**	1.250	**0.632**	0.642	New interval of uncertainty is $[a_3, \mu_3]$.
4	0.750	1.250	1.000	**1.000+ε**	0.632	**0.634**	To obtain a separation in the last iteration we add a small number ε to μ_k. Here we choose $\varepsilon = 0.10$.
5	0.750	1.100					This is $[a_4, \mu_4+\varepsilon]$.

The final interval of uncertainty is $[0.750, 1.100]$. For the minimum point we obtain $\bar{\lambda} = 0.925 \pm 0.175$, and $F_{\min} \approx 0.633$.

Answer: $F_{\min} = 0.633$. Uncertainty interval: a) $[0.944,\ 1.236]$, b) $[0.750,\ 1.100]$.

Note: The number of iterations in this example is perhaps unrealistically low. — In this case the exact minimum is easily found by elementary methods: $\bar{\lambda} = 1$.

2.7 The derivatives are

$$F'(\lambda) = 2\lambda e^{\lambda^2} - 2, \qquad F''(\lambda) = (2 + 4\lambda^2)e^{\lambda^2}.$$

The results of the successive iterations are

k	1	2	3	4	5	6	7
λ_k	0	1	0.7892	0.6767	0.6537	0.6529	0.6529

Answer: 0.6529.

Chapter 3

3.1 Let $F(\lambda) = q(\boldsymbol{x}_k + \lambda \boldsymbol{d}_k)$. Then by the chain rule

$$F'(\lambda) = \nabla q(\boldsymbol{x}_k + \lambda \boldsymbol{d}_k)^T \boldsymbol{d}_k,$$

where in this case $\nabla q(\boldsymbol{x}) = \boldsymbol{H}\boldsymbol{x} + \boldsymbol{c}$. (Cf. exercise 1.2.) Hence

$$\begin{aligned} F'(\lambda) = 0 \quad &\iff \quad \big(\boldsymbol{H}(\boldsymbol{x}_k + \lambda \boldsymbol{d}_k) + \boldsymbol{c}\big)^T \boldsymbol{d}_k = 0 \quad \iff \\ &\iff \quad \boldsymbol{d}_k^T \boldsymbol{H}\boldsymbol{x}_k + \lambda \boldsymbol{d}_k^T \boldsymbol{H}\boldsymbol{d}_k + \boldsymbol{d}_k^T \boldsymbol{c} = 0 \quad \iff \\ &\iff \quad \lambda = -\frac{\boldsymbol{d}_k^T(\boldsymbol{H}\boldsymbol{x}_k + \boldsymbol{c})}{\boldsymbol{d}_k^T \boldsymbol{H}\boldsymbol{d}_k}. \end{aligned}$$

Note that $\boldsymbol{d}_k^T \boldsymbol{H}\boldsymbol{d}_k > 0$ since $\boldsymbol{H}$ is positive definite.

Answer: $\lambda_k = -\dfrac{\boldsymbol{d}_k^T(\boldsymbol{H}\boldsymbol{x}_k + \boldsymbol{c})}{\boldsymbol{d}_k^T \boldsymbol{H}\boldsymbol{d}_k}$.

Note: The answer can also be written as $\lambda_k = -\dfrac{\boldsymbol{d}_k^T \nabla q(\boldsymbol{x}_k)}{\boldsymbol{d}_k^T \boldsymbol{H}\boldsymbol{d}_k}$.

Note: An alternative solution is given in the proof of lemma 3, page 71.

3.2 a) E.g. as in exercise 3.1 we have

$$\lambda_k = -\frac{\boldsymbol{d}_k^T(\boldsymbol{H}\boldsymbol{x}_k + \boldsymbol{c})}{\boldsymbol{d}_k^T \boldsymbol{H}\boldsymbol{d}_k}.$$

In the steepest descent case $\boldsymbol{d}_k = -\nabla q(\boldsymbol{x}_k) = -\boldsymbol{g}_k$, and $\boldsymbol{g}_k = \boldsymbol{H}\boldsymbol{x}_k + \boldsymbol{c}$. Hence

$$\lambda_k = \frac{\boldsymbol{g}_k^T \boldsymbol{g}_k}{\boldsymbol{g}_k^T \boldsymbol{H}\boldsymbol{g}_k}.$$

b) Using the iteration formula in SD we obtain the following iteration formula for the gradient:

$$\begin{aligned} \boldsymbol{g}_{k+1} &= \boldsymbol{H}\boldsymbol{x}_{k+1} + \boldsymbol{c} = \boldsymbol{H}(\boldsymbol{x}_k - \lambda_k \boldsymbol{g}_k) + \boldsymbol{c} = \\ &= \boldsymbol{H}\boldsymbol{x}_k + \boldsymbol{c} - \lambda_k \boldsymbol{H}\boldsymbol{g}_k = \boldsymbol{g}_k - \lambda_k \boldsymbol{H}\boldsymbol{g}_k. \end{aligned}$$

Now, using the formula for λ_k above we get

$$\boldsymbol{g}_{k+1}^T \boldsymbol{g}_k = \boldsymbol{g}_k^T \boldsymbol{g}_k - \lambda_k \boldsymbol{g}_k^T \boldsymbol{H} \boldsymbol{g}_k = \boldsymbol{g}_k^T \boldsymbol{g}_k - \boldsymbol{g}_k^T \boldsymbol{g}_k = 0.$$

c) In the first iteration step $\boldsymbol{g}_2 \perp \boldsymbol{g}_1$. In the second step, where $\boldsymbol{g}_3 \perp \boldsymbol{g}_2$, in two dimensions it follows that $\boldsymbol{g}_3 \parallel \boldsymbol{g}_1$. In the next step $\boldsymbol{g}_4 \parallel \boldsymbol{g}_2$, etc. There are only two search directions, determined by the first chosen direction, $-\boldsymbol{g}_1 = -\boldsymbol{\nabla} q(\boldsymbol{x}_1)$.

3.11 At the point $\boldsymbol{x} = \boldsymbol{0}$ the Hessian is

$$\boldsymbol{H}(\boldsymbol{0}) = \begin{pmatrix} 0 & 1 \\ 1 & 2 \end{pmatrix}$$

which is indefinite. The search direction in Newton's method is

$$\boldsymbol{d} = -\boldsymbol{H}(\boldsymbol{0})^{-1} \boldsymbol{\nabla} f(\boldsymbol{0}) = \begin{pmatrix} -2 \\ 0 \end{pmatrix},$$

which is orthogonal to $\boldsymbol{\nabla} f(\boldsymbol{0}) = (0, 2)$. This implies that the slope in the $\pm\boldsymbol{d}$-directions is equal to 0.

The matrix

$$\boldsymbol{H}(\boldsymbol{0}) + \varepsilon \boldsymbol{I} = \begin{pmatrix} \varepsilon & 1 \\ 1 & 2+\varepsilon \end{pmatrix}$$

is positive definite if and only if $\varepsilon > 0$ and the determinant $\varepsilon^2 + 2\varepsilon - 1 > 0$, that is, if and only if $\varepsilon > \sqrt{2} - 1$. Choosing $\varepsilon = 1$ yields

$$\boldsymbol{d}_1 = -\left(\boldsymbol{H}(\boldsymbol{0}) + \boldsymbol{I}\right)^{-1} \boldsymbol{\nabla} f(\boldsymbol{0}) = \begin{pmatrix} 1 \\ -1 \end{pmatrix}.$$

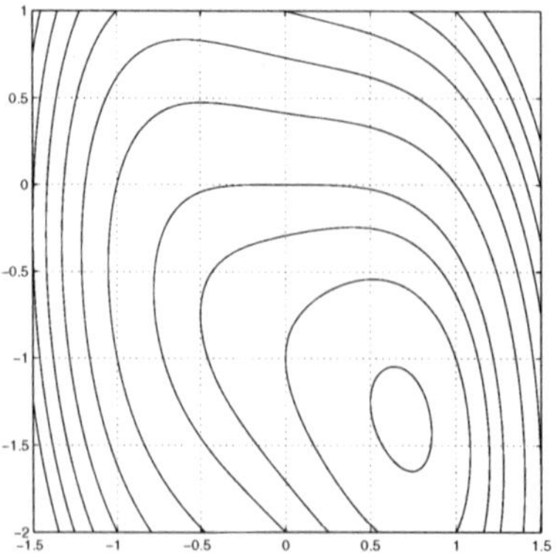

Exercise 3.11

Answer: $\varepsilon > \sqrt{2} - 1$. If $\varepsilon = 1$ then $\boldsymbol{d}_1 = (1, -1)$.

Remark. The level curves of f are shown in the figure on the previous page.

3.14 By definition the vectors are linearly independent if

$$\sum_1^n \lambda_k \boldsymbol{d}_k = \boldsymbol{0} \quad \Longrightarrow \quad \lambda_k = 0 \ \text{ for all } k.$$

Thus, assume that $\sum_1^n \lambda_k \boldsymbol{d}_k = \boldsymbol{0}$ and multiply by $\boldsymbol{d}_l^T \boldsymbol{H}$. This gives

$$\sum_{k=1}^n \lambda_k \boldsymbol{d}_l^T \boldsymbol{H} \boldsymbol{d}_k = 0.$$

By assumption $\boldsymbol{d}_l^T \boldsymbol{H} \boldsymbol{d}_k = 0$ if $k \neq l$. Consequently, all that remains of the sum is one term, i.e., $\lambda_l \boldsymbol{d}_l^T \boldsymbol{H} \boldsymbol{d}_l = 0$. Since $\boldsymbol{d}_l^T \boldsymbol{H} \boldsymbol{d}_l \neq 0$ ($\boldsymbol{H}$ is positive definite and $\boldsymbol{d}_l \neq \boldsymbol{0}$) we obtain $\lambda_l = 0$. This being true for all l, the assertion is proved.

3.17 a) With $f(\boldsymbol{x}) = x_1^2 + 4x_2^2 - 4x_1 - 8x_2$ we have $\boldsymbol{\nabla} f = \begin{pmatrix} 2x_1 - 4 \\ 8x_2 - 8 \end{pmatrix}$. We take $\boldsymbol{y}_1 = \boldsymbol{x}_1 = \boldsymbol{0}$ and choose

$$\boldsymbol{d}_1 = -\boldsymbol{\nabla} f(\boldsymbol{0}) = 4 \begin{pmatrix} 1 \\ 2 \end{pmatrix}.$$

Then $\boldsymbol{y}_2 = \boldsymbol{y}_1 + \lambda_1 \boldsymbol{d}_1$, where λ_1 solves $\min_\lambda f(\boldsymbol{y}_1 + \lambda \boldsymbol{d}_1)$. Since

$$f(\boldsymbol{y}_1 + \lambda \boldsymbol{d}_1) = f(4\lambda, 8\lambda) = 272\lambda^2 - 80\lambda$$

we find that $\lambda_1 = \frac{5}{34}$ and $\boldsymbol{y}_2 = \frac{10}{17} \begin{pmatrix} 1 \\ 2 \end{pmatrix}$. In the next iteration step, to compute $\boldsymbol{y}_3$, we must first find the new direction $\boldsymbol{d}_2$. It is given by

$$\boldsymbol{d}_2 = -\boldsymbol{\nabla} f(\boldsymbol{y}_2) + \beta_1 \boldsymbol{d}_1.$$

Here

$$\beta_1 = \frac{\|\boldsymbol{\nabla} f(\boldsymbol{y}_2)\|^2}{\|\boldsymbol{\nabla} f(\boldsymbol{y}_1)\|^2} = \frac{\|\frac{24}{17}(-2, 1)\|^2}{\|4\,(1, 2)\|^2} = \frac{36}{289},$$

thus

$$\boldsymbol{d}_2 = -\frac{24}{17} \begin{pmatrix} -2 \\ 1 \end{pmatrix} + \frac{36 \cdot 4}{289} \begin{pmatrix} 1 \\ 2 \end{pmatrix} = \frac{120}{289} \begin{pmatrix} 8 \\ -1 \end{pmatrix}.$$

Now, $\boldsymbol{y}_3 = \boldsymbol{y}_2 + \lambda_2 \boldsymbol{d}_2$, where λ_2 solves $\min_\lambda f(\boldsymbol{y}_2 + \lambda \boldsymbol{d}_2)$. Inserting

$$\boldsymbol{y}_2 + \lambda \boldsymbol{d}_2 = \frac{10}{17}\begin{pmatrix} 1 \\ 2 \end{pmatrix} + \lambda \frac{120}{289}\begin{pmatrix} 8 \\ -1 \end{pmatrix}$$

in f and minimizing (or using ready-made formulas, exercise 3.1) we find $\lambda_2 = \frac{17}{40}$ and

$$\boldsymbol{y}_3 = \begin{pmatrix} 2 \\ 1 \end{pmatrix}.$$

Finally let $\boldsymbol{x}_2 = \boldsymbol{y}_3 = (2, 1)$. We have found the exact minimum point in one full iteration. (Are you surprised?)

Answer: The function has minimum at $\boldsymbol{x}_2 = (2, 1)$.

b) We verify that $\boldsymbol{d}_1$ and $\boldsymbol{d}_2$ are $\boldsymbol{H}$-conjugate by computing $\boldsymbol{d}_1^T \boldsymbol{H} \boldsymbol{d}_2$, which turns out to be equal to 0, as it should. To check that $\boldsymbol{d}_1$ is a descent direction at $\boldsymbol{y}_1$ we compute

$$\boldsymbol{\nabla} f(\boldsymbol{y}_1)^T \boldsymbol{d}_1 = (-4 \quad -8)\, 4 \begin{pmatrix} 1 \\ 2 \end{pmatrix} = -80,$$

which is less than zero. Similarly we find that $\boldsymbol{\nabla} f(\boldsymbol{y}_2)^T \boldsymbol{d}_2 < 0$.

3.18 a) With $f(\boldsymbol{x}) = x_1^2 + 4x_2^2 - 4x_1 - 8x_2$ we have $\boldsymbol{\nabla} f = \begin{pmatrix} 2x_1 - 4 \\ 8x_2 - 8 \end{pmatrix}$. We take $\boldsymbol{y}_1 = \boldsymbol{x}_1 = \boldsymbol{0}$ and choose

$$\boldsymbol{d}_1 = -\boldsymbol{\nabla} f(\boldsymbol{0}) = 4 \begin{pmatrix} 1 \\ 2 \end{pmatrix}.$$

Then $\boldsymbol{y}_2 = \boldsymbol{y}_1 + \lambda_1 \boldsymbol{d}_1$, where λ_1 solves $\min_\lambda f(\boldsymbol{y}_1 + \lambda \boldsymbol{d}_1)$. Since

$$f(\boldsymbol{y}_1 + \lambda \boldsymbol{d}_1) = f(4\lambda, 8\lambda) = 272\lambda^2 - 80\lambda$$

we find that $\lambda_1 = \frac{5}{34}$ and

$$\boldsymbol{y}_2 = \frac{10}{17}\begin{pmatrix} 1 \\ 2 \end{pmatrix}.$$

In the next iteration, to compute $\boldsymbol{y}_3$, we must first find the new direction $\boldsymbol{d}_2$. It is given by

$$\boldsymbol{d}_2 = -\boldsymbol{D}_2 \boldsymbol{\nabla} f(\boldsymbol{y}_2).$$

Here, since $\boldsymbol{D}_1 = \boldsymbol{I}$,

$$\boldsymbol{D}_2 = \boldsymbol{I} + \frac{\boldsymbol{p}_1\boldsymbol{p}_1^T}{\boldsymbol{p}_1^T\boldsymbol{q}_1} - \frac{\boldsymbol{q}_1\boldsymbol{q}_1^T}{\boldsymbol{q}_1^T\boldsymbol{q}_1},$$

where

$$\boldsymbol{p}_1 = \lambda_1\boldsymbol{d}_1 = \frac{10}{17}\begin{pmatrix} 1 \\ 2 \end{pmatrix}, \qquad \boldsymbol{q}_1 = \nabla f(\boldsymbol{y}_2) - \nabla f(\boldsymbol{y}_1) = \ldots = \frac{20}{17}\begin{pmatrix} 1 \\ 8 \end{pmatrix}.$$

Hence,

$$\boldsymbol{D}_2 = \boldsymbol{I} + \frac{1}{34}\begin{pmatrix} 1 & 2 \\ 2 & 4 \end{pmatrix} - \frac{1}{65}\begin{pmatrix} 1 & 8 \\ 8 & 64 \end{pmatrix} = \frac{1}{34 \cdot 65}\begin{pmatrix} 2241 & -142 \\ -142 & 294 \end{pmatrix}.$$

This gives

$$\boldsymbol{d}_2 = -\frac{1}{34 \cdot 65}\begin{pmatrix} 2241 & -142 \\ -142 & 294 \end{pmatrix}\frac{24}{17}\begin{pmatrix} -2 \\ 1 \end{pmatrix} = \frac{24}{65}\begin{pmatrix} 8 \\ -1 \end{pmatrix}.$$

We now have $\boldsymbol{y}_3 = \boldsymbol{y}_2 + \lambda_2\boldsymbol{d}_2$, where λ_2 solves $\min_\lambda f(\boldsymbol{y}_2 + \lambda\boldsymbol{d}_2)$. Inserting (we put $\mu = \lambda \cdot \dfrac{24}{65}$)

$$\boldsymbol{y}_2 + \lambda\boldsymbol{d}_2 = \frac{10}{17}\begin{pmatrix} 1 \\ 2 \end{pmatrix} + \mu\begin{pmatrix} 8 \\ -1 \end{pmatrix}$$

in f and minimizing (or using ready-made formulas, exercise 3.1) we find $\lambda_2 = \dfrac{65}{8 \cdot 17}$ and

$$\boldsymbol{y}_3 = \begin{pmatrix} 2 \\ 1 \end{pmatrix}.$$

Finally let $\boldsymbol{x}_2 = \boldsymbol{y}_3 = (2, 1)$. We have found the exact minimum point in one full iteration.

Answer: The function has minimum at $\boldsymbol{x}_2 = (2, 1)$.

b) See the solution to exercise 3.17 b).

3.19 We provide some computational results ($\boldsymbol{g} = \nabla f$):

$$\boldsymbol{g}_1 = \boldsymbol{c}, \quad \boldsymbol{d}_1 = -\boldsymbol{c}, \quad \lambda_1 = \frac{1}{2}, \quad \boldsymbol{x}_2 = -\lambda_1\boldsymbol{c} = \begin{pmatrix} -1 \\ -\frac{1}{2} \\ -\frac{1}{2} \\ -1 \end{pmatrix},$$

$$\boldsymbol{g}_2 = \begin{pmatrix} -\frac{1}{2} \\ 1 \\ 1 \\ -\frac{1}{2} \end{pmatrix}, \quad \beta_2 = \frac{1}{4}, \quad \boldsymbol{d}_2 = \begin{pmatrix} 0 \\ -\frac{5}{4} \\ -\frac{5}{4} \\ 0 \end{pmatrix}, \quad \lambda_2 = \frac{2}{5}, \quad \boldsymbol{x}_3 = \begin{pmatrix} -1 \\ -1 \\ -1 \\ -1 \end{pmatrix},$$

$$\boldsymbol{x}_3 = \bar{\boldsymbol{x}} = -\boldsymbol{H}^{-1}\boldsymbol{c}.$$

3.24 By the recursion formula and the fact that $f(x^*) = x^*$ we have

$$x^{(k+1)} - x^* = f(x^{(k)}) - f(x^*) = f'(\xi_k)\,(x^{(k)} - x^*).$$

Here ξ_k is some number between $x^{(k)}$ and x^*. (We used the mean value theorem of differential calculus.) Taking norms we obtain

$$|x^{(k+1)} - x^*| = |f'(\xi_k)|\,|x^{(k)} - x^*|.$$

By assumption $|f'(x^*)| < 1$, which by continuity implies that there is a neighbourhood I of x^* in which $|f'(x)| < c$ for some constant $c < 1$. Assuming that $x^{(1)}$ and $x^{(2)}$ both belong to I it follows from the inequality above that all $x^{(k)}$ belong to I, and that

$$|x^{(k+1)} - x^*| < c\,|x^{(k)} - x^*|, \qquad k = 1, 2, \ldots .$$

This shows that the convergence is (at least) linear, with convergence ratio less than or equal to $c \approx |f'(x^*)|$.

Chapter 4

4.3 This statement is not as obvious as it may seem, since taking the closure of a set S might introduce some boundary point of the closure which is not a boundary point of S. We have to prove that this is not the case.

Let $\bar{\boldsymbol{x}} \in \partial(\mathrm{cl}(S))$. We must prove that $\bar{\boldsymbol{x}} \in \mathrm{cl}(S)$. Assume otherwise, that is, assume that $\bar{\boldsymbol{x}}$ is an outer point of S. Then there is a ball

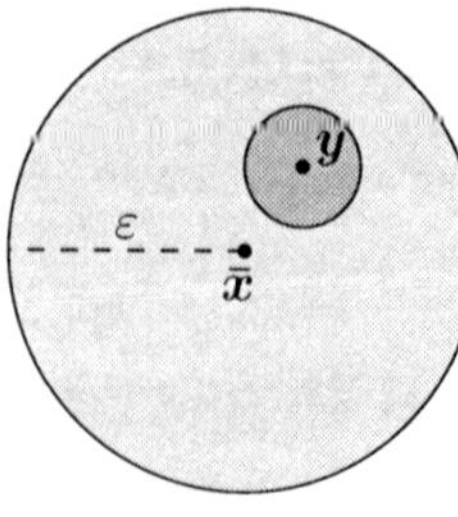

$\|\boldsymbol{x}-\bar{\boldsymbol{x}}\| < \varepsilon$ with center $\bar{\boldsymbol{x}}$ in which no point belongs to S. Any $\boldsymbol{y}$ in this ball has of course the property that there is a ball with center $\boldsymbol{y}$ with no point belonging to S. Hence any such $\boldsymbol{y}$ is an outer point of S and consequently does not belong to $\mathrm{cl}(S)$. This is a contradiction to the assumption that $\bar{\boldsymbol{x}} \in \partial(\mathrm{cl}(S))$.

Thus the statement is proved.

4.6 Assume that $\boldsymbol{y}_1 \in \boldsymbol{A}S$ and $\boldsymbol{y}_2 \in \boldsymbol{A}S$, i.e.,

$$\boldsymbol{y}_1 = \boldsymbol{A}\boldsymbol{x}_1, \quad \boldsymbol{y}_2 = \boldsymbol{A}\boldsymbol{x}_2, \qquad \text{where} \quad \boldsymbol{x}_1, \boldsymbol{x}_2 \in S.$$

Let $0 \le \lambda \le 1$. Then

$$\lambda \boldsymbol{y}_1 + (1-\lambda)\boldsymbol{y}_2 = \boldsymbol{A}(\lambda \boldsymbol{x}_1 + (1-\lambda)\boldsymbol{x}_2).$$

Since S is convex the element $\lambda \boldsymbol{x}_1 + (1-\lambda)\boldsymbol{x}_2$ belongs to S. Thus,

$$\lambda \boldsymbol{y}_1 + (1-\lambda)\boldsymbol{y}_2 \in \boldsymbol{A}S.$$

We have hereby shown that $\boldsymbol{A}S$ is convex.

4.14 That $\boldsymbol{A}\boldsymbol{x} \le 0 \Rightarrow \boldsymbol{c}^T\boldsymbol{x} \le 0$ is equivalent to the system $\boldsymbol{A}\boldsymbol{x} \le 0$, $\boldsymbol{c}^T\boldsymbol{x} > 0$ having no solution. According to Farkas' theorem this is equivalent to the existence of a solution to $\boldsymbol{A}^T\boldsymbol{y} = \boldsymbol{c}$, $\boldsymbol{y} \ge \boldsymbol{0}$. In our case $\boldsymbol{A}^T\boldsymbol{y} = \boldsymbol{c}$ means

$$\begin{cases} y_1 + 3y_2 - 3y_3 = -1 \\ y_1 + 4y_2 - 4y_3 = -2 \\ 2y_1 - y_2 + ay_3 = 4 \end{cases} \iff \begin{cases} y_1 + 3y_2 - 3y_3 = -1 \\ y_2 - y_3 = -1 \\ (a-1)y_3 = -1. \end{cases}$$

If $a = 1$ there is no solution. Thus in this case the given implication is false. If $a \ne 1$ we get the unique solution

$$\boldsymbol{y} = \left(2, \frac{-a}{a-1}, \frac{-1}{a-1}\right).$$

The implication is true exactly when $\boldsymbol{y} \ge \boldsymbol{0}$, i.e., when all coordinates are ≥ 0. The third coordinate requires that $a < 1$, and then the second one gives $a \ge 0$.

Answer: $0 \le a < 1$.

4.18 We have

$$\begin{cases} \boldsymbol{A}\boldsymbol{x} \le \boldsymbol{0} \\ \boldsymbol{x} \le \boldsymbol{0} \end{cases} \iff \mathcal{A}\boldsymbol{x} \le \boldsymbol{0} \qquad \text{where } \mathcal{A} = \begin{pmatrix} \boldsymbol{A} \\ \boldsymbol{I} \end{pmatrix}.$$

Let the first alternative in Farkas' theorem be

$$\mathcal{A}\boldsymbol{x} \leq 0, \quad \boldsymbol{c}^T\boldsymbol{x} > 0.$$

Then the second alternative becomes

$$\mathcal{A}^T \begin{pmatrix} \boldsymbol{y} \\ \boldsymbol{z} \end{pmatrix} = \boldsymbol{c}, \quad \begin{pmatrix} \boldsymbol{y} \\ \boldsymbol{z} \end{pmatrix} \geq \boldsymbol{0}.$$

In this case $\mathcal{A}^T = (\boldsymbol{A}^T \ \ \boldsymbol{I})$, so that

$$\mathcal{A}^T \begin{pmatrix} \boldsymbol{y} \\ \boldsymbol{z} \end{pmatrix} = \boldsymbol{c} \quad \Longleftrightarrow \quad \boldsymbol{A}^T\boldsymbol{y} + \boldsymbol{z} = \boldsymbol{c} \quad \Longleftrightarrow \quad \boldsymbol{A}^T\boldsymbol{y} = \boldsymbol{c} - \boldsymbol{z}.$$

This system of equations is solvable with solutions $\boldsymbol{y}, \boldsymbol{z} \geq \boldsymbol{0}$ if and only if the system of inequalities

$$\boldsymbol{A}^T\boldsymbol{y} \leq \boldsymbol{c}, \quad \boldsymbol{y} \geq \boldsymbol{0}$$

is solvable. This proves the assertion.

4.20 First, assume that C is a convex cone. Then

$$\begin{cases} \boldsymbol{x}_1, \boldsymbol{x}_2 \in C \\ \lambda_1, \lambda_2 \geq 0 \end{cases} \overset{C \text{ convex}}{\Longrightarrow} \frac{\lambda_1}{\lambda_1+\lambda_2}\boldsymbol{x}_1 + \frac{\lambda_2}{\lambda_1+\lambda_2}\boldsymbol{x}_2 \in C \quad \Longrightarrow$$

$$\overset{C \text{ cone}}{\Longrightarrow} \quad (\lambda_1+\lambda_2)\left(\frac{\lambda_1}{\lambda_1+\lambda_2}\boldsymbol{x}_1 + \frac{\lambda_2}{\lambda_1+\lambda_2}\boldsymbol{x}_2\right) = \lambda_1\boldsymbol{x}_1 + \lambda_2\boldsymbol{x}_2 \in C.$$

(This reasoning is not valid in the case $\lambda_1 = \lambda_2 = 0$, but then the result is obvious.)

For the converse, assume that

$$\begin{cases} \boldsymbol{x}_1, \boldsymbol{x}_2 \in C \\ \lambda_1, \lambda_2 \geq 0 \end{cases} \quad \Longrightarrow \quad \lambda_1\boldsymbol{x}_1 + \lambda_2\boldsymbol{x}_2 \in C.$$

Then C is a cone (take $\lambda_2 = 0$), and C is convex (assume $\lambda_1+\lambda_2 = 1$).

Chapter 5

5.6 Since x_1, x_5 are the basic variables we must begin by expressing everything in x_2, x_3, x_4, using Gaussian elimination.

* x_1	x_2	x_3	x_4	* x_5	
-1	-6	7	-1	-5	0
5	-4	13	-2	1	20
1	-1	5	-1	1	8

Produce $\begin{pmatrix} 0 \\ 1 \end{pmatrix}$ in the x_5-column

*				*	
4	-11	32	-6	0	40
4	-3	8	-1	0	12
1	-1	5	-1	1	8

Produce $\begin{pmatrix} 1 \\ 0 \end{pmatrix}$ in the x_1-column

*		↓		*			
0	-8	24	-5	0	28		
1	$-\frac{3}{4}$	(2)	$-\frac{1}{4}$	0	3	$\frac{3}{2}$	←
0	$-\frac{1}{4}$	3	$-\frac{3}{4}$	1	5	$\frac{5}{3}$	

Here the simplex algorithm starts. Basic variables are marked with a $*$. 2 is the pivot element. New basic variables are x_3 and x_5

	↓	*		*			
-12	1	0	-2	0	-8		
$\frac{1}{2}$	$-\frac{3}{8}$	1	$-\frac{1}{8}$	0	$\frac{3}{2}$	< 0 no restriction	
$-\frac{3}{2}$	($\frac{7}{8}$)	0	$-\frac{3}{8}$	1	$\frac{1}{2}$	$\frac{1/2}{7/8} = \frac{4}{7}$	←

Pivot element $\frac{7}{8}$. New basic variables are x_2 and x_3.

	*	*			
$-\frac{72}{7}$	0	0	$-\frac{11}{7}$	$-\frac{8}{7}$	$-\frac{60}{7}$
$-\frac{1}{7}$	0	1	$-\frac{2}{7}$	$\frac{3}{7}$	$\frac{12}{7}$
$-\frac{12}{7}$	1	0	$-\frac{3}{7}$	$\frac{8}{7}$	$\frac{4}{7}$

End. All coefficients in the objective function row are negative.

Thus we find that the objective function has minimum equal to $-60/7$ when $x_1 = x_4 = x_5 = 0$, $x_2 = 4/7$, $x_3 = 12/7$.

Answer: Minimum is $-\frac{60}{7}$ for $\boldsymbol{x} = (0, \frac{4}{7}, \frac{12}{7}, 0, 0)$.

5.11 $\boldsymbol{x} = (1\ 9\ 0)^T$ is a basic feasible solution with $f(\boldsymbol{x}) = -4$, and a change in x_3 does not affect the function value $f = -4$, because when x_1 and x_2 are eliminated the coefficient of x_3 in f is 0. Now, x_3 can be increased until x_1 becomes 0. Then another basic feasible solution is obtained: $\hat{\boldsymbol{x}} = (0\ 10\ 2)^T$ with $f(\hat{\boldsymbol{x}}) = -4$. Thus there must be infinitely many optimal solutions along the straight line between $\boldsymbol{x}$ and $\hat{\boldsymbol{x}}$. A point $\boldsymbol{y}$ on this line can be written

$$\boldsymbol{y} = \lambda \boldsymbol{x} + (1-\lambda)\hat{\boldsymbol{x}} = \begin{pmatrix} \lambda \\ 10-\lambda \\ 2-2\lambda \end{pmatrix} \qquad \text{for some } \lambda \in [0,1].$$

Verify that $f(\boldsymbol{y}) = -4y_1 - 2y_3 = \ldots = -4$ independent of λ!

Generally, a solution is non-unique if, when the simplex method has terminated, some coefficient of a non-basic variable is equal to 0, and if it is possible to obtain a positive value of this variable by performing another simplex step.

Note: Non-uniqueness can also be checked via the dual problem. See e.g. exercise 5.14.

5.12 First we check that the right hand side $\boldsymbol{b}$ is non-negative. Then we introduce slack variables with appropriate signs, turning all inequalities into equalities. Finally we introduce artifical variables u_1, u_2, u_3. The problem to solve is then to minimize

$$u_1 + u_2 + u_3$$

subject to

$$\begin{aligned}
5x_1 + 5x_2 - \ \ 3x_3 + s_1 \qquad\qquad\quad + u_1 \qquad\qquad &= 9 \\
-9x_1 - 8x_2 + 10x_3 \qquad - s_2 \qquad\qquad + u_2 \qquad &= 7 \\
-3x_1 + 3x_2 - \ \ 7x_3 \qquad\qquad - s_3 \qquad\qquad + u_3 &= 2 \\
x_1, x_2, x_3,\ s_1, s_2, s_3, ; u_1, u_2, u_3 \geq 0. &
\end{aligned}$$

The initial solution is $x_i = 0$, $i = 1, 2, 3$, $s_j = 0$, $j = 1, 2, 3$ and $u_1 = 9$,

$u_2 = 7$, $u_3 = 2$. The initial tableaux is

						*	*	*	
0	0	0	0	0	0	−1	−1	−1	18
5	5	−3	1	0	0	1	0	0	9
−9	−8	10	0	−1	0	0	1	0	7
−3	3	−7	0	0	−1	0	0	1	2

After two iteration steps the final simplex tableaux is

		*	*					*	
−9.3	−2.6	0	0	−0.7	−1	−1	−0.3	0	6.9
2.3	2.6	0	1	−0.3	0	1	0.3	0	11.1
−0.9	−0.8	1	0	−0.1	0	0	0.1	0	0.7
−9.3	−2.6	0	0	−0.7	−1	0	0.7	1	6.9

The minimum value is 6.9, which is strictly positive. Hence the set is empty.

5.14 The objective function has value 76 at $\hat{\boldsymbol{y}}$.

The dual to the given problem (or, if you prefer to consider the given problem as dual, the primal problem) is

$$\begin{aligned} \text{Minimize} \quad & x_1 + 2x_2 + 9x_3 + 9x_4 \\ \text{subject to} \quad & x_1 + 3x_2 - 2x_3 + 5x_4 = 4 \\ & 2x_1 - x_2 + 4x_3 + x_4 = 29 \\ & -x_1 - x_2 + 3x_3 - x_4 = 13 \\ & x_1, x_2, x_3, x_4 \geq 0. \end{aligned}$$

The solution $\hat{\boldsymbol{y}}$ is feasible to the given problem, gives equality in the three first constraints and strict inequality in the fourth. Hence, by the complementary slackness principle, $\hat{\boldsymbol{y}}$ is an optimal solution if and only if the dual has a feasible solution $\boldsymbol{x}$ with $x_4 = 0$. Insertion of this condition in the constraints gives

$$\left\{ \begin{aligned} x_1 + 3x_2 - 2x_3 &= 4 \\ 2x_1 - x_2 + 4x_3 &= 29 \\ -x_1 - x_2 + 3x_3 &= 13 \end{aligned} \right. \iff \left\{ \begin{aligned} x_1 &= 3 \\ x_2 &= 5 \\ x_3 &= 7. \end{aligned} \right.$$

The solution $\hat{\boldsymbol{x}} = (3, 5, 7, 0)$ is feasible to the dual problem ($\hat{\boldsymbol{x}} \geq \boldsymbol{0}$). Thus $\hat{\boldsymbol{x}}$ is optimal for the dual and $\hat{\boldsymbol{y}}$ is optimal for the primal problem.

As a further check we note that $\hat{\boldsymbol{x}}$ gives the minimum value 76 for the dual problem, equal to the maximum value of the primal problem.

Chapter 6

6.7 Assume $\boldsymbol{a}, \boldsymbol{b} \in \mathbf{R}^n$ and $0 < \lambda < 1$. By the convexity of h

$$h(\lambda \boldsymbol{a} + (1-\lambda)\boldsymbol{b}) \le \lambda h(\boldsymbol{a}) + (1-\lambda)h(\boldsymbol{b}).$$

From the fact that g is nondecreasing we now conclude that

$$g\big(h(\lambda \boldsymbol{a} + (1-\lambda)\boldsymbol{b})\big) \le g\big(\lambda h(\boldsymbol{a}) + (1-\lambda)h(\boldsymbol{b})\big).$$

Using the convexity of g on the right hand side it follows that

$$g\big(h(\lambda \boldsymbol{a} + (1-\lambda)\boldsymbol{b})\big) \le \lambda g\big(h(\boldsymbol{a})\big) + (1-\lambda)g\big(h(\boldsymbol{b})\big),$$

which was to be proven.

6.11 a) The function $x \mapsto e^x$ of one variable is convex. As a consequence the function of three variables

$$(x, y, z) \mapsto e^x$$

is convex. By the same means we find that the other two terms are convex function. We could also argue in the following way: $x+2y+3z$ is a convex (affine) function, the exponential function is convex and nondecreasing, hence the composite function

$$(x, y, z) \mapsto e^{x+2y+3z}$$

is convex.

In conclusion, the function f is a sum of three convex functions, and is consequently convex.

b) If we confine ourselves to the line $(x, y, z) = (t, t, t)$ we obtain the function

$$t \longmapsto t^3 + 3t^2.$$

This function is neither convex nor concave, since its second derivative $6t + 6$ changes sign. Then the same goes for g. — As an alternative you could use Sylvester's criterion.

Answer: a) Convex, b) neither convex nor concave.

6.17 The following functions are convex:

$(x, y) \longmapsto e^x,$	since $x \mapsto e^x$ is convex
$(x, y) \longmapsto x^2 - 2xy + 2y^2 = (x-y)^2 + y^2,$	positive definite, hence convex
$(x, y) \longmapsto x - 4y + 8,$	affine, hence convex.

Since a sum of convex functions is convex, f is convex. Furthermore, the point $(0,1)$ is a stationary point for f since

$$f'_x(0,1) = f'_y(0,1) = 0.$$

Thus f has a minimum at $(0,1)$.

6.21 a) We have

$$f(\boldsymbol{x}) = \|\boldsymbol{A}\boldsymbol{x} - \boldsymbol{b}\|^2 = (\boldsymbol{A}\boldsymbol{x} - \boldsymbol{b})^T(\boldsymbol{A}\boldsymbol{x} - \boldsymbol{b}) = \boldsymbol{x}^T\boldsymbol{A}^T\boldsymbol{A}\boldsymbol{x} - 2\boldsymbol{b}^T\boldsymbol{A}\boldsymbol{x} + \boldsymbol{b}^T\boldsymbol{b}.$$

Then

$$\nabla f(\boldsymbol{x}) = 2\boldsymbol{A}^T\boldsymbol{A}\boldsymbol{x} - 2\boldsymbol{A}^T\boldsymbol{b}$$

and

$$\nabla f(\boldsymbol{x}) = \boldsymbol{0} \quad \Longleftrightarrow \quad \boldsymbol{A}^T\boldsymbol{A}\boldsymbol{x} = \boldsymbol{A}^T\boldsymbol{b}.$$

We have now found a necessary condition for f to have a minimum at $\boldsymbol{x}$. But this condition is also sufficient, since f is convex. Namely, the matrix $\boldsymbol{A}^T\boldsymbol{A}$ is positive semidefinite, since

$$\boldsymbol{x}^T\boldsymbol{A}^T\boldsymbol{A}\boldsymbol{x} = (\boldsymbol{A}\boldsymbol{x})^T\boldsymbol{A}\boldsymbol{x} = \|\boldsymbol{A}\boldsymbol{x}\|^2 \geq 0 \qquad \text{for all } \boldsymbol{x}.$$

b) The solution to the system $\boldsymbol{A}^T\boldsymbol{A}\boldsymbol{x} = \boldsymbol{A}^T\boldsymbol{b}$ is unique if and only if $\boldsymbol{A}^T\boldsymbol{A}$ is invertible, i.e., if and only if the rank of $\boldsymbol{A}$ equals m. (However, $\boldsymbol{A}\boldsymbol{x}$ is of course always uniquely determined.) Assuming this to be the case the formula for the solution is

$$\boldsymbol{x} = (\boldsymbol{A}^T\boldsymbol{A})^{-1}\boldsymbol{A}^T\boldsymbol{b}.$$

6.25 Put $F(\lambda) = f(\boldsymbol{a} + \lambda\boldsymbol{d})$ for $\lambda \geq 0$. Then $F'_+(0) = f'(\boldsymbol{a}; \boldsymbol{d})$. The fact that f is convex implies that $F(\lambda)$ is a convex function of λ, and thus that the difference quotient

$$\frac{F(\lambda) - F(0)}{\lambda - 0}$$

is an increasing function of $\lambda > 0$.

Assume that $\boldsymbol{d}$ is a direction of descent. Then $F(\lambda) < F(0)$ for all sufficiently small λ. This implies that the difference quotient of F is negative, and hence in the limit $\lambda \to +0$ that $f'(\boldsymbol{a}; \boldsymbol{d}) < 0$, where the inequality is strict since we take the limit from the right of a negative and increasing function.

To prove the converse statement, assume that $f'(\boldsymbol{a}; \boldsymbol{d}) < 0$. Then the difference quotient $\dfrac{F(\lambda) - F(0)}{\lambda - 0}$ is < 0 when λ is positive and close enough to 0. This implies $F(\lambda) < F(0)$, and hence that $\boldsymbol{d}$ is a direction of descent.

Without the assumption of convexity, the statement is true in the last direction only. For example, in one variable $f(x) = -x^2$ at 0 has ± 1 as directions of descent, yet $f'_+(0) = f'_-(0) = 0$.

Chapter 7

7.1 Let $g_1(\boldsymbol{x}) = -x_1 + 2x_2$ and $g_2(\boldsymbol{x}) = -3x_1 + x_2$. By theorem 2 the gradient of f can be written

$$\nabla f(\boldsymbol{0}) = -u_1 \nabla g_1(\boldsymbol{0}) - u_2 \nabla g_2(\boldsymbol{0}) = u_1 \begin{pmatrix} 1 \\ -2 \end{pmatrix} + u_2 \begin{pmatrix} 3 \\ -1 \end{pmatrix}, \qquad u_1, u_2 \geq 0.$$

Hence $\nabla f(\boldsymbol{0})$ is a positive linear combination of $(1, -2)$ och $(3, -1)$, and thus belongs to a sector bounded by the normals to the lines bounding the domain of f.

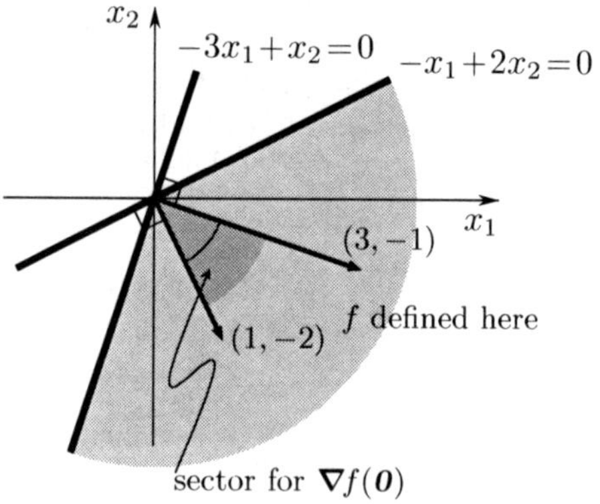

7.3 The Lagrange function is

$$L(\boldsymbol{x}; \boldsymbol{u}) = 6ax_1 - 4x_2 + u_1(x_1x_2 - x_2 - 10) + u_2(6x_1 + 4x_2 - 38).$$

Note that the constraints $x_1 \geq 0$ and $x_2 \geq 0$ are not active at $(3, 5)$ and hence need not be included in the Lagrange function. The other two constraints are active there. The derivatives of L are

$$\begin{cases} \dfrac{\partial L}{\partial x_1}(\boldsymbol{x}; \boldsymbol{u}) = 6a + u_1 x_2 + 6u_2 \\[2ex] \dfrac{\partial L}{\partial x_2}(\boldsymbol{x}; \boldsymbol{u}) = -4 + u_1 x_1 - u_1 + 4u_2. \end{cases}$$

The point $(3,5)$ is a KKT point if and only if $\nabla_{\boldsymbol{x}} L(3,5;u_1,u_2) = \boldsymbol{0}$ for some multipliers $u_1, u_2 \geq 0$. Setting the gradient equal to zero gives

$$\begin{cases} 6a + 5u_1 + 6u_2 = 0 \\ -4 + 2u_1 + 4u_2 = 0 \end{cases} \iff \begin{cases} u_1 = -3a - 3 \\ u_2 = \dfrac{5+3a}{2}. \end{cases}$$

Now $-3a - 3 \geq 0 \Leftrightarrow a \leq -1$ and $\frac{5+3a}{2} \geq 0 \Leftrightarrow a \geq -\frac{5}{3}$. Thus $(3,5)$ is a KKT point if and only if $-\frac{5}{3} \leq a \leq -1$.

Answer: $-\frac{5}{3} \leq a \leq -1$.

7.7 Since the positivity conditions can be disregarded the Lagrange function is

$$L(x,y,z;v) = x + y + z + v\Big(\frac{a}{x} + \frac{b}{y} + \frac{c}{z} - 1\Big),$$

defined for $x, y, z > 0$ and all v. The CQ is easily seen to be fulfilled everywhere in this case. The KKT conditions are:

$$\begin{cases} \dfrac{\partial L}{\partial x} = 1 - \dfrac{va}{x^2} = 0 \\ \dfrac{\partial L}{\partial y} = 1 - \dfrac{vb}{y^2} = 0 \\ \dfrac{\partial L}{\partial z} = 1 - \dfrac{vc}{z^2} = 0 \\ \dfrac{a}{x} + \dfrac{b}{y} + \dfrac{c}{z} = 1. \end{cases}$$

The first three equations give

$$\frac{a}{x} = \sqrt{\frac{a}{v}}, \qquad \frac{b}{y} = \sqrt{\frac{b}{v}}, \qquad \frac{c}{z} = \sqrt{\frac{c}{v}}$$

respectively, which inserted into the fourth one yields

$$\sqrt{v} = \sqrt{a} + \sqrt{b} + \sqrt{c}.$$

Hence the only KKT point $(\bar{x}, \bar{y}, \bar{z})$ is given by

$$\begin{cases} \bar{x} = a + \sqrt{ab} + \sqrt{ac} \\ \bar{y} = b + \sqrt{ab} + \sqrt{bc} \\ \bar{z} = c + \sqrt{ac} + \sqrt{bc}. \end{cases}$$

Since we are searching for a minimum we can certainly confine ourselves to some compact set $\alpha \leq x \leq A$, $\beta \leq y \leq A$, $\gamma \leq z \leq A$. Thus the minimum exists, and the only possibility is $(\bar{x}, \bar{y}, \bar{z})$. The minimal value is $\bar{x} + \bar{y} + \bar{z} = a + b + c + 2\sqrt{ab} + 2\sqrt{ac} + 2\sqrt{bc}$.

Answer: The least value is $a + b + c + 2\sqrt{ab} + 2\sqrt{ac} + 2\sqrt{bc}$.

7.11

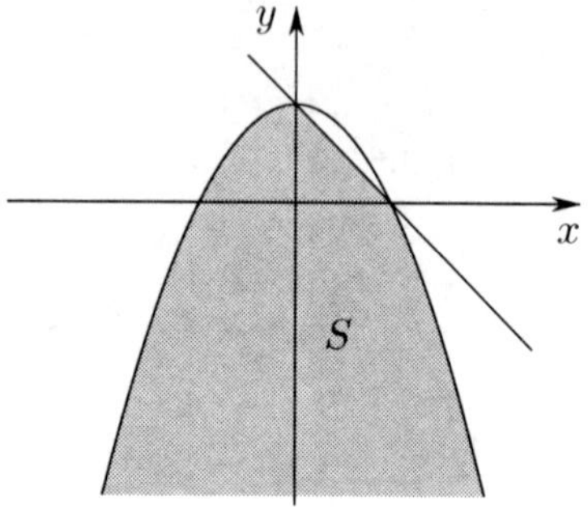

The Lagrange function is

$$L(x, y; u_1, u_2) =$$

$$= x^2 + 2xy + 2y^2 - 5x - 5y + u_1(x + y - 1) + u_2(x^2 + y - 1),$$

$$u_1, u_2 \geq 0.$$

- The KKT conditions are:

$$\begin{cases} \dfrac{\partial L}{\partial x} = 2x + 2y - 5 + u_1 + 2u_2 x = 0 \\ \dfrac{\partial L}{\partial y} = 2x + 4y - 5 + u_1 + u_2 = 0. \end{cases}$$

- The complementary slackness conditions are:

$$\begin{cases} u_1(x + y - 1) = 0, \\ u_2(x^2 + y - 1) = 0. \end{cases}$$

- Feasibility conditions:

$$\begin{cases} x + y \leq 1, \\ y \leq 1 - x^2. \end{cases}$$

There are four cases to consider:

Case 1: $u_1 = 0$, $u_2 = 0$.

This implies that $x = \frac{5}{2}$, $y = 0$. However, this point lies outside S, that is, it is not feasible.

Case 2: $u_1 > 0$, $u_2 = 0$.

We obtain the following system of equations:

$$\begin{cases} 2x + 2y - 5 + u_1 = 0 \\ 2x + 4y - 5 + u_1 = 0 \\ x + \ y - 1 \qquad = 0 \end{cases} \iff \begin{cases} x = 1 \\ y = 0 \\ u_1 = 3, \end{cases}$$

with $u_1 > 0$ and $(x, y) = (1, 0) \in S$, i.e. feasible.

Since f is convex (it is the sum of a positive definite quadratic form and an affine function) and the constraint functions g_1 and g_2 are convex, the sufficient conditions are fulfilled. Hence we have found the minimum point, $(x, y) = (1, 0)$, and there is no need to investigate the remaining two cases! (Also, we don't have to worry about the CQ.)

Answer: Minimum at $(1, 0)$.

7.14 The global minimizer is most easily obtained by solving the equivalent problem of one variable: minimize $x_1(x_1 - 1)^3$ over $\mathbf{R}$. The minimum occurs at $x_1 = \frac{1}{4}$. Thus $\boldsymbol{x}^* = (\frac{1}{4}, -\frac{27}{64})$.

Putting $h(x_1, x_2) = x_2 - (x_1 - 1)^3$ we have

$$\nabla h(x_1, x_2) = \begin{pmatrix} -3(x_1 - 1)^2 \\ 1 \end{pmatrix} = \frac{1}{16} \begin{pmatrix} -27 \\ 16 \end{pmatrix} \quad \text{at } \boldsymbol{x}^*.$$

The feasible directions at $\boldsymbol{x}^*$ are all $\boldsymbol{d} \neq \boldsymbol{0}$ orthogonal to $\nabla h(\boldsymbol{x}^*)$, that is, all

$$\boldsymbol{d} = t \begin{pmatrix} 16 \\ 27 \end{pmatrix}, \qquad t \neq 0.$$

Since $L(\boldsymbol{x}; v) = x_1 x_2 + v(x_2 - (x_1 - 1)^3)$ the KKT conditions are

$$\frac{\partial L}{\partial x_1} = x_2 - 3v(x_1 - 1)^2 = 0$$
$$\frac{\partial L}{\partial x_2} = x_1 + v = 0.$$

At $\boldsymbol{x}^*$ we have $v^* = -x_1^* = -\frac{1}{4}$. The Hessian is

$$\nabla^2_{\boldsymbol{xx}} L(\boldsymbol{x}; v) = \begin{pmatrix} -6v(x_1 - 1) & 1 \\ 1 & 0 \end{pmatrix},$$

which at $(\boldsymbol{x}^*, v^*)$ becomes $\begin{pmatrix} -9/8 & 1 \\ 1 & 0 \end{pmatrix}$. This matrix $\boldsymbol{W}$ is indefinite. However,

$$\boldsymbol{d}^T \boldsymbol{W} \boldsymbol{d} = 576t^2 > 0 \quad \text{for all feasible directions } \boldsymbol{d} = t \begin{pmatrix} 16 \\ 27 \end{pmatrix}, \quad t \neq 0.$$

Thus the sufficiency conditions are satisfied at $\boldsymbol{x}^*$.

Answer: Local minimum at $(\frac{1}{4}, -\frac{27}{64})$.

Chapter 8

8.1 a) The Lagrange function is

$$L(\boldsymbol{x}; u) = x_1^2 + 2x_2^2 + u(-x_1 - x_2 + 1), \qquad \boldsymbol{x} \in X,\ u \geq 0.$$

The dual function $\theta(u)$ is found by minimizing L over X. Completing squares gives

$$L(\boldsymbol{x}; u) = \left(x_1 - \frac{u}{2}\right)^2 + 2\left(x_2 - \frac{u}{4}\right)^2 - \frac{3}{8}u^2 + u.$$

Thus we have minimum for $x_1 = \dfrac{u}{2}$, $x_2 = \dfrac{u}{4}$, and the minimal value is

$$\theta(u) = -\frac{3}{8}u^2 + u, \qquad u \geq 0.$$

b) By use of the derivative we easily find that the maximum of $\theta(u)$ is equal to $\dfrac{2}{3}$ and occurs at $u = \dfrac{4}{3}$.

c) Let

$$\bar{u} = \frac{4}{3}, \qquad \bar{x}_1 = \frac{\bar{u}}{2} = \frac{2}{3}, \qquad \bar{x}_2 = \frac{\bar{u}}{4} = \frac{1}{3}.$$

Then

$$f(\bar{x}_1, \bar{x}_2) = \frac{2}{3} = \theta(\bar{u}).$$

Consequently there is no duality gap, and $(\bar{x}_1, \bar{x}_2;\ \bar{u})$ is a saddle point by theorem 3. It follows from theorem 1 that $(\bar{x}_1, \bar{x}_2)$ solves the primal problem. Thus the minimum value is $\dfrac{2}{3}$.

Answer: a) $\theta(u) = -\dfrac{3}{8}u^2 + u, \quad u \geq 0.$ b) Maximum is $\dfrac{2}{3}$ at $u = \dfrac{4}{3}$.

c) Minimum is $\dfrac{2}{3}$ at $x_1 = \dfrac{2}{3}$, $x_2 = \dfrac{1}{3}$.

Chapter 9

9.1 a) By drawing a figure of the domain one easily finds the minimum point $(0, 1)$.

b) Let

$$F(x,\mu) = x_1^2 + x_2^2 + \mu\left(\max(0,\ 2x_1 + x_2 - 2)^2 + \max(0,\ -x_2 + 1)^2\right).$$

More explicitly,

$$F(x,\mu) =$$

$$\begin{cases} x_1^2 + x_2^2 & \text{when } 2x_1{+}x_2{-}2 \le 0, \quad -x_2{+}1 \le 0 \\ x_1^2 + x_2^2 + \mu(2x_1{+}x_2{-}2)^2 & \text{when } 2x_1{+}x_2{-}2 \ge 0, \quad -x_2{+}1 \le 0 \\ x_1^2 + x_2^2 + \mu(-x_2{+}1)^2 & \text{when } 2x_1{+}x_2{-}2 \le 0, \quad -x_2{+}1 \ge 0 \\ x_1^2 + x_2^2 + \mu(2x_1{+}x_2{-}2)^2 + \mu(-x_2{+}1)^2 & \text{when } 2x_1{+}x_2{-}2 \ge 0, \quad -x_2{+}1 \ge 0. \end{cases}$$

c) Newton's method makes use of the iteration formula

$$\boldsymbol{x}_{k+1} = \boldsymbol{x}_k - \boldsymbol{H}(\boldsymbol{x}_k)^{-1}\boldsymbol{d}_k,$$

where $\boldsymbol{d}_k = \nabla f(\boldsymbol{x}_k)$. With $\mu = 1$ the results of the computations are as follows.

k	$\boldsymbol{x}_k$	$\boldsymbol{d}_k$	$\boldsymbol{H}(\boldsymbol{x}_k)$
0	$(2, 6)$	$(36, 28)$	$\begin{pmatrix} 10 & 4 \\ 4 & 4 \end{pmatrix}$
1	$(0.67, 0.33)$	$(1.33, -0.67)$	$\begin{pmatrix} 2 & 0 \\ 0 & 4 \end{pmatrix}$
2	$(0, 0.5)$	$(0, 0)$	

We don't get any further than $(0, 0.5)$ when $\mu = 1$. (Why?)

d) Letting $\mu = 10$ and starting at $(0, 0.5)$ we get

k	$\boldsymbol{x}_k$	$\boldsymbol{d}_k$	$\boldsymbol{H}(\boldsymbol{x}_k)$
0	$(0, 0.5)$	$(0, -9)$	$\begin{pmatrix} 2 & 0 \\ 0 & 22 \end{pmatrix}$
1	$(0, 0.91)$	$(0, 0)$	

Answer: $(0, 0.91)$.

Chapter 10

10.1 a) $\mathbf{A}$ is a point-to-set map from $\mathbf{R}$ to $\mathbf{R}$. (See the figure.) To show that $\mathbf{A}$ is closed means to show that

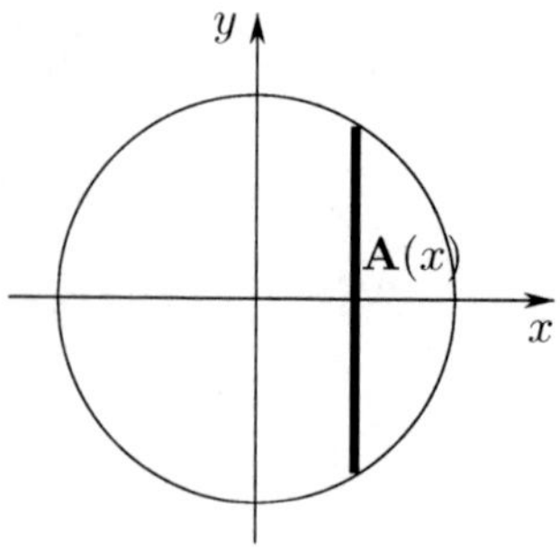

$$\left.\begin{cases} x_k \to x \text{ as } k \to \infty \\ y_k \to y \text{ as } k \to \infty \\ y_k \in \mathbf{A}(x_k) \end{cases}\right. \implies y \in \mathbf{A}(x).$$

In the present case we must show that

$$x_k^2 + y_k^2 \le 1 \quad \text{for all } k \quad \implies \quad x^2 + y^2 \le 1.$$

This, however, is obvious.

b) Also $\mathbf{B}$ is a point-to-set map from $\mathbf{R}$ to $\mathbf{R}$. Consider the sequences

$$x_k = \frac{1}{k}, \quad y_k = \frac{1}{2}, \qquad k = 2, 3, \dots .$$

Here $y_k \in \mathbf{B}(x_k)$, since $\frac{1}{k^2} + \frac{1}{2^2} \le 1$ for $k \ge 2$. However, the limits are $x = 0$, $y = \frac{1}{2}$, and here $y \notin \mathbf{B}(x)$ since $\frac{1}{2} \notin [-1, 0]$. This example shows that $\mathbf{B}$ is not closed.

Answer: a) $\mathbf{A}$ is closed. b) $\mathbf{B}$ is not closed.

10.3 The point-to-set mapping $\mathbf{A}$ is given by

$$\begin{cases} y_1 = 2x_1 - 1 \\ y_2 \in [x_2 - 1, x_2 + 2]. \end{cases}$$

Assume that $\left(\boldsymbol{x}^{(k)}\right)_1^\infty$ and $\left(\boldsymbol{y}^{(k)}\right)_1^\infty$ are sequences in $\mathbf{R}^2$ such that

$$\begin{aligned} \boldsymbol{x}^{(k)} &= (x_1^{(k)}, x_2^{(k)}) \to \boldsymbol{x} = (x_1, x_2) \\ \boldsymbol{y}^{(k)} &= (y_1^{(k)}, y_2^{(k)}) \to \boldsymbol{y} = (y_1, y_2) \end{aligned} \qquad \text{as } k \to \infty,$$

and assume further that $\boldsymbol{y}^{(k)} \in \mathbf{A}(\boldsymbol{x}^{(k)})$ for all k. Then in particular $y_1^{(k)} = 2x_1^{(k)} - 1$, which in the limit implies that $y_1 = 2x_1 - 1$. Furthermore, for all k

$$x_2^{(k)} - 1 \le y_2^{(k)} \le x_2^{(k)} + 2,$$

which implies that $x_2 - 1 \le y_2 \le x_2 + 2$. Thus $\boldsymbol{y} \in \mathbf{A}(\boldsymbol{x})$, and the proof is finished.

10.6 Consider any convergent sequence $\big((\boldsymbol{x}^{(k)},\, \boldsymbol{y}^{(k)})\big)_1^\infty$ of elements in Z. Obviously

$$Z \ni (\boldsymbol{x}^{(k)},\, \boldsymbol{y}^{(k)}) \to (\boldsymbol{x}, \boldsymbol{y}) \qquad \Longleftrightarrow \qquad \begin{cases} \boldsymbol{x}^{(k)} \to \boldsymbol{x} \\ \boldsymbol{y}^{(k)} \to \boldsymbol{y} \\ \boldsymbol{y}^{(k)} \in \mathbf{A}(\boldsymbol{x}^{(k)}). \end{cases}$$

We now have

$$Z \text{ closed} \iff (\boldsymbol{x}, \boldsymbol{y}) \in Z \iff \boldsymbol{y} \in \mathbf{A}(\boldsymbol{x}) \iff \mathbf{A} \text{ closed}.$$

Thus Z is closed if and only if the map $\mathbf{A}$ is closed.

Note: The set Z is a generalized version of the notion of the graph of a function.

Appendix A

A.1 That $\boldsymbol{H}$ is invertible follows e.g. from

$$\boldsymbol{H}\boldsymbol{x} = \boldsymbol{0} \quad \Longrightarrow \quad \boldsymbol{x}^T\boldsymbol{H}\boldsymbol{x} = 0 \quad \Longrightarrow \quad \boldsymbol{x} = \boldsymbol{0},$$

where the last implication is due to $\boldsymbol{H}$ being positive definite.

Putting $\boldsymbol{y} = \boldsymbol{H}\boldsymbol{x}$, since $\boldsymbol{H}$ is symmetric we have

$$\boldsymbol{y}^T\boldsymbol{H}^{-1}\boldsymbol{y} = \boldsymbol{x}^T\boldsymbol{H}\boldsymbol{H}^{-1}\boldsymbol{H}\boldsymbol{x} = \boldsymbol{x}^T\boldsymbol{H}\boldsymbol{x}.$$

From this we conclude that

$$\boldsymbol{x}^T\boldsymbol{H}\boldsymbol{x} > 0 \text{ for all } \boldsymbol{x} \neq \boldsymbol{0} \quad \Longleftrightarrow \quad \boldsymbol{y}^T\boldsymbol{H}^{-1}\boldsymbol{y} > 0 \text{ for all } \boldsymbol{y} \neq \boldsymbol{0}.$$

Hence $\boldsymbol{H}^{-1}$ is positive definite if and only if $\boldsymbol{H}$ is positive definite.

A.4 a) Gaussian elimination gives

$$\begin{pmatrix} 3 & 2 & 0 \\ 2 & 6 & 1 \\ 0 & 1 & 2 \end{pmatrix} \xrightarrow{d_1=3} \begin{pmatrix} 1 & \frac{2}{3} & 0 \\ 0 & \frac{14}{3} & 1 \\ 0 & 1 & 2 \end{pmatrix} \xrightarrow{d_2=\frac{14}{3}}$$

$$\longrightarrow \begin{pmatrix} 1 & \frac{2}{3} & 0 \\ 0 & 1 & \frac{3}{14} \\ 0 & 0 & \frac{25}{14} \end{pmatrix} \xrightarrow{d_3=\frac{25}{14}} \begin{pmatrix} 1 & \frac{2}{3} & 0 \\ 0 & 1 & \frac{3}{14} \\ 0 & 0 & 1 \end{pmatrix}.$$

The pivot elements are $d_1 = 3$, $d_2 = \dfrac{14}{3}$, $d_3 = \dfrac{25}{14}$. All are positive, hence the matrix is positive definite.

b) The subdeterminants to be considered are

$$|3| = 3, \qquad \begin{vmatrix} 3 & 2 \\ 2 & 6 \end{vmatrix} = 14, \qquad \begin{vmatrix} 3 & 2 & 0 \\ 2 & 6 & 1 \\ 0 & 1 & 2 \end{vmatrix} = 25.$$

Since all three are strictly positive $\boldsymbol{H}$ is positive definite by Sylvester's criterion.

c) A Cholesky factorization is obtained by putting $\boldsymbol{H} = \boldsymbol{L}\boldsymbol{L}^T$ with

$$\boldsymbol{L} = \begin{pmatrix} l_{11} & 0 & 0 \\ l_{21} & l_{22} & 0 \\ l_{31} & l_{32} & l_{33} \end{pmatrix}.$$

The elements of $\boldsymbol{L}$ can be computed by multiplying through, which for our $\boldsymbol{H}$ gives

$$\begin{cases} l_{11}^2 & = 3 \\ l_{11}l_{21} & = 2 \\ l_{11}l_{31} & = 0 \\ l_{21}^2 + l_{22}^2 & = 6 \\ l_{21}l_{31} + l_{22}l_{32} & = 1 \\ l_{31}^2 + l_{32}^2 + l_{33}^2 & = 2. \end{cases}$$

The first equation gives $l_{11} = \pm\sqrt{3}$, where we choose the plus sign. Now the second and third equations give $l_{21} = 2/\sqrt{3}$ and $l_{31} = 0$. Thereafter we get $l_{22} = \sqrt{14/3}$ from the fourth equation, and finally we obtain $l_{32} = \sqrt{3/14}$, $l_{33} = 5/\sqrt{14}$. To summarize, we have

$$\boldsymbol{L} = \begin{pmatrix} \sqrt{3} & 0 & 0 \\ 2/\sqrt{3} & \sqrt{14/3} & 0 \\ 0 & \sqrt{3/14} & 5/\sqrt{14} \end{pmatrix}.$$

The existence of a Cholesky factorization implies that $\boldsymbol{H}$ is positive definite.

d) The eigenvalues (computed numerically) are $\approx 2.3691, 1.4746, 7.1563$, all positive. Hence $\boldsymbol{H}$ is positive definite.

A.6 a) The Jacobian is

$$\nabla f(\boldsymbol{x}) = \begin{pmatrix} -400x_1(x_2 - x_1^2) - 2(1 - x_1) \\ 200(x_2 - x_1^2) \end{pmatrix}.$$

We see that $\nabla f(\boldsymbol{x}^*) = \boldsymbol{0}$, making $\boldsymbol{x}^*$ a stationary point. The Hessian is

$$\boldsymbol{H}(\boldsymbol{x}) = \begin{pmatrix} 1200x_1^2 - 400x_2 + 2 & -400x_1 \\ \\ -400x_1 & 200 \end{pmatrix}$$

and at the stationary point

$$\boldsymbol{H}(\boldsymbol{x}^*) = \begin{pmatrix} 802 & -400 \\ -400 & 200 \end{pmatrix}.$$

This matrix is positive definite, since $802 > 0$ and $\det \boldsymbol{H}(\boldsymbol{x}^*) = 400 > 0$ (Sylvester's criterion).

b) The condition that $\det \boldsymbol{H}(\boldsymbol{x}) = 0$ is

$$200(1200x_1^2 - 400x_2 + 2) - (-400x_1)^2 = 0 \quad \Longleftrightarrow \quad x_2 - x_1^2 = 0.005.$$

We now use Sylvester's criterion in the version of exercise A.5.b). If $f(\boldsymbol{x}) < 0.0025$ then in particular

$$100(x_2 - x_1^2)^2 < 0.0025 \quad \Longrightarrow \quad |x_2 - x_1^2| < 0.005 \quad \Longrightarrow \quad \det \boldsymbol{H}(\boldsymbol{x}) \neq 0.$$

Then for reasons of continuity $\det \boldsymbol{H}(\boldsymbol{x}) > 0$ for these $\boldsymbol{x}$, and since also $200 > 0$ the Hessian is positive definite.

Index